W9-DAH-409

Microorganisms and Bioterrorism

INFECTIOUS AGENTS AND PATHOGENESIS

Series Editors: Mauro Bendinelli, *University of Pisa*
Herman Friedman, *University of South Florida College of Medicine*

Recent volumes in this series:

CHLAMYDIA PNEUMONIAE
Infection and Disease
Edited by Herman Friedman, Yoshimasa Yamamoto, and Mauro Bendinelli

DNA TUMOR VIRUSES
Oncogenic Mechanisms
Edited by Giuseppe Barbanti-Brodano, Mauro Bendinelli, and Herman Friedman

ENTERIC INFECTIONS AND IMMUNITY
Edited by Lois J. Paradise, Mauro Bendinelli, and Herman Friedman

HELICOBACTER PYLORI INFECTION AND IMMUNITY
Edited by Yoshimasa Yamamoto, Herman Friedman, and Paul S. Hoffman

HERPESVIRUSES AND IMMUNITY
Edited by Peter G. Medveczky, Herman Friedman, and Mauro Bendinelli

HUMAN RETROVIRAL INFECTIONS
Immunological and Therapeutic Control
Edited by Kenneth E. Ugen, Mauro Bendinelli, and Herman Friedman

INFECTIOUS DISEASES AND SUBSTANCE ABUSE
Edited by Herman Friedman, Catherine Newton, and Thomas W. Klein

IN VIVO MODELS OF HIV DISEASE AND CONTROL
Edited by Herman Friedman, Steven Specter, and Mauro Bendinelli

MICROORGANISMS AND AUTOIMMUNE DISEASES
Edited by Herman Friedman, Noel R. Rose, and Mauro Bendinelli

MICROORGANISMS AND BIOTERRORISM
Edited by Burt Anderson, Herman Friedman, and Mauro Bendinelli

OPPORTUNISTIC INTRACELLULAR BACTERIA AND IMMUNITY
Edited by Lois J. Paradise, Herman Friedman, and Mauro Bendinelli

PULMONARY INFECTIONS AND IMMUNITY
Edited by Herman Chmel, Mauro Bendinelli, and Herman Friedman

RAPID DETECTION OF INFECTIOUS AGENTS
Edited by Steven Specter, Mauro Bendinelli, and Herman Friedman

---

A Continuation Order Plan is available for this series. A continuation order will bring delivery of each new volume immediately upon publication. Volumes are billed only upon actual shipment. For further information please contact the publisher.

Microorganisms and Bioterrorism

Edited by

Burt Anderson
University of South Florida
Tampa, Florida, USA

Herman Friedman
University of South Florida
Tampa, Florida, USA

and

Mauro Bendinelli
University of Pisa
Pisa, Italy

Springer

Burt Anderson
University of South Florida College
of Medicine
Tampa, Florida 33612-4799
USA
banderso@hsc.usf.edu

Herman Friedman
University of South Florida College
of Medicine
Tampa, Florida 3612-4799
USA
hfriedma@hsc.usf.edu

Mauro Bendinelli
Universita di Pisa
Via San Zeno 37
I- 56127 Pisa
Italy
bendinelli@biomed.unipi.it

Library of Congress Control Number: 2005930722

ISBN-10: 0-387-28156-8

ISBN-13: 978-0387-28156-8

Printed in the United States of America (BS/DH)

9 8 7 6 5 4 3 2 1

springeronline.com

Contributors

KEN ALIBEK • The National Center for Biodefense, George Mason University, Manassas, VA 20910

PHILIP AMUSO • Center for Biological Defense, University of South Florida, Tampa, FL

BURT ANDERSON • Department of Medical Microbiology and Immunology, University of South Florida College of Medicine, Tampa, FL

THOMAS E. BLANK • Headquarters, U.S. Army Medical Research Institute of Infectious Diseases, Ft. Detrick, Frederick, MD 21702

DONALD H. BOUYER • Department of Pathology and Center for Biodefense and Emerging Infectious Diseases, The University of Texas Medical Branch at Galveston, Galveston, TX 77555-0609

JOEL A. BOZUE • Bacteriology Division, U.S. Army Medical Research Institute of Infectious Diseases, Ft. Detrick, Frederick, MD 21702

ANDREW CANNONS • Center for Biological Defense, University of South Florida, Tampa, FL

DONALD J. CHABOT • Headquarters, U.S. Army Medical Research Institute of Infectious Diseases, Ft. Detrick, Frederick, MD 21702

CHRISTOPHER K. COTE • Headquarters, U.S. Army Medical Research Institute of Infectious Diseases, Ft. Detrick, Frederick, MD 21702

WILLIAM A. DAY • Headquarters, U.S. Army Medical Research Institute of Infectious Diseases, Ft. Detrick, Frederick, MD 21702

DAVID L. ERICKSON • Laboratory of Human Bacterial Pathogenesis, Rocky Mountain Laboratories, National Institute of Allergy and Infectious Diseases, National Institutes of Health, Hamilton, MT

SANDRA G. GOMPF • Department of Infectious and Tropical Medicine and Allergy/Immunology Section Chief, James A. Haley Veterans Hospital, Tampa, FL

KATE F. GRIFFIN • Defence Science and Technology Laboratory, Porton Down, Salisbury, Wiltshire, SP4 0JQ, United Kingdom

LAURA R. HENDRIX • Department of Medical Microbiology and Immunology, Texas A&M University System Health Science Center, College Station, TX 77843

B. JOSEPH HINNEBUSCH • Laboratory of Human Bacterial Pathogenesis, Rocky Mountain Laboratories, National Institute of Allergy and Infectious Diseases, National Institutes of Health, Hamilton, MT

MICHELLE KUTZLER • Department of Pathology and Laboratory Medicine, University of Pennsylvania School of Medicine, Philadelphia, PA

JORDAN LEWIS • Florida Infectious Disease Institute and University of South Florida Infectious and Tropical Medicine Division, Tampa, FL

CATHERINE LOBANOVA • The National Center for Biodefense, George Mason University, Manassas, VA 20910

STEPHEN L. MICHELL • Defence Science and Technology Laboratory, Porton Down, Salisbury, Wiltshire, SP4 0JQ, United Kingdom

J. D. MILLER • Viral and Rickettsial Zoonoses Branch, Division of Viral and Rickettsial Diseases, Center for Disease Control, Atlanta, GA

STEPHEN A. MORSE • Bioterrorism Preparedness and Response Program, Centers for Disease Control and Prevention, Atlanta, GA

DAVID B. MORTON • Centre for Biomedical Ethics, University of Birmingham, B152TT, United Kingdom

EKNATH NAIK • Department Global Health, Epidemiology and Statistics, University of South Florida, Tampa, FL

R. MARTIN ROOP II • Department of Microbiology and Immunology, E. Carolina University School of Medicine, Greenville, NC 27858-4354

KASI RUSSELL • Department of Medical Microbiology and Immunology, Texas A&M University System Health Science Center, College Station, TX 77843

JAMES E. SAMUEL • Department of Medical Microbiology and Immunology, Texas A&M University System Health Science Center, College Station, TX 77843

ANGELO SCORPIO • Headquarters, U.S. Army Medical Research Institute of Infectious Diseases, Ft. Detrick, Frederick, MD 21702

E. I. SHAW • Viral and Rickettsial Zooroses Branch, Division of Viral and Rickettsial Diseases, Center for Disease Control, Atlanta, GA

KALEY TASH • Harvard University, Boston, MA

H. A. THOMPSON • Viral and Rickettsial Zoonoses Branch, Division of Viral and Rickettsial Diseases, Center for Disease Control, Atlanta, GA

RICHARD W. TITBALL • Defence Science and Technology Laboratory, Porton Down, Salisbury, Wiltshire, SP4 0JQ, United Kingdom

KENNETH E. UGEN • Department Medical Microbiology and Immunology, University of South Florida College of Medicine, Tampa, FL

MICHELLE WRIGHT VALDERAS • Department of Microbiology and Immunology, E. Carolina University School of Medicine, Greenville, NC 27858-4354
DAVID H. WALKER • Department of Pathology and Center for Biodefense and Emerging Infectious Diseases, The University of Texas Medical Branch at Galveston, Galveston, TX 77555-0609
DAVID B. WEINER • Department of Pathology and Laboratory Medicine, University of Pennsylvania School of Medicine, Philadelphia, PA
SUSAN L. WELKOS • Headquarters, U.S. Army Medical Research Institute of Infectious Diseases, Ft. Detrick, Frederick, MD 21702
GUOQUAN ZHANG • Department of Medical Microbiology and Immunology, Texas A&M University System Health Science Center, College Station, TX 77843

Preface to the Series

The mechanisms of disease production by infectious agents are presently the focus of an unprecedented flowering of studies. The field has undoubtedly received impetus from the considerable advances recently made in the understanding of the structure, biochemistry, and biology of viruses, bacteria, fungi, and other parasites. Another contributing factor is our improved knowledge of immune responses and other adaptive or constitutive mechanisms by which hosts react to infection. Furthermore, recombinant DNA technology, monoclonal antibodies, and other newer methodologies have provided the technical tools for examining questions previously considered too complex to be successfully tackled. The most important incentive is probably the regenerated idea that infection might be the initiating event in many clinical entities, presently classified as idiopathic or of uncertain origin.

Infectious pathogenesis research holds great promise. As more information is uncovered, it is becoming increasingly apparent that our present knowledge of the pathogenic potential of infectious agents is often limited to the most noticeable effects, which sometimes represents only the tip of the iceberg. For example, it is now well appreciated that pathologic processes caused by infectious agents may emerge clinically after an incubation of decades, and may result from genetic, immunologic, and other indirect routes more than from the infecting agent itself. Thus, there is a general expectation that continued investigation will lead to the isolation of new agents of infection, the identification of hitherto unsuspected etiologic correlations, and, eventually, more effective approaches to prevention and therapy.

Studies on the mechanisms of disease caused by infectious agents demand a breadth of understanding across many specialized areas, as well as much cooperation between clinicians and experimentalists. The series *Infectious Agents and Pathogenesis* is intended not only to document the state of the art in this fascinating and challenging field, but also to help lay bridges among diverse areas and people.

M. Bendinelli
H. Friedman

Introduction

The threat of bioterrorism has become a major challenge for the twenty-first century. However, the potentials of infectious agents as bioweapons have been recognized for centuries. Throughout history there have been attempts to initiate infectious disease outbreaks and epidemics during warfare. In the last decade the attention of the biomedical community, as well as governments and the United Nations, has increasingly focused on the threat of bioterrorism, especially the use of biological and/or chemical weapons against military and civilian populations. As an example, there is now much interest concerning microbial infection and bioterrorism in the medical microbiology and immunology communities. This volume addresses such concerns and emphasizes both basic and clinical concepts, as well as problematic implications of infection by various microbes now recognized as potential bioterrorism agents.

The first chapter by Drs. Andrew Canons, Philip Amuso, and Burt Anderson from the University of South Florida is an overview of the biotechnology of bioterrorism both in the public health response to possible acts of bioterrorism, as well as for the concerns about the misuse of biotechnology. The second chapter is a historical perspective of microbial bioterrorism by Dr. Steven Morse, Director of the Bioterrorism Division at the Center for Disease Control and Prevention in Atlanta, GA. This chapter describes in detail historical aspects concerning the early use of biological agents in warfare, development and international conventions to prohibit the use of such weapons, and a brief description of important incidents of infectious agents as bioterrorist agents and use during the last few centuries. The next chapter by Dr. Sandra Gompf from the University of South Florida discusses the role of public health physicians and infectious diseases specialists in the control of microbial bioterrorism.

Dr. Ken Alibek and associates from the National Center for Biodefense at George Mason University then present a detailed chapter concerning the role of innate immunity in protection against biological weapons, especially antimicrobial agents and vaccines to such agents. This chapter describes in detail various components of the immune response system that has a role in host defense, especially of the pulmonary system, which would be the first system involved in an aerosol bioterrorist attack. This chapter also discusses in detail the possibility that the immune system may be nonspecifically stimulated in an antigen-independent manner by nonspecific immune modulators, both synthetic and natural, to enhance resistance to microbial bioterrorist agents. This chapter also discusses the possibility of immunotherapeutics to neutralize or lyse pathogens by interfering with microbial gene expression and growth by enhancing and activating immune cells or a combination of both. The possible use of cytokines, including interferons and interferon inducers to stimulate resistance against infectious diseases, which could be used as biological weapons, is discussed. The possibility of inhalation administration of cytokines to ameliorate aerosol spread of a bioterrorist agent is also discussed.

The next chapter by biomedical scientists from the University of Pennsylvania and University of South Florida discusses host responses relevant to vaccine-induced immunity and therapeutic strategies to treat bioterrorist agents, especially viruses. Dr. Chris Coté and colleagues from Frederick, MD, then discuss in detail Bacillus anthracis as a possible bioterrorist agent, especially pathogenicity and infection caused by this organism. Drs. Richard Titball and Steven Michell from the United Kingdom then describe pathogenesis and immunity to Tularemia, an organism considered a high priority as a possible bioterrorist agent. Drs. Valderas and Roop from E. Carolina University then discuss Brucella as a possible bioterrorism agent, including pathogenicity and immunity involved in infection by this organism.

Drs. Joseph Hinnebush and David Ericson from Hamilton, MT, discuss Pasteurella pestis, an important organism causing pneumonic plague, as a possible bioterrorist agent. Drs. J. Miller, E. Shaw, and H. Thompson from the Centers for Disease Control in Atlanta discuss Q fever and Coxciella burnetti as possible bioterrorist agents. Dr. Laura Hendrix and colleagues from University of Texas at College Station discuss further Q fever as a possible bioterrorist agent from the viewpoint of genomic and proteomic approaches. Drs. David Walker and D. Bouyer from the University of Texas Medical Branch in Galveston discuss Spotted Fever Rickettsia as a potential bioterrorist agent.

Because it is assumed that bioterrorist microbes may cause epidemic disease and panic among civilian populations during a possible terrorist attack or even warfare, there is widespread concern on this issue by various governments and organizations. Thus it is widely acknowledged and recognized that much emphasis is now being focused on laboratory and clinical studies concerning basic research as well as clinical and epidemiologic investigations about microbes that can be considered possible bioterrorist weapons. Although there have been several documented probable bioterrorist attacks in the United States in the past decade, these are considered to be due to individuals rather than governments or terrorist organizations. Nevertheless, there is now much concern about the possibility that microbial agents will be used in attacks against civilian populations, or the environment, by terrorists. The editors of this volume, as well as the authors of individual chapters, are encouraged by recent advances and new knowledge about microorganisms that have been till now considered mainly esoteric until the threat of bioterrorism became a national concern and encouraged new and precise information being accumulated.

There is little doubt that the rapidly emerging interest in understanding the nature and pathogenic mechanisms of agents that may be used as bioweapons provides increasing impetus for further research of the infectious diseases that they may induce. This has resulted in further detailed understanding of the nature and mechanisms whereby opportunistic pathogens that may be used as bioweapons cause disease. The chapters in this book by experts in the field will be of value, the editors believe, for laboratory investigators and biomedical scientists in general, as well as for clinicians and health professionals.

The editors wish to express their gratitude to Ms. Ilona Friedman who served as an outstanding editorial assistant for this volume as well as for all the books in this series.

Burt Anderson
Herman Friedman
Mauro Bendinelli

Contents

1

Biotechnology and the Public Health Response to Bioterrorism

ANDREW CANNONS, PHILIP AMUSO, and BURT ANDERSON

1. INTRODUCTION

The tragic events of September 11, 2001 have forced a reconsideration of international security. The role of relatively small, but well organized, terrorists groups in inflicting mass casualties must be considered a real threat. In October 2001, *Bacillus anthracis* spores were intentionally released through the mailing of contaminated letters distributed through the United States postal service. Again, apparently small groups, or perhaps an individual, were able to utilize a biological agent as a weapon to cause death and affect relatively large areas of the United States, including the District of Columbia, Florida, Connecticut, New Jersey, and New York.[1] Furthermore, the resulting hysteria and public response has focused recent attention on the use of biological agents as potential weapons.

The use of the US postal system as an efficient "artificial mechanical vector" has served to remind us that the dispersion and transmission of intentional infections may share few similarities with naturally acquired infections caused by the same agent. Characterization of the intentionally released *B. anthracis* spores revealed that they were related to the Ames strain—a relatively well-studied laboratory strain. Much media attention has been focused on the use

ANDREW CANNONS and PHILIP AMUSO • Center for Biological Defense, University of South Florida Tampa, FL 33612. BURT ANDERSON • Department of Medical Microbiology and Immunology, College of Medicine, University of South Florida, Tampa, FL 33612.

of agents other than *B. anthracis* that might be used as biological weapons. The identification of such agents and the establishment of the ABC priority category list is largely based on a historical perspective of offensive biological weapons programs dating to the cold war and earlier.[2,3] The impact of molecular biology, gene cloning, and biotechnology on not only the control and prevention, but also the development of biological weapons must remain clearly in focus.

The entire genome sequence for at least one representative strain for all category A agents has been determined. Most of these sequences have already been deposited in public access databases. The opportunity to use this information to enhance the detection, surveillance, and diagnosis of infections caused by bioterrorism (BT) agents is great. Conversely, the potential for misuse to engineer pathogens for intentional release must also be considered. Such open public access to genome sequences and other scientific reports describing details about mechanisms of pathogenesis, antibiotic resistance, etc. raises questions about use, misuse, and the dissemination of such information.

This chapter will describe the public health infrastructure that has been put in place to respond in the event of acts of bioterrorism as well as the role of biotechnology in the control and prevention of infections caused by the intentional release of agents of bioterrorism. Specifically, we describe how the laboratory response network (LRN) successfully responded to the index case of anthrax in Florida. We also describe the overall operational structure of the LRN and its role to detect, monitor, and characterize microbial agents that may have been used as weapons. In addition, we have summarized the potential for the misuse of biotechnology for the development of future biological weapons. We have also attempted to describe how the concern for such misuse may have a profound impact on how information is exchanged within the scientific community and among public health officials. Finally, we summarize the biodefense research funding trends and the reaction to these trends among the scientific community, as well as public concern about possible risks associated with the establishment of both regional and national biocontainment laboratory facilities.

2. US PUBLIC HEALTH SERVICE INFRASTRUCTURE—THE LABORATORY RESPONSE NETWORK

On October 4, 2001, an index case of inhalational anthrax that occurred in Boca Raton, Florida, was confirmed when a Florida LRN reference laboratory discovered the presence of *B. anthracis* in a clinical specimen. This discovery triggered a cascade of events that nearly crippled the public health and emergency response systems throughout the United States. "White powder" instantly became a household phrase synonymous with "*Bacillus anthracis* spores", causing fear and panic throughout the nation.

Although LRN personnel had planned and trained for the "real thing" for nearly two years, they were unable to predict just how the system would respond to an actual BT event. Few responsible officials had predicted that a mere handful of intentionally inflicted anthrax cases would bring the public health system to its knees under a deluge of "suspicious" samples coming from a terrified public. Across the nation, LRN laboratories tested 125,000 samples by the time the investigation was completed. This amounted to more than 1 million separate tests. Law enforcement units, HazMat units, fire rescue units, and public health personnel were occupied around the clock for over a month.

The LRN was established in 1999 when the US Congress appropriated funds to address the threat of terrorism. Previously, Congress passed the Response to Weapons of Mass Destruction Act of 1997 (commonly known as the Nunn-Lugar-Domenici legislation), which focused on providing training and equipment for emergency first-responder personnel. The 120 largest US cities were specified in this legislation. The Department of Health and Human Services (DHHS) sponsors community-level medical and public health preparedness and response associated with emergencies resulting from terrorist incidents. DHHS, through the Office of Emergency Preparedness (OEP), lead a national effort to develop Metropolitan Medical Response Systems (MMRS) to enhance the local ability to respond to the health and medical consequences of a terrorist incident.

DHHS, through the Centers for Disease Control and Prevention (CDC) program announcement 99051: "public health preparedness and response for bioterrorism", initiated an effort to upgrade national public health capability to counter bioterrorism. Toward this effort, CDC initiated this cooperative agreement program for States, US Territories, and major local public health departments to help upgrade their capabilities. The LRN is a multi-level system designed to link front-line clinical microbiology laboratories in hospitals and other institutions, to state and local public health laboratories, and supporting advanced capacity public health, military, veterinary, agricultural, water and food testing laboratories at the federal level. LRN testing is performed according to consensus protocols using reagents provided by the CDC, exclusively available to LRN reference and national laboratories (see Fig. 1).

2.1. Threat Agents by Category

The CDC has divided the threat agents into three categories: A, B, and C, based on their potential for dissemination and/or for person to person transmission, the expected mortality/morbidity, and their potential to cause public panic and social disruption.

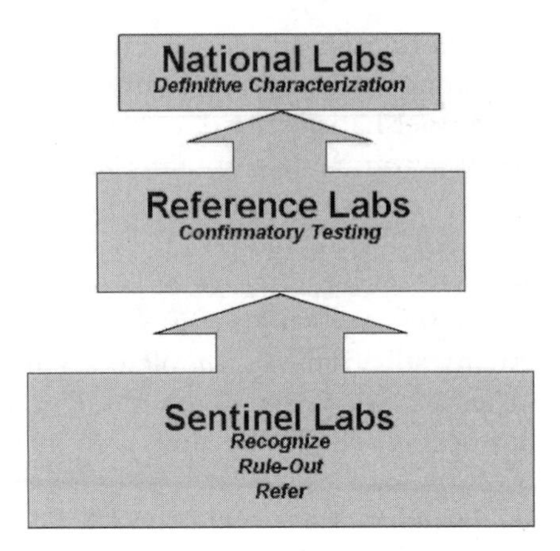

FIGURE 1. The CDC Laboratory Response Network: A multi-level system linking clinical, public health and national labs.

Category A

- Can be easily disseminated or transmitted person to person
- Cause high mortality, with the potential for major public health impact
- Might cause public panic and social disruption, and
- Require special action for public health preparedness

Category A agents include:

- *Bacillus anthracis* (anthrax)
- *Yersinia pestis* (plague)
- *Francisella tularensis* (tularemia)
- Variola major (small pox)
- *Clostridium botulinum* toxin
- Hemorrhagic fever viruses

Category B

- Moderately easy to disseminate
- Cause moderate morbidity and low mortality
- Require specific enhancements of CDC's diagnostic capacity and enhanced disease surveillance

Category B agents include:

- *Coxiella burnetti* (Q fever)
- *Burkholderia mallei* (glanders)
- *Staphylococcus* enterotoxin B
- Epsilon toxin of *Clostridium perfringens*

- Ricin toxin (from castor beans)
- *Brucella* species (brucellosis)
- Encephalitis viruses

Category C

This could be engineered for mass dissemination in the future because of:

- Availability
- Ease of production and dissemination; and
- Potential for high morbidity and mortality and major health impact

Category C agents include:

- Nipah virus (encephalitis with high mortality)
- Hantavirus (pulmonary syndrome)
- Yellow fever virus
- Multidrug-resistant tuberculosis

3. DETECTION, MONITORING, AND IDENTIFICATION OF BT AGENTS

Level A LRN protocols are used by sentinel laboratories (formerly known as level A laboratories) and are available in the public domain via the Intranet at www.bt.cdc.gov. These protocols consist of "rule out" procedures used by hospital laboratories and commercial laboratories, to screen for category A bioterrorism threat agents. If a sentinel laboratory is unable to rule out a threat agent the specimen is immediately transferred to an LRN reference laboratory for further analysis. Since these protocols are subject to change by CDC at any time and without notice, it would be pointless to include them in this chapter. Rather, we recommend that the readers frequently check the web site (www.bt.cdc.gov) for updates.

In stark contrast, LRN reference laboratories (formerly known as level B and level C laboratories) use protocols that are not available in the public domain. Instead, these protocols are available at a secure password-protected Intranet web site accessible only by LRN reference laboratory members. For security reasons we are not allowed to provide specific, detailed information about the LRN reference protocols. However, we can say that they include the use of advanced molecular biology methods in conjunction with the reagents mentioned previously, which are available only to LRN reference laboratories.

The detection of pathogenic microorganisms in clinical and environmental samples requires essentially the same parameters. These include, ease of use, rapidity, high sensitivity, high specificity, and the lowest possible cost.

Additional desirable attributes are multiplexing, to detect more than one organism, and the ability to discriminate between naturally occurring pathogens, pathogens resulting from a BT incident, and hoaxes. Currently, no system in use incorporates all these parameters. Public health laboratories currently rely on both culture and molecular procedures for the detection of BT pathogens. Samples for analysis can be powders used to intentionally deliver agents, such as talc, cornstarch, and flour; soil; air; clinical (tissues and body fluids); water; and food. Reference laboratories must be able to process all of these matrices. Culture, still considered to be the gold standard for testing, has the advantages of being able to detect more than one agent at a time, does not require huge investments in instrumentation, high sensitivity, and initial processing can be rapid.[4] However, growth in culture can be slow, requiring 24–72 hours of incubation and further analysis must be conducted by trained microbiologists. On the other hand, nucleic-acid-based detection systems have great potential since they are sensitive, specific, and rapid. The polymerase chain reaction (PCR) assay is widely used by clinical laboratories for the identification of infectious disease agents.[5] Real-time PCR, utilizing patented chemistries (e.g., Applied Biosystems Taqman®, Roche LightCycler®), improves the time of detection to hours compared to days. However, most PCR-based detection systems are limited since they can only target one agent in a single run, the primers and probes for some of the BT agents may not be available, the equipment and instrumentation can be expensive, and interpretation of data may require a very high level of technical acumen. Additionally, sample processing for recovery of PCR-competent nucleic acid can be extensive for some complex matrices such as environmental powders and food.

Powders and environmental samples are the most common non-clinical samples submitted to LRN reference laboratories for analysis. Processing of these samples for real time PCR analysis for *B. anthracis* can be difficult since the powder may become aerosolized, thus requiring biosafety level 3 conditions for safe manipulation. Additionally, the bacterial DNA has to be extracted from what could be a complex and PCR inhibitory matrix. Methods have been developed that utilize sonication[6] or autoclaving[7] for DNA extraction. A procedure that incorporates sonication, germination, and autoclaving has been published, providing a safe and rapid method (6 hours) for sample preparation and PCR detection of greater than 10 spores of *B. anthracis* in various powders.[8] Should the reference laboratories be deluged with samples for testing, procedures such as this could help alleviate bottlenecks and decrease sample processing times.

Food is also a carrier that could be used to purposely distribute pathogens. Enteric pathogens have been used as BT agents, including the use of *Salmonella enterica* in salad bars[9] and the use of *Shigella* contaminated pastries to sicken co-workers.[10] Conventional methods to identify bacterial pathogens in foods are time-consuming and laborious. Real-time PCR detection offers a rapid alternative,[11] although not without problems. The major challenge comes with sample preparation and extraction of the pathogen DNA from the food. There are many different food matrices that need to be considered, some of which can interfere with extraction and/or inhibit the nucleic acid amplification process (http://www.cfsan.fda.gov/~ebam/bam-1.html). Strategies for template preparation directly from food have been evaluated[12] with detection limits of 10^3CFU/g of food. Increased sensitivity is achieved with initial enrichment, whereby even five hours growth can increase the detections limits of pathogens 10–100 fold.

The food emergency response network (FERN) is a counter- terrorism program formed by the Food and Drug Administration (FDA) and modeled after the LRN. This is a network of state and federal laboratories committed to analyzing food samples in the event of a biological, chemical, or radiological terrorist attack. FERN is developing protocols to be used by state and federal laboratories for detecting pathogens in food samples. Again, these protocols will not be available to the public domain and will be maintained on a secure password-protected web site.

Emergency responders need pathogen detection systems that will yield results that meet the requirements of laboratory-based detection systems, as well as the requirements that they be rugged and mobile. Currently, the ideal system is not available. Most field-use systems rely on immunological detection methods such as lateral flow assays. These assays are generally rapid to use, taking approximately 15 minutes, require minimal sample processing and only limited technical expertise. However, immunoassays are typically 1000-fold less sensitive than nucleic acid amplification and culture procedures, with detection limits in the range of 10^5 CFU/ml.[13] In addition, specificity can be limited. For example, close relatives of *B. anthracis,* which are common contaminants of the environment, can be picked up using some *B. anthracis* specific assays.[13] Low sensitivity and specificity lead to false negatives and false positives, respectively, and can result in an incorrect emergency response. PCR-based systems for emergency responders are also being developed. These include instrumentation such as the Handheld Advanced Nucleic Acid Analyzer (HANAA), a fully automated sample processor that uses TaqMan®-based PCR for the detection and identification of BT agents and the Ruggedized Advanced Pathogen Identification Device (RAPID®), which is a field-deployable thermal cycler using LightCycler PCR technology.[14] However, use of PCR in the field is currently limited because of its complexity, requirement for sophisticated and sensitive equipment, and the need for trained personal for operation and interpretation of results.

Presently, all BT samples collected by emergency responders are transported to an LRN reference laboratory for characterization and evaluation using LRN protocols. The use of accurate field detection instruments would improve triage activities and consequently reduce the workload for reference laboratories. However, this can only happen when a suitable and well-characterized field detection system is available.

Continual monitoring can be an effective method for early detection and alleviation of a BT event and thus can supplement and complement the work of the LRN and emergency responders. This monitoring can be both environmental and/or epidemiological, including surveillance of syndromic data.[15] Environmental monitoring can be important for pathogen detection immediately after an event/exposure but before symptoms are recognized. As with pathogen detection in the lab, environmental surveillance requires the development of specialized equipment. Ideally, these monitoring systems should be automated and be able to discriminate between naturally occurring agents and those used in a BT event. Presently, few such surveillance systems exist, such as the Biohazard Detection System (BDS), a PCR-based technology developed by Cepheid[16] and employed in the US postal system; the Biological Aerosol Sentry and Information System (BASIS), an environmental monitoring system used during the 2002 Winter Olympics utilizing sampling stations to collect aerosol samples, which were then brought to a laboratory for pathogen analysis;[15] the nationwide BioWatch program that has been implemented to monitor the air in several US cities,[15] and was modeled on BASIS; and the Autonomous Pathogen Detection System (APDS), an automatic aerosol monitoring system for areas of high risk.[14] It is likely that additional technology will be developed over the next few years that will increase and improve our capacity for environmental monitoring of pathogenic microorganisms. The progress in microarray technology and integration of automation will increase the accuracy and variety of agent detection, while reducing analysis time and reagent consumption.

4. THE POTENTIAL FOR MISUSE OF BIOTECHOLOGY

The biotechnology revolution has generated a range of methodologies that have been exploited for the development of diagnostics, antimicrobials and vaccines. In particular, the availability of full genome sequences for most recognized agents of bioterrorism has provided the tools to design diagnostic reagents as well as new generation vaccines. Clearly, such information has allowed for the improvement of protocols used by the LRN and enabled many areas of both basic and applied research on these agents. Many different approaches have been used for the development of recombinant vaccines for anthrax (see Chapter 6) with some impressive successes. The ability to mine the *B. anthracis* genome for additional genes that may ultimately constitute new generation vaccines will likely result in the development of additional vaccine strategies based on biotechnology.

Since the first gene cloning experiments approximately 30 years ago, countless molecular biology tools and reagents have been developed. These techniques allow us to detect, amplify, and express virtually any gene in a matter of days. Such techniques have rapidly accelerated our ability to study the virulence of bacterial and viral pathogens and have even resulted in the identification of many new agents of disease. Ultimately, the long-term goal resulting from the application of modern molecular biology techniques to infectious disease research is to improve diagnosis and prevention of the diseases that they cause. However, just as these techniques have proven invaluable in the control and prevention of infectious disease, the same powerful techniques could be used for the development of enhanced biological weapons. Creating chimeric pathogens with increased virulence, multidrug resistance or strains that are not targeted by standard vaccines, are now possible.(17) Even modestly competent molecular biology laboratories have individuals who may be able to modify many of the "traditional agents" associated with biological warfare.

The possibility of using biotechnology to engineer a new class of agents, termed advance biological warfare (ABW) agents, must also be considered. In addition, biotechnology may have applications supporting weaponization, dissemination, and delivery as well.(18) Thus, the genetic modification of "traditional agents" associated with biological warfare must be viewed as only a part of the picture. The form of such ABW agents can only be the subject of conjecture. It has been suggested that ABW agents could be engineered to target specific human biological systems such as the cardiovascular, gastrointestinal, neurological, or immunological systems.(18) Transgenic plants and animals could be engineered to produce large quantities of bioregulatory or toxin proteins. Transgenic insects, such as bees, wasps, or mosquitoes, could be developed to produce and deliver biological toxins. For instance, a mosquito could be genetically altered to produce and secrete a biological toxin into its saliva. This same mosquito would then serve as the vector to deliver the toxin during its feeding process. Despite the potential to produce toxins that might be effective at low doses and deliver these toxins, such transgenic insects would likely go unnoticed. Many of the counterproliferation, detection, and medical countermeasures that have been developed for "traditional agents" will be ineffective for ABW agents such as protein-based transgenics.(18) Five important attributes of a biological warfare (BW) agent have been described:(19)

- High virulence coupled with high host specificity
- High degree of controllability
- Lack of timely countermeasures to the attacked population
- Ability to camouflage the BW agent with relative ease
- High degree of resistance to adverse environmental forces

Daly notes that of these five, the last attribute is the most difficult to genetically engineer into an organism. Accordingly, he suggests that it may be simpler to engineer BW attributes into organisms that are naturally resistant to environmental forces. This raises the possibility of engineering extremophiles for use as BT agents. Among these are agents that are resistant to environmental factors including desiccation, ultraviolet radiation, high temperatures and pressures, or decontamination compounds. *Deinococcus radiodurans* is one such bacterium that is resistant to multiple environmental factors including radiation, as the species designation implies. This Gram-positive coccus displays remarkable resistance to even Megarad doses of radiation. Furthermore, availability of the complete *D. radiodurans* genome sequence, together with genetic systems to express foreign genes, makes this bacterium amenable to genetic engineering. Daly has suggested that it may be possible to synthesize and store viruses within microorganisms such as *D. radiodurans.* The use of extremophiles such as *D. radiodurans* may seem to be a complicated approach to the development of BT agents. Furthermore, the ability of such agents to establish efficient infections in the human population may, in itself, require extensive genetic engineering. Despite these facts the use of extremophiles as BT agents should be considered as a threat that must not be ignored.

5. BIOTECHNOLOGY, PUBLIC HEALTH INTEREST AND THE EXCHANGE OF SCIENTIFIC INFORMATION

The potential to misuse the so-called dual-use technology (civil and military) is clearly illustrated in genomics research.(20) Genome sequences are available for most recognized BT agents and countless numbers of bacterial and viral pathogens. Although this information has undoubtedly enhanced our ability to detect, prevent, and treat infections caused by BT agents, experts in the field of genomics recognize the possibility of "tailoring" classical BW agents to make them harder to detect, diagnose, and treat.(21) How then can we make pathogen genome sequences, or for that matter scientific information related to possible BT agents, available for valid scientific research? What types of research constitute studies that should not be published and accessible to all? These are some of the questions that must be addressed to ensure that biodefense research findings accessible to the public domain are not used for nefarious purposes.(22)

In a study from 2001, it was shown the recombinant mousepox virus that expresses IL-4 suppressed natural killer cell and cytotoxic T-lymphocyte responses.(23) The natural genetic resistance of some mice strains to mousepox virus was overcome. More importantly, recently immunized genetically resistant mice were shown to be susceptible to infection with this virus. Thus, not only does the virus-encoded IL-4 suppress primary antiviral cell-mediated immune responses, but it also can inhibit the expression of immune memory responses. The finding that a poxvirus can circumvent immune memory has potential implications for smallpox vaccination efforts, as well as providing a tool that could be misused to manipulate the immune system. Not surprisingly, this publication created quite a topic for discussion among both scientists and government officials about the potential for misuse of information in such publications. In another study, infectious poliovirus was generated from a synthetic cDNA template, showing that it is possible to synthesize infectious agents *in vitro* from genome sequence data.[24] Again, this report sounded an alarm and raised questions from the public and some in the scientific community, about the possibility of synthesizing larger and more complicated viruses such as HIV, Ebola or even smallpox from genome sequence information. The Patriot Act (2001) and the Public Health Security and Bioterrorism Preparedness and Response Act of 2002 imposed new regulations on the conduct of research involving select agents. Concern has been raised in the scientific community that such regulations will have a "chilling effect upon legitimate scientific inquiry".[25] The need for scientific self-governance has been suggested and the essential components of such a system have been proposed.[25]

5.1. Public Perception of Biodefense Research

Over the next decade, the Unites States will spend billions of dollars to develop countermeasures against biological and chemical weapons.[26] In the past, most of this type of research was conducted at a few government facilities. However, it is likely that much of the research conducted in the coming decade will be performed in academic settings. Advocacy groups such as the Sunshine Project of Austin, Texas (http://www.sunshine-project.org) and the Council on Responsible Genetics in Boston, Mass (http://gene-watch.org), are opposed to high-security containment laboratories and have taken to monitoring compliance issues at academic research facilities. Citizen and community opposition to maximum containment biosafety level 4 facilities has been encountered in response to proposed facilities in both Boston[27] and Montana.[28] Recently, three laboratory workers at a Boston University laboratory apparently contracted tularemia from mishandling a virulent strain of *Francisella tularensis.*[29] This further raised community concern about biosafety in the Boston University laboratories, which was recently awarded the National Institutes of Health funding to build a National Biocontainment Biosafety Level 4 facility.[30] Again, public concern and a lawsuit by local residents have resulted.

In addition to biosafety concerns in the public and scientific community, a backlash reaction to the large spending increases in federal research dollars on biodefense has recently surfaced. In a recent letter to the Director of NIH, over 700 microbiologists indicated their concern about the redirection of NIH grant funds from other projects to biodefense-related projects (accessible at: http://waksman.rutgers.edu/NIH-MBC_BM/current/). In that letter it was pointed out that funding for prioritized BW agents has increased by 1500% from 1996 to 2005. In contrast, research on pathogens that are non-biodefense associated has decreased by 27%. It was suggested that the greatest threat to public health are existing and emerging infectious diseases. Basic research on many of these agents is currently under funded in comparison to the prioritized BW agents. Given the large number of cases, as well as the morbidity and mortality associated with some of the organisms that are not identified as priority agents, as compared to those that are, it is difficult to refute this argument.

SUMMARY

The events of September 11, 2001, as well as the subsequent intentional release of anthrax spores in contaminated letters, have changed the way both the public and the scientific community view the use of biotechnology in biodefense research. The enhancement of our public health infrastructure has greatly aided our ability to respond to potential future acts of bioterrorism, and to a certain extent, emerging and re-emerging infectious diseases The Laboratory Response Network is an example of infrastructure that was already in place prior to September 11, 2001, but that has been strengthened considerably after that date. The possibility of governmental control over access to scientific information that might be considered dual use remains an active topic for debate. Backlash, from the public due primarily to safety concerns and among scientists over concerns about prioritization of biodefense projects, are more recent developments related to the government funding for biodefense. The need to balance our obligation to respond to acts of bioterrorism, along with all other public health threats, will undoubtedly continue to be a matter of contention for several years to come.

REFERENCES

1. Jernigan, J. A., and Stephens, D. S., *et al.*, 2001, Bioterrorism-related inhalational anthrax: the first 10 cases reported in the United States, *Emerg. Infect. Dis.* **7**:933–944.
2. Christopher, G. W., and Cieslak, T. J., *et al.*, 1997, Biological warfare. A historical perspective, *JAMA.* **278**:412–417.
3. Marty, A., 2001, History of the development and use of biological weapons. Laboratory aspects of biowarfare., *Clinics Lab. Med.* **21**:421–434.
4. Peruski, L. F., and Peruski, A. H., 2003, Rapid diagnostic assays in the genomic biology era: detection and identification of infectious disease and biological weapon agents, *Biotechniques.* **35**:840–46.
5. Whelan, A. C., and Persing, D. H., 1996, The role of nucleic acid amplification and detection in the clinical microbiology laboratory, *Ann. Rev. Microbiol.* **50**:349–73.
6. Belgrader, P., and Hansford, D., 1999, A minisonicator to rapidly disrupt bacterial spores for DNA analysis, *Anal. Chem.* **71**:4232–4236.
7. Dang, J. L., Heroux., K., 2001, Bacillus spore inactivation methods affect detection assays, *Appl. Environ. Microbiol.* **67**:3665–3670.
8. Luna, V. A., and King, D., 2003, Novel sample preparation method for safe and rapid detection of *Bacillus anthracis* spores in environmental powders and nasal swabs, *J. Clin. Micro.* **41**:1252–1255.
9. Torok, T. J., and Tauxe, R. V., 1997, A large community outbreak of *salmonellosis* caused by intentional contamination of restaurant salad bars, *JAMA.* **278**:389–395.
10. Kolavic, S. A., and Kimura, A., 1997, An outbreak of *Shigella dysenteriae* type-2 among laboratory workers due to intentional food contamination, *JAMA.* **278**:396–398.

11. Jordan, J. A., 2000, Real-time detection of PCR products and microbiology, *New technologies for life sciences: a trends guide.* **6**:61–66.
12. Heller, L. C., and Davis, C. R., 2003, Comparison of methods for DNA isolation from food samples for detection of shiga toxin-producing *Escherichia coli* by real-time PCR, *App. Environ. Micro.* **69**:1844–1846.
13. King, D., and Luna, V., 2003, Performance assessment of three commercial assays for direct detection of *Bacillus anthracis* spores, *J. Clin. Micro.* **41**:3454–3455.
14. Ivnitski, D., O'Neil, D. J. O., 2003, Nucleic acid approaches for detection and identification of biological warfare and infectious disease agents, *BioTechniques.* **35**:862–869.
15. Fitch, J. P., and Raber, E., *et al.*, 2003, Technology challenges in responding to biological or chemical attacks in the civilian sector, *Science,* **302**:1350–1354.
16. Meehan, P. J., and Rosenstein, N. E., 2004, Responding to detection of aerosolized *Bacillus anthracis* by autonomous detection systems in the workplace, *MMWR.* **53**:1–12.
17. Greenfield, R. A., Lutz, B. D., Huycke, M. M., and Gilmore, M. S., 2002, Unconventional biological threats and the molecular biological response to biological threats, *Am. J. Med. Sci.* **323**:350–357.
18. Petro, J. B., Plasse, T. R., and McNulty, J. A., 2003, Biotechnology: impact on biological warfare and biodefense, *Biosecurity and bioterrorism: biodefense strategy, practice and science.* **1**:161–168.
19. Daly, M. J., 2001, The emerging impact of genomics on the development of biological weapons–threats and benefits posed by engineered extremophiles, *Clinics Lab. Med.* **21**:619–629.
20. Black, J. L., 2003, Genome projects and gene therapy: gateways to next generation biological weapons, *Military Med.* **168**:864–871.
21. Fraser, C. M., and Dando, M. R., 2001, Genomics and future biological weapons: the need for preventive action by the biomedical community, *Nature Genetics.* **29**:253–256.
22. Atlas, R. M., 2002, National security and the biological research community, *Science.* **298:**753–754.
23. Jackson, R. J., Ramsay, A. J., *et al.*, 2001, Expression of mouse interleukin-4 by a recombinant ectromelia virus suppresses cytolytic lymphocyte responses and overcomes genetic resistance to mousepox, *J. Virol.* **75**:1205–1209.
24. Cello, J., Paul, A. V., and Wimmer, E., 2002, Chemical synthesis of poliovirus cDNA: generation of infectious virus in the absence of natural template, *Science.* **297**:1016–1018.
25. Kwik, G., Fitzgerald, J., Inglesby, T. V., and O'Toole, T., 2003, Biosecurity: responsible stewardship of bioscience in an age of catastrophic terrorism. *Biosecurity and bioterrorism: biodefense strategy, practice and science.* **1**:27–35.
26. Kahn, L., 2004, Biodefense research: can secrecy and safety co-exist? *Biosecurity and bioterrorism: biodefense strategy, practice and science.* **2**:81–85.
27. Lawler, A., 2004, Boston weighs a ban on biodefense studies, *Science (News).* **304**:665.
28. Kaiser, J., 2004, Citizens sue to block Montana biodefense lab., *Science (News).* **305**:1088.
29. Dalton, R., 2005, Infections scare inflames fight against biodefense network, *Nature.* **433**:344.
30. Lawler, A., 2005, Boston University under fire for pathogen mishap, *Science.* **307**:501.

2

Historical Perspectives of Microbial Bioterrorism

STEPHEN A. MORSE

1. INTRODUCTION

A number of events over the last decade have served to focus attention on the threat of terrorism and the use of biological or chemical weapons against military and civilian populations for the purpose of causing illness or death. It is increasingly recognized that agricultural animals and plants also present a vulnerable target to terrorists.[1,2] Most significantly, the threat of terrorism has attracted the attention of policy makers in all levels of government in the United States. However, policy makers and analysts have differed in their assessment of the threat of bioterrorism. Many authorities believed that the threat of bioterrorism was growing, particularly from non-state sponsored groups.[3] Some of them contended that it was only a matter of time before a terrorist used biologic agents to cause mass casualties, while others argued that the historical record provided no basis for concern. Moreover, some even questioned the wisdom of funding preparedness efforts.[4] However, the situation changed in October 2001 when an individual or individuals sent spores of *Bacillus anthracis* to media companies in New York City and Boca Raton, Florida[2] resulting in five deaths and considerable panic throughout the country.

1.1. Definitions

For the purposes of this article, the working definition of a biological agent is "a microorganism (or a toxin derived from it) which causes disease in man, plants, or animals or causes deterioration of material."[6] In this context, the biological agents are normally divided into three categories: anti-personnel,

STEPHEN A. MORSE • Centers for Disease Control and Prevention, Atlanta, GA, 30333.

anti-animal, and anti-plant. In addition, the use of biological agents is often classified by the manner in which they are used. For example: *biological warfare* has been defined as a specialized type of warfare conducted by a government against a target; *bioterrorism* has been defined as the threat or use of biological agents (or toxins) by individuals or groups motivated by political, religious, ecological, or other ideological objectives.[7] Terrorists can be distinguished from other types of criminals by their motivation and objective; criminals may also be driven by psychological pathologies and may use biological agents. When criminals use biological agents for murder, extortion, or revenge it is called a *biocrime.*[7]

1.2. Development and Prohibition of Biological Weapons

In November 1918, an armistice ended World War I in which eight million soldiers and nearly as many civilians were killed. However, that armistice could not halt the even greater ravages of an influenza pandemic. In the course of a single year beginning in the spring of 1918, the virus spread globally killing more than 20 million people. No one thought that this influenza pandemic was a deliberate act of war; however, the magnitude of the impact of this epidemic apparently impressed the statesmen of the era.[8] When the Geneva Protocol was issued in 1925 to ban, in warfare, the use of asphyxiating, poisonous, or other gases, which had been responsible for about one million casualties during World War I,[6] the provision was extended to include bacteriological agents as well.[9] The Geneva Protocol affirmed that chemical and biological weapons were "justly condemned by the general opinion of the civilized world."[8]

In 1972, the convention on the prohibition of the development, production, and stockpiling of bacteriological (biological) and toxin weapons and on their destruction (referred to as the Biological Weapons Convention or BWC) was opened for signature. Since it entered into force in 1975, the BWC has been signed and ratified by 141 countries, signed but not ratified by 18 countries, and observed by the Government of Taiwan. The BWC prohibits the development, production, stockpiling, or acquisition of microbial or other biological agents or toxins of types and in quantities that have no justification for prophylactic, protective, or other peaceful purposes.[9] The BWC also prohibits the weapons, equipment, or means of delivery designed to use such agents or toxins for hostile purposes or in armed conflict.[9] In addition, it requires that each State Party (i.e., those countries that both signed and ratified the BWC) destroy, or divert to peaceful purposes, all agents, toxins, weapons, equipment, and means of delivery which are in its possession or under its jurisdiction or control. Each State Party also agreed not to transfer any of the agents, toxins, weapons, equipment, or means of delivery to any recipient, or induce any State to manufacture or otherwise acquire such organisms or equipment for non-peaceful purposes.[9] Unfortunately, the BWC has no verification provisions, and there have been significant difficulties in determining the existence or status of State programs.

The use of biological agents for the purpose of warfare has been associated with State programs for the development of biological weapons. Some of these programs were very large, employing thousands of people. Leitenberg[10] has recently reviewed the activities of several of these programs. Although there are inherent differences between terrorist and criminal use of biological agents, the criminal faces many of the same obstacles as the terrorist. Both must acquire, develop, and employ biological weapons, so the technical constraints will probably be similar.

The purpose of this chapter will be to provide a historical account of the use of biological agents for warfare, terrorism, or criminal purposes. However, in doing so one must appreciate that there are problems in assessing the historical use of biological agents because of: (1) difficulties in verifying an alleged or attempted biological attack; (2) the use of allegations of biological attacks for propaganda purposes; (3) a lack of pertinent microbiological or epidemiological data; (4) the presence of naturally occurring endemic or epidemic diseases during hostilities; and (5) the secrecy surrounding biological weapons programs.[11] This review relies heavily on the extensive research of Carus,[7] who identified most of the events and attempts to use biological agents since 1900. This review will not cover hoaxes or attempts to acquire biological agents for nebulous purposes (e.g., Larry Wayne Harris incident[7]).

2. EARLY USE OF BIOLOGICAL AGENTS IN WARFARE

An examination of the way in which biological agents were used for warfare in the fourteenth through the middle of the nineteenth centuries demonstrates a correlation with the prevailing theories of infectious diseases.

2.1. Early Theories of Infectious Disease

A number of theories have evolved over time to explain the origin of epidemics. Two of these have been referred to as *miasma* and *contagion.* In the theory of miasma, epidemics were thought to be due to an atmospheric poison generated in external nature.[12] Disease was thought to be a consequence of "bad air" resulting from extensive decomposition. An example of how this theory might have played a role occurred at the siege of Thun l'eveque (Table I) where catapults were used to hurl dead horses over the wall into the castle. Years later and based on eyewitness testimony, the chronicler Froissart wrote that the "stynke and ayre was so abomydable, that they considered howe that finally they coude nat long endure" (as cited in ref 13). Miasma originated from the first-hand observations of Hippocrates (460–360 B.C.) of the effects of climate, season, and locality on outbreaks of disease.[12] However, it was Galen

(130–200 A.D.) who developed the idea of "miasmic corruption of the air", which figured prominently in the way medicine viewed epidemics in the Middle Ages. Galen thought that epidemic disease resulted from "the inhalation of air fouled by putrid exhalations that might come from sources far and near, such as masses of unburied corpses of the slain in battle, or from swamps and stagnant water in summertime, or from the excessive heat of foul air in close and in unventilated hovels".[(12)] By the fourteenth century, the idea that the immediate cause of epidemics was some sort of corruption in the air was widely accepted. It was believed that this corrupted air could gain entrance to the body by way of the lungs or through wide-open pores in the skin as a result of excesses, bathing, or heat.

Also in the fourteenth century, additional prominence was given to the idea of contagion. In the theory of contagion, the "poison" was originally generated in man himself and spread person-to-person by contact with the sick or dead, or with their personal effects (fomites). The fact that some diseases were infectious and could be transmitted person-to-person had been long recognized. There are numerous biblical references in the Books of Leviticus, Numbers, and Deuteronomy to the "infectivity" of lepers. Thucydides thought that the "plague" of Athens in 430 B.C. was exceedingly infectious.[(12)] In the fourteenth century, smallpox and measles were added to the list of infectious diseases.[(12)]

In the sixteenth century, Girolamo Fracastoro (1478–1553) recognized and defined three kinds of contagion. In his treatise *De contagione et contagionis morbis et eorum curatione*, Fracastoro attributed certain diseases including plague, smallpox, and measles, to specific tiny seeds (*seminaria*) and stated that these specific contagions could be spread directly from person-to-person, indirectly via infected clothing, wooden objects, or other fomites, and even at a distance, such as through the air.[(14)]

At various times, both theories were combined. Some adherents to the theory of contagion argued that the sick could radiate infection through the air in their immediate vicinity (a local miasma), while some adherents to the theory of miasma admitted a limited degree of contagion at the peak of a severe epidemic.

2.2. Selected Incidents from the Fourteenth to the Nineteenth Centuries

Table I lists a number of substantiated and unsubstantiated incidents occurring between 1340 and 1863 in which biological agents were used for the purpose of warfare. If we assume that there was intent to spread disease in order to weaken or defeat the enemy, it is possible to correlate the methods used to deliver the disease with the prevailing theories of miasma and contagion.

The claim that biological warfare was used at the 1346 siege of Caffa by the Mongols deserves special mention because of its association with the spread of the Black Death, which devastated Europe, the Near East, and North Africa

TABLE I
Selected Substantiated and Unsubstantiated Incidents of Biological Warfare, 1340–1863

Period	Incident	Reference
1340	Jean, Duke of Normandy besieged the castle of Thun l'eveque, which had been captured by the Englishman, Sir Walter of Manny. Catapults were used to cast dead horses over the wall into the castle. Chronicled years later by Jean Froissart, based on eyewitness testimony of participants from both sides.	13
1346	Plague hits Mongol forces besieging the Genoese city of Caffa (now Feodosija, Ukraine) on the Crimean coast. Fresh corpses of plague victims were lobbed into the city. Plague breaks out in the city; 85,000 plague deaths in the region. Mongols abandon siege. Event chronicled in 1348–1349 by the Italian, Gabriele de' Mussi.	13,15
1422	At the siege of Karlstein, machines were used to catapult corpses of those who died in battle, and manure or garbage, into the city. The incident was described 250 years after the event and is not considered credible.	13
1500	Pizarro presented the indigenous peoples of South America with variola-contaminated clothing.	16
1710	Siege of Reval (now Tallin, Estonia), Sweden. Russians were said to have hurled corpses of plague victims into the besieged city, following which plague broke out in the city. There is no documentation to support this claim. The event is referred to in a Swedish military document.	13
1763	During the French and Indian War (1754–1767), Sir Jeffrey Amherst, commander of British forces in North America, suggested the deliberate use of smallpox to "reduce" native American tribes hostile to the British. Captain Ecuyer (one of Amherst's subordinates), fearing an attack on Ft. Pitt from native Americans, acquired two variola-contaminated blankets and a handkerchief from a smallpox hospital and in a false gesture of good will distributed them to the native Americans. Several outbreaks of smallpox occurred in various tribes in the Ohio River valley.	13
1775	In Boston, British attempted to spread smallpox among the continental forces by inoculating (variolation) civilians fleeing the city. In the south, there is evidence that the British were going to distribute slaves who had escaped during hostilities, and were sick with smallpox, back to the rebel plantations in order to spread the disease.	13,14
1861–1863	General W.T. Sherman complained that retreating confederate troops were deliberately shooting farm animals in ponds so that their "stinking carcasses" would contaminate water supplies for the Union forces, resulting in troops weakened and demoralized by gastrointestinal disease. Allegations that Dr. Luke Blackburn, a future governor of Kentucky, attempted to infect clothing with variola and then sell it to unsuspecting Union troops could not be substantiated.	6

in the mid-fourteenth century. Wheelis[13] believes that the hurling of plague-infected cadavers into the besieged city of Caffa was not only plausible, technically feasible, and consistent with contemporary notions of disease causation, but that it provided the best explanation of the entry of plague into the city. The attack itself appeared to be successful as it produced casualties within the city; however, it was of no strategic importance as the city remained in Italian hands and the Mongols abandoned the siege. These facts not withstanding, Wheelis[15] provides a convincing argument that this incident did not have a decisive role in the spread of plague to Europe.

3. THE GERM THEORY AND BIOTERRORISM

The seminal work of Robert Koch (1843–1910) provided the basis for the development of a new generation of biological weapons. Although others provided indirect evidence for the importance of microorganisms in causing human diseases, it was Koch who clearly conceptualized and provided experimental support for the germ theory of disease. In his early work on anthrax, he used microscopy to demonstrate that the blood of diseased animals contained large numbers of a spore-forming bacterium. He also showed that the bacteria could be cultured outside the animal body in nutrient fluids. However, to link a specific microorganism to a specific disease, the organism must first be isolated in pure culture. Toward this end, Koch developed several ingenious methods, e.g., use of nutrient medium solidified with gelatin or agar, to isolate microorganisms in pure culture. Once microorganisms were available in pure culture he was able to formulate the criteria, now called Koch's postulates, for proving that a specific type of microorganism causes a specific disease. The effects of these discoveries can be seen in the types of incidents that occurred since 1900.

3.1. Selected Confirmed Incidents, 1900–2003

Since 1900, there have been a number of incidents in which the use of a biological agent was suspected. Carus[7] used specific criteria to confirm that a biological agent was used in a criminal, terrorist, or state-sponsored event. Most of the incidents in which the use of a biological agent has been confirmed are listed in Table II. Of these 29 confirmed incidents, the majority ($N = 19$) was of the criminal type. Of the remainder, five were state-sponsored and four were considered to be terrorist events. The anthrax attack of 2001, which is currently under investigation, is listed as unknown.

Among these 29 confirmed incidents, 15 involved the use of bacteria, 10 toxins, 5 viruses, and 1 used the ova of a parasitic roundworm. In three of the confirmed incidents, biological agents or toxins were used against livestock or other animals; the remainder involved the use of biological agents or toxins

TABLE II
Selected Confirmed Incidents Where Biological Agents Were Used, 1900–2003

Date	Incident(s)	Type	Reference
1910	Patrick O'Brien de Lacy and Vladimir Pantchenko (a physician) were convicted in St. Petersburg, Russia, of murdering Captain Vassilli Buturlin (de Lacy's brother in law) by injection with diphtheria toxin	Criminal	17
1909–1918	Henri Girard used *S. typhi* and poisonous mushrooms to murder people to whom he sold insurance policies in order to obtain the death benefits. He was responsible for killing two people; six others recovered after being infected or poisoned. Girard was studying bacteriology at the time of the first murder	Criminal	7,17
1913	Karl Hopf infected his third wife with *V. cholerae* and typhus organisms. He murdered his father, two of his children and his first wife with arsenic. He was also accused of attempting to poison his second and third wives and his mother. Hopf had training in handling drugs. He was convicted in a German court	Criminal	7
1915–1918	The German Secret Service instituted a covert biological warfare campaign in the United States during the early part of World War I, while the United States was still neutral. They used *B. mallei* (glanders) and *B. anthracis* (anthrax) to infect horses and mules that Allies purchased in the United States for use by their forces in Europe. In Romania, they infected sheep bound for Russia with glanders and anthrax. In Argentina, they infected sheep, cattle, and horses with glanders and anthrax that were being shipped to Britain and to the Indian army. They purportedly used *B. mallei* and *V. cholerae* against allied forces during the German retreat	State	18
1916	Arthur Warren Waite was a dentist who had made serious attempts to acquire virulent pathogens. He killed his mother-in-law by putting pathogenic microorganisms in her food. When an attempt to kill his father-in-law using pathogens was unsuccessful, he poisoned him with arsenic	Criminal	7
1932	The League of Nation's General Assembly established the Lytton Commission in December, 1931, to investigate Japan's conquest of Manchuria. When commissioners visited Manchuria in 1932, the Japanese served them fruit "laced" with *V. cholerae*. No one became ill	State	7
1933	Dr. Taranath Bhatacharyna was a physician with training in bacteriology. He and Benoyendra Chandra Pandey murdered 20-year-old Amarendra Pandey (half brother of Benoyendra) with a lethal dose of *Y. pestis* after a feud over the division of their father's estate	Criminal	7,19

(*continued*)

TABLE II
(*Continued*)

Date	Incident(s)	Type	Reference
1936	Dr. Tei-Sabro Takahashi was a Japanese physician who used food contaminated with *S. typhi* to infect 17 people, including three who subsequently died. The incidents involved competing physicians, their families, and his wife	Criminal	7
1939	Dr. Kikuko Hirose, a Japanese physician, gave pastries contaminated with *S. typhi* and *S. paratyphi* to her former husband who, in turn, shared them with others. Twelve became ill and one died	Criminal	7
1932–1945	Japan conducted biological weapons research at facilities in China (e.g., Unit 731). Prisoners were infected with pathogens including *B. anthracis, N. meningitidis, V. cholerae, Y. pestis,* and *Shigella* spp. Between 1932 and 1945, more than 10,000 prisoners died as a result of experimental infection or execution following experimentation. At least 11 Chinese cities were attacked with biological weapons sprayed from aircrafts or introduced into water supplies or food items. Plague-infected fleas were released from aircraft over Chinese cities to initiate plague epidemics	State	20
1952	The Mau Mau used the plant toxin from the African milk bush (*Synadenium grantii*) to kill livestock in what is now Kenya.	Terrorist	7
1964	Dr. Mitsuru Suzuki, a Japanese physician with training in bacteriology, was arrested for infecting four colleagues with a sponge cake contaminated with dysentery. He was subsequently linked to a series of typhoid fever and dysentery outbreaks involving approximately 200 people, including four deaths. Prosecutors claimed he did this to complete his dissertation, which involved studies of *S. typhi* recovered from numerous persons. A culture of *S. typhi* was stolen from Japan's NIH; another culture was isolated from an infected patient	Criminal	7
1970	Eric Kranz, a postgraduate student in parasitology at MacDonald College, infected four of his room-mates using food contaminated with large numbers of embryonated ova of *Ascaris suum,* a parasitic roundworm found in pigs. The infected individuals presented with symptoms and signs of lower respiratory tract disease, the more severely ill being in acute respiratory failure	Criminal	21
1977	Arnfinn Nesset, who ran a nursing home for the elderly in Norway, was convicted of murdering 22 of his patients by injecting them with curacit, which is derived from curare	Criminal	7

TABLE II
(*Continued*)

Date	Incident(s)	Type	Reference
1978	The Bulgarian secret police attempted to assassinate Vladimir Kostov, a Bulgarian defector who had served as a news correspondent and was also a major in the D.S. (Bulgarian equivalent to the K.G.B.). A small metal pellet containing ricin was injected into Kostov who subsequently became ill but did not die	State	7,22
1978	In London, the Bulgarian secret police assassinated Georgi Markov, a Bulgarian dissident and announcer for Radio Free Europe. He was killed by ricin contained in a small platinum-iridium pellet that was injected into the back of his thigh by means of a modified umbrella tip. He died 4 days later	State	7
1981	A group calling itself "Dark Harvest" was responsible for leaving a package of soil on the grounds of the Chemical Defense Establishment located at Porton Down, England. By doing so, they claimed they were returning "seeds of death" to their source. The group claimed the soil was a part of a larger quantity (300 pounds) removed from Gruinard Island where tests of anthrax bombs were conducted in 1941. They also stated that microbiologists from two universities and locals were involved in removing the soil. Analysis showed that the soil contained *B. anthracis* (approx. 10 organisms/gram of soil)	Terrorist	7
1984	The Rajneeshees, a religious cult, employed biological agents against inhabitants of The Dalles, Oregon in an attempt to influence the local government. In the first incident, two county commissioners visiting the commune were given drinking water contaminated with *S.* Typhimurium–both became sick. Later, members of the cult contaminated salad bars, salad dressing, and coffee creamers in local restraunts with *S.*Typhimurium. As a result, 751 people became sick. Attempts were also made to contaminate the water system. Bactrol discs containing *S.*Typhimurium were legitimately obtained from VWR Scientific in Seattle for use in their medical clinic's state-licensed laboratory. It was later removed to a clandestine laboratory where large quantities were grown	Terrorist	7,23
1990–1995	The Aum Shinrikyo is a religious cult that was responsible for the 1995 dissemination of sarin gas in the Tokyo subway system. The cult claimed they had 10,000 members and assets of $\geq$\$300 million dollars. The cult was also involved in biological warfare activity involving botulinum neurotoxin, *B. anthracis*, *C. burnetii*, and attempted to obtain Ebola virus from Zaire. The group attempted to use aerosolized biological agents against nine targets including:		

(continued)

TABLE II
(*Continued*)

Date	Incident(s)	Type	Reference
	botulinum toxin (Japan Parliament, Narita International airport, downtown Tokyo, and Tokyo subway); and anthrax (sprayer on the roof of the Aum building in East Tokyo, a truck sprayer around the Diet in Central Tokyo, Imperial palace, Yokohama, and the U.S. Naval base at Yokosuka). None of these attacks were successful due to the selection of the wrong strain or to the conscience of the individual responsible for filling the dissemination devices	Terrorist	7,24,25
1990	Graham Farlow was an asymptomatic HIV-positive inmate at a prison in New South Wales, Australia. He injected a guard (Geoffrey Pearce) with HIV-contaminated blood. The guard became infected with HIV. Farlow died of AIDS	Criminal	26
1992	Brian T. Stewart worked as a phlebotomist at a St. Louis, MO hospital. He injected his 11-month-old son with HIV-contaminated blood during a fight over payment of child support	Criminal	7
1993	Iwan E was a Dutch man who injected his former girlfriend (Gina O) with 2.5 ml of HIV-contaminated blood after she broke up with him	Criminal	7,27
1994	Dr. Richard J. Schmidt, a married Louisiana gastroenterologist, injected a former lover with HIV-contaminated blood. Laboratory tests demonstrated that she contracted the same strain of HIV as found in one of Dr. Schmidt's patients	Criminal	7
1995	Dr. Debora Green, an oncologist, attempted on three occasions to kill her estranged cardiologist husband (Dr. Michael Farrar) by putting ricin in his food. When these attempts failed, she set fire to her house killing two of her three children. Green was a heavy drinker and appeared to suffer from a severe psychiatric disorder	Criminal	7,28
1996	Diane Thompson worked in the laboratory at St. Paul Medical Center hospital in Dallas, TX. She contaminated pastries (blueberry muffins and doughnuts) with *S. dysenteriae* type 2, placed them in a break room, and sent an email to laboratory personnel that food was available in the break room. Twelve people who worked in the laboratory became sick after eating the contaminated food; another person became ill after consuming pastry brought home by one of the laboratory workers. Four of the people were sufficiently sick to require hospitalization. A year earlier, she infected her boy friend (John P. Richy) with same organism. Thompson then falsified laboratory test results so that physicians would not learn of his infection. She infected him again after his release from the hospital, and a third time by injecting him with microorganisms while purporting to take a blood specimen	Criminal	7,29

TABLE II
(Continued)

Date	Incident(s)	Type	Reference
1997	Unknown farmers deliberately and illegally introduced rabbit hemorrhagic disease (a calicivirus) into the south island of New Zealand as an animal control tool to kill feral rabbits	Criminal	7
2001	Shortly after 9/11 someone mailed letters containing spores of *B. anthracis* to media companies and governmental officials resulting in 22 cases of anthrax (11 inhalational and 11 cutaneous). Five of those with inhalational anthrax died	Unknown	5,30
2002	Chen Zhengping spiked food in a rival's pastry shop in Tangshan, near Nanjing, with tetramine, a toxin from the red whelk. Up to 300 people fell sick and 38 people died	Criminal	31
2003	A letter signed "Fallen Angel" complaining about new federal trucking regulations and a threat to use ricin was enclosed in a package with a vial that contained ricin. Also, another letter addressed to the White House and signed "Fallen Angel" was intercepted at an off-site mail sorting facility. The letter contained low potency ricin	Criminal	32

against humans. The largest non-state sponsored event involved 751 people who were deliberately infected by *Salmonella* Typhimurium (Table II). There were no confirmed incidents involving the use of biological agents against crops. Most of the incidents (4/5) involving viruses occurred in the 1990s, were of a criminal nature, and involved the injection of the human immunodeficiency virus (HIV). Ricin was the most common toxin used ($n = 4$) in those incidents where toxins were used in assassination (state-sponsored) or for murder (criminal). Bacteria were used in 11/24 non-state sponsored incidents. The majority of the bacterial species that were used belonged to the Enterobacteriaceae, which is reflective of the method used to deliver the bacteria.

The method used to disseminate the agent was not known in every case. However, in those cases where it was known, ingestion of contaminated food ($N = 15$), injection ($N = 10$), and inhalation ($N = 2$) were most often used.

It was also interesting to note that a significant proportion of the non-state sponsored incidents (16/22; 73%) were perpetrated by individuals with scientific or medical training.

3.2. Probable or Possible use of Biological Agents, 1900–2003

In addition to those confirmed instances in which biological agents have been used for criminal, terrorist, or military purposes, there have been a number of instances in which it is likely, though not confirmed, that a biological agent(s) was used. The incidents described in Table III have been selected from those described by Carus.[6]

TABLE III
Probable or Possible Use of Biological Agents, 1900–2003

Date	Incident(s)	Reference
1900	It has been claimed that castor beans (ricin) were used in Malawi to kill unwanted offspring by inclusion of the seed in food	7
1909	Dr. Bennett Clark Hyde, a surgeon living in Kansas City, MO, was indicted for the alleged murder of Colonel Thomas H. Swope. He was never convicted nor acquitted of the charge. However, he likely used *S. typhi* to infect several individuals of whom one died	7
1910	In Mexico, supporters of Pancho Villa were thought to have used botulinum toxin against Mexican Federal troops	7
1917	In the United States, there were reports that German sympathizers possibly contaminated certain brands of courtplaster (an adhesive plaster used to cover small wounds) with *C. tetani*	7
1940s	An Egyptian gangster purportedly used stolen culture of *S. typhi* as part of a plot to murder insured victims	7
1942	Polish Resistance purportedly used typhus against German forces. Also, the staff at the Institute of Hygiene in Warsaw reportedly contaminated letters with spores of *B. anthracis* and sent them to the Gestapo so that they would be reluctant to open other letters naming resistance fighters	7
1947	During the 1947 war, "Zionists" might have contaminated wells around Gaza with *S. dysenteriae* and *S. typhi*	7
1969	Dr. John R. Hill, a Houston plastic surgeon, purportedly killed his wife (Joan Robinson Hill) using an injection of a bacterial mixture	33
1971	The KGB attempted to assassinate Alexander Solzhenitsyn with ricin in Novocherkassk, USSR	7
1976	The Rhodesian Central Intelligence Organization (CIO) used the services of Robert Symington (Professor of Anatomy, University of Rhodesia) who, in turn, recruited other faculty members and students into his program to develop chemical and biological agents. In 1975, researchers tested some of the agents on detainees. Members of the Selous Scouts were used to disseminate *V. cholerae* in the Ruya river and water supply of the town of Cochemane in Mozambique. Deaths that were attributed to cholera occurred in both areas. *B. anthracis* was introduced into rural areas of western Zimbabwe resulting in several hundred human deaths	7,34
1970–1980	Eastern Bloc agents may have attempted to assassinate Stefan Bankov on a flight between Seattle and London in 1974. In 1981, Boris Korczak, a CIA double agent, was killed with a platinum-iridium pellet containing ricin	7
1987–1990	Dr. David Acer, a Florida dentist infected with HIV, transmitted the disease to six of his patients. Intentional infection of these patients is a possibility although there is no direct evidence. The source of their infection remains controversial	7,35

TABLE III
(Continued)

Date	Incident(s)	Reference
1989	In South Africa, a covert operation group (Civilian Cooperation Bureau [CCB]) employed biological agents against the South West Africa People's Organization (SWAPO). Dr. Wouter Basson was the head of Project Coast, codename for South Africa's covert biological and chemical weapons program. Roodeplaat Research Laboratories, which was associated with the program, produced approx. 500 products including 32 bottles containing *V. cholerae* cultures, chocolate and cigarettes laced with *B. anthracis*, beer containing botulinum toxin, and sugar containing *Salmonella* spp. It was claimed that they infected three Russian advisors to the ANC and one died	7,34
1990	Nine cases of diarrheal disease due to *G. lamblia* occurred in residents of an apartment building in Edinburgh, Scotland. Investigators discovered that one of the tanks that supplied water to the building had been broken into and contaminated with *Giardia*-containing fecal material	6,36
1992	Canadian authorities in Alberta alleged that Marilyn Tan deliberately injected Con Boland with HIV-contaminated blood during a sexual encounter. There was insufficient evidence to convict Tan	7

4. CONCLUDING REMARKS AND PERSPECTIVES

For centuries, biological agents have been used either for warfare, terrorist, or criminal activities. Some of the perpetrators have been governments (state-sponsored), non-governmental groups (e.g., religious cults), or individuals. Today, the threat of mass casualties from bioterrorism is real; however, from a historical perspective there have been relatively few confirmed instances in which a terrorist has used a biological agent. Most of the confirmed instances involving the use of a biological agent have been for criminal purposes. Current efforts to prepare the public health system in the United States for bioterrorism through the enhancement of surveillance, epidemiology, and laboratory capacity should lead to the early detection of the event and minimize the number of casualties.[37]

REFERENCES

1. Wilkening D. A., 1999, in: *The New Terror. Facing the Threat of Biological and Chemical Weapons* (S.D. Drell, A. D.Soafer, and G. D. Wilson eds.), Hoover Institution Press, Stanford, CA., pp. 76–114.
2. Horn F., 2000, in: *Hype or Reality? The "New Terrorism" and Mass Casuality Attacks* (B. Roberts ed.), The Chemical and Biological Arms Control Institute, Alexandria, VA, pp. 109–115.

3. Carus W.S., 1998, Biological warfare threats in perspective, *Crit. Rev. Microbiol.* **24**:149–155.
4. Cohen H. W., Gould R. M., and Sidel V. W., 1999, Bioterrorism initiatives: public health in reverse, *Am. J. Pub. Health.* **89**:1629–1631.
5. Jernigan D. B., Raghunathan P. L., Bell B. P., Brechner R., Bresnitz E. A., Butler J. C., *et al.*, 2002, Investigation of bioterrorism-related anthrax, United States, 2001: epidemiological findings, *Emerg. Infect. Dis.* **8**:1019–1028.
6. Smart J. K., 1997, in: *Textbook of Military Medicine, Medical Aspects of Chemical and Biological Warfare* (F. R. Sidell, E. T. Takafuji, and D. R. Franz, eds.), Office of the Surgeon General, Washington D.C., pp. 9–86.
7. Carus W. S., 2002, *Bioterrorism and Biocrimes. The Illicit Use of Biological Agents Since 1900*, Fredonia Books, Amsterdam.
8. Steinbruner J. D., 1997–98. Biological weapons: a plague upon all houses, *Foreign Policy, Winter 1997–98*, pp. 85–96.
9. Atlas R. M., 1998, Biological weapons pose challenge for microbiology community, *ASM News* **64**:1–7.10.
10. Leitenberg M., 2001, Biological weapons in the twentieth century: a review and analysis, *Crit. Rev. Microbiol.* **27**:267–320.
11. Christopher G. W., Cieslak T. J., Parvin J. A., and Eitzen E. M. Jr., 1997, Biological warfare: a historical perspective, *JAMA* **278**:412–417.
12. Hirst L. F., 1999, *The Conquest of Plague. A Study of the Evolution of Epidemiology*, Oxford University Press, London.
13. Wheelis M., in: *SIPRI Chemical & Biological Warfare Studies, 18. Biological and Toxin Weapons: Research, Development and Use From the Middle Ages to 1945* (E. Geissler and J. E. van Courtland Moon eds.), Oxford University, Oxford pp. 8–34.
14. Hopkins D. R., 1983, *Princes and Peasants. Smallpox in History*, University of Chicago Press, Chicago.
15. Wheelis M., 2002, Biological warfare at the 1346 siege of Caffa, *Emerg. Infect. Dis.* **8**:971–975.
16. Noah D. L., Huebner K. D., Darling R. G., and Waeckerle J. F., 2002, The history and threat of biological warfare and terrorism, Emerg. *Med. Clin. N. Am.* **20**:255–271.
17. Thompson C. J. S., 1940, *Poisons and Poisoners: With Historical Accounts of Some Famous Mysteries in Ancient and Modern Times*, Harold Shayler, London.
18. Wheelis M., 1999, in: *SIPRI Chemical & Biological Warfare Studies, 18. Biological and Toxin Weapons: Research, Development and Use from the Middle Ages to 1945*, (E. Geissler and J. E. van Courtland Moon eds.), Oxford University Press, Oxford, pp. 35–62.
19. Lambert D. P., 1937, The Pakur murder, *Medico-legal Criminol Rev.* **5**:297–302.
20. Harris S., 1999, in: *SIPRI Chemical & Biological Warfare Studies, 18. Biological and Toxin Weapons; Research, Development and Use From the Middle Ages to 1945* (E. Geissler and J. E. van Courtland Moon, eds.), Oxford University Press, Oxford, pp. 127–152.
21. Phils J. A., Harold A. J., Whiteman G. V., and Perelmutter L., 1972, Pulmonary infiltrates, asthma and eosinophilia due to *Ascaris suum* infestation in man, *N. Engl. J. Med.* **18**:965–970.
22. Kostov V., 1988, *The Bulgarian Umbrella: The Soviet Direction and Operation of the Bulgarian Secret Service in Europe*, St. Martins Press, New York.
23. Török T. J., Tauxe R. V., Wise R. P., Livengood J. R., Sokolow R., Mauvais S., *et al.*, 1997, A large community outbreak of salmonellosis caused by intentional contamination of salad bars, *JAMA* **278**:389–395.
24. Olsen K. B., 1999, Aum Shinrikyo: once and future threat? *Emerg. Infect. Dis.* **5**:513–516.
25. Kaplan D. E. and Marshall A., 1996, *The Cult at the End of the World*, Crown Publishers, New York.
26. Jones P. D., 1991, HIV transmission by stabbing despite zidovudine prophylaxis, *Lancet* **338**:884.

27. Veenstra J., Schurrman R., Cornelissen M., van't Wout A. B., Boucher C. A. B., Schuitemaker H., *et al.*, 1995, Transmission of zidovudine-resistant human immunodeficiency virus type 1 variants following deliberate injection of blood from a patient with AIDS: characteristics and natural history of the virus, *Clin. Infect. Dis.* **21**:556–560.
28. Rule A., 1997, *Bitter Harvest: A Woman's Fury, A Mother's Sacrifice*, Simon & Schuster, New York.
29. Kolavic S. A., Kimura A., Simons S. L., Slutsker S. L., Barth S., and Haley C. E., 1997, An outbreak of *Shigella dysenteriae* type 2 among laboratory workers due to intentional food contamination, *JAMA*, **278**:396–398.
30. Hsu V. P., Lukacs S. L., Handzel T., Hayslett J., Harper S., Hales T., *et al.*, 2002, Opening a Bacillus anthracis-containing envelope, Capitol Hill, Washington, D.C.: the public health response, *Emerg. Infect. Dis.* **8**:1039–1043.
31. *New York Times*, Man admits poisoning food in rival's shop, killing 38 in China, September 18, 2002.
32. ABC News, New York, February 3, 2004, http://abcnews.go.com/sections/US/WorldNewsTonight/ricin_white_house_040203.html.
33. Thompson T., 1976, *Blood and Money*, Doubleday and Co., Garden City, NY.
34. Mangold T. and Goldberg J., 1999, *Plague Wars. A True Story of Biological Warfare*, St. Martin's Press, New York.
35. Rom M. C., 1997, *Fatal Extraction: The Story Behind the Florida Dentist Accused of Infecting His Patients with HIV and Poisoning Public Health*, Jossey-Bass Publishers, San Francisco, CA.
36. Ramsey C. N., and Marsh J., 1990, Giardiasis due to deliberate contamination of water supply, *Lancet* **336**:880–881.
37. Kahn A. S., Morse S., and Lillibridge S., 2000, Public-health preparedness for biological terrorism in the U.S.A., *Lancet* **356**:1179–1182.

3

The Infectious Disease Physician and Microbial Bioterrorism

SANDRA G. GOMPF, JORDAN LEWIS, EKNATH NAIK, and KALEY TASH

1. INTRODUCTION

On the morning of September 11, 2001, and later in the weeks that chronicled the spread of anthrax through the U.S. mail, our global consciousness of the terrorist threat was altered. We had awakened to a nightmare. Microbes are a perfect metaphor for our fears: our world seemed *infected* with terrorists, unlimited in virulence, waiting to emerge from dormancy. The metaphor had become real. Although the atmosphere evokes cold-war fears, the world of this century is more complex than that of the McCarthy-era. The infectious disease physician's role in bioterrorism response must be framed in this context.

2. THE EVOLUTION OF THE GLOBAL COMMUNITY, INFECTION, AND BIOTERRORISM

Modern bioterrorism attacks at the level of the individual, but its origins are global, and we must acknowledge its roots in the sociopolitical and ecologic changes of the last half-century. The decline of colonial empires and the breakup of the Soviet block have left a power vacuum in many parts of the

SANDRA G. GOMPF • University of South Florida College of Medicine, Tampa, Florida. JORDAN LEWIS • Florida Infectious Disease Institute, Tampa, Florida. EKNATH NAIK • University of South Florida College of Medicine, Tampa, Florida. KALEY TASH • Harvard University, Boston, Massachussetts.

world, and both repressive governments and radical dissidents have sometimes risen to fill it. Meanwhile, poverty, political oppression, and cultural inequalities inflame the disaffected, providing terrorist organizations with a steady flow of manpower. Decades of violence—especially in the Middle East, the African continent, and the former USSR—propagate the very conditions that first lead to militancy. Perhaps, most disturbingly, first-world governments often send a mixed message to fledgling nations. With no viable means of enforcing international bans on biological warfare, many countries continue to pursue biological weapons programs.

Although social and political instability fuel terrorist movements, microbial evolution arms them with new weapons. Human forces have affected microbial ecology in several ways. First, the use of pesticides and antimicrobials in agriculture and animal husbandry has selected for resistance. In the late 1970s, fresh from the victory against smallpox, many experts fully expected that infectious disease was a dead science: the microbial threat was to be conquered by the end of the twentieth century. But germs fought back: the adaptability of biological agents to human assaults has become a demonstration project of sorts for evolutionary science—sometimes with adverse consequences for national defense. At the time of writing this, for example, multidrug-resistant *Acinetobacter* (now endemic to many areas of the Middle East) has unexpectedly spread to the U.S. military and veterans—care facilities, via injured active duty personnel who have acquired this pathogen during warfare.(1,2) Second, the spread of humans into new environments permits exposure to new pathogens, particularly from animal hosts. Many pathogens originating from wild animals are potential bioterrorist agents—including anthrax from deer, monkeypox from illegally imported giant Gambian rats,(3) Ebola hemorrhagic fever from game animals, and SARS-associated metapneumovirus from palm civet cats. The increasing ease of human travel can quickly spread these infections from remote geographic areas. Third, human-driven environmental influences may impact microbial spread and evolution. Global warming may change migratory patterns and local biodiversity, allowing microbes to infect new hosts and reservoirs, including humans.Pesticide resistance in the *Anopheles* mosquito and favorable climactic change has recently complicated control of malaria in the Amazon.(4)

Although sociopolitical and ecological factors help us understand the proliferation of both terrorists and germs, how have the two become connected—and how has bioterrorism reached our hospitals? A look at the global communications network may offer insight. Thanks to the internet and other forms of digital communication, access to information on weapons manufacture and microbiology is readily available and virtually impossible to regulate. Meanwhile, as radio and television networks provide 24-hour coverage of world events, terrorists rely upon these institutions to propagate fear. Finally, the global media makes Western affluence and liberalism increasingly visible among impoverished or oppressed populations. The West has become an easy scapegoat for radical movements. But while global communications may facilitate bioterrorism, it also strengthens our response. International e-mail listserves, such as

ProMED,[5] the Federation of American Scientists Program for Monitoring Infectious Diseases, allow rapid notification of potential outbreaks throughout the world. Information may prove both a poison and an antidote.

3. THE EVOLVING PRACTICE OF INFECTIOUS DISEASE

Largely in response to the trends discussed above, the practice of infectious diseases has evolved from a fairly esoteric subspecialty of internal medicine to a broad and diverse field involving both academic and community-based clinicians. Infectious diseases practice today may encompass patient care, direction of public health initiatives, epidemiologic monitoring, infection control in inpatient facilities, and management of HIV infection (on the verge of becoming a subspecialty in itself), among many others. The many roles of the infectious disease physician have not always been recognized by the public health and military sectors. Effective involvement of infectious disease physicians in preparedness-planning requires communications between physicians and public agencies. In spite of the many directions in which they are pulled, infectious disease physicians are deeply interested and concerned about preparedness against all microbial threats and have organized on the regional and national level (most notably via the Infectious Diseases Society of America) to effectively advocate for supportive public policy, as well as for the inclusion of infectious disease and other civilians in preparedness planning.[6]

Though infectious disease was once a largely academic subspecialty, preparedness requires extensive involvement of community-based physicians. While academic infectious disease physicians are likely to provide leadership in the integration with public health, it must be recognized that a community-based physician was the first to detect the case of anthrax in the attacks of 2001 and the West Nile virus cluster in New York City.[7,8] This said, it must be noted that the infectious disease physician in private practice is exceedingly busy and subject to increasing economic constraints; thus rapid contact with the community physician is likely to be difficult. Coordination of efforts between public health facilities, and community physicians with diverse missions and economic constraints may be challenging. From a practical standpoint, once recognition occurs in an institution, the immediate response will be implicitly deferred to infection control staff and local infectious disease physicians.[9]

4. INTEGRATING THE INFECTIOUS DISEASE PHYSICIAN WITH PUBLIC HEALTH RESPONSE

From a public health perspective, the infectious disease physician will interface with bioterrorism response primarily in three ways, among some others (Fig. 1). First, early case recognition and communications with the public health system will likely fall on the shoulders of infectious disease physicians and first line providers. Second, infectious disease physicians must

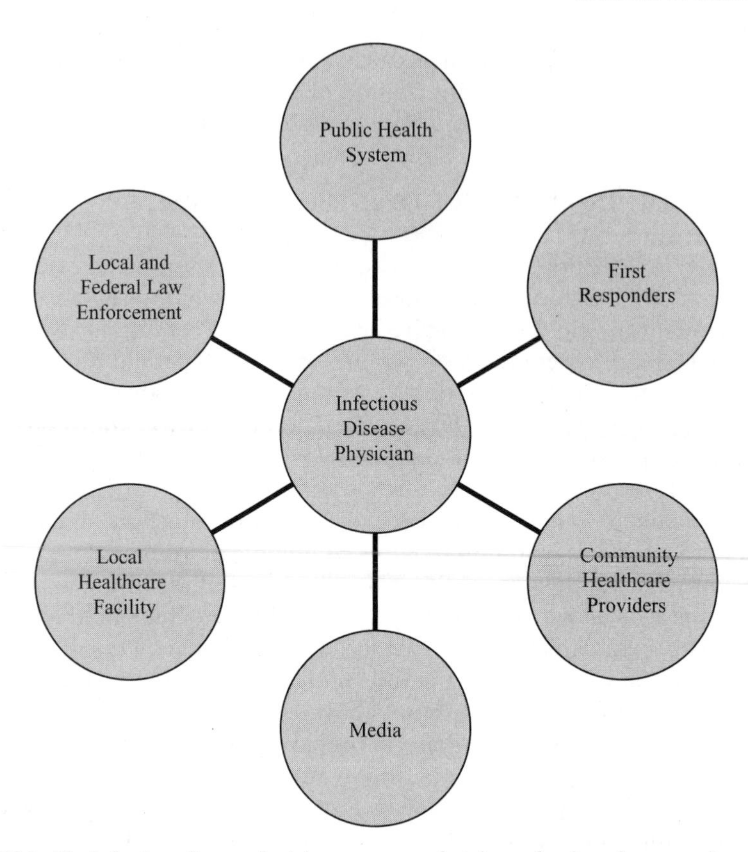

FIGURE 1. The infectious disease physician wears many hats in academic and community practice. While the infectious disease physician's response to a bioterrorist incident may be affected by many factors, he/she will likely be involved at the site of patient care, and will communicate with other health care providers, first responders, involved health care facilities, the public health system, local and federal law enforcement, and the media. Infectious disease physicians will likely serve in the education of the health care community as well.

educate other first line providers (such as emergency and primary care physicians, physician extenders, and nurses) and the public and media. Third, the communications stream between infectious disease physicians, epidemiologists, community health providers, the media, and the public is a complex issue that must be carefully considered in developing a preparedness program.

5. PREVENTION, EARLY RECOGNITION, AND THE INFECTIOUS DISEASE PHYSICIAN

Initial prevention remains in the hands of law enforcement agencies whose role is to gather and interpret intelligence and threats and act prior to the release of a disease-causing agent. However, the development of new vaccines and ongoing research into more effective antibiotics are a critical role of the infectious disease researcher. Antibiotic overuse and rising antibiotic resistance strains our ability to efficiently treat infectious disease, and amplifies the burden on health care. Prevention of antibiotic resistance and research focused on vaccine development is critical to reducing the health-care burden and to improving bioterrorism preparedness.

Once an attack has occurred, the first indication of a biological incident will be an increase in persons seeking care from community physicians and emergency departments. The ability of clinicians to recognize unusual disease symptoms, order appropriate diagnostics, and notify the local health department will largely determine the final impact on public morbidity and mortality. The rapid detection of any type of pathogenic outbreak is a fundamental challenge for the current communicable diseases surveillance system. The collaboration of public health and infectious disease physicians will expedite early diagnosis, treatment, and control.[10] However, passive disease reporting systems lack the speed and force to rapidly implement control measures such as vaccination, prophylaxis and quarantine. A variety of approaches are currently being explored to provide a more rapid identification of an outbreak occurrence. Briefly, many of these involve the monitoring of syndromic symptoms in patients. Many potential bioweapons produce nonspecific clinical symptoms, and screening for syndromes may facilitate early detection. Therefore syndromic surveillance detects a rise in reported syndromes suggestive of epidemic pathogens, such as meningitis, rash with fever, and unexplained death, and allows the public health agency to follow up on any significant deviations from the norm, typically on a per hospital basis or a pooled community. Syndromic surveillance provides an opportunity for more timely intervention with hopefully minimal overall cost and labor. Local infectious disease physicians and infection control practitioners may then be contacted for rapid institution of treatment, transmission control, and to direct investigation of other cases.[11,12]

What constitutes an optimal syndromic surveillance system is currently under study. Cost issues appear to be low, but differ among the settings in which it has been instituted. An unforeseen benefit of surveillance that has been noted in practice is that individual emergency department physicians, however observant, usually are not on serial shifts and may not readily detect rising trends; however, using syndromic surveillance and generating daily reports has provided an early warning system for seasonal events such as rotavirus and influenza epidemics. Armed with such forewarning, providers may update themselves on current treatment recommendations, plan for infection control needs, and avoid unnecessary diagnostic tests (along with their attendant costs and discomforts to patients).[13] Unfortunately, the Health Insurance Portability and Accountability Act (HIPAA) of 1996, which regulates the dissemination of personal medical information, has arguably been a hindrance to disclosure in many municipalities using syndromic surveillance.[9]

6. EDUCATION, THE INFECTIOUS DISEASE PHYSICIAN, AND PREPAREDNESS

The need for provider and public education became evident after the anthrax attacks and the resulting hysteria in the fall of 2001. From a practical perspective, an infectious disease physician's response to an act of bioterrorism and the response to any naturally occurring infectious outbreak follow a similar template. Hantavirus, West Nile virus, SARS, and avian influenza are recently emerged pathogens to which the response has required rapid recognition, heightened case surveillance, aggressive control measures, efficient communications, and prompt education of the media, the public, and the health care infrastructure. These outbreaks have certainly served an educational purpose in response planning, and have acted as important trial runs of preparedness programs, highlighting best practices and areas for improvement. Educating the clinicians and infectious disease physicians will be paramount in the early recognition and treatment of rare and difficult-to-diagnosis illnesses. Studies have shown many health professionals are not well informed about CDC class A pathogens and have poor self-perceived abilities to diagnose and manage victims of a bioterrorism attack. Clearly, improving the knowledge of health care professionals in hospitals and the community is needed to permit early recognition and treatment of victims. Additionally, health care workers are not immune to fear and hysteria; and SARS proved that health care workers, who accounted for up to one-fifth of all cases, may be directly in the line of fire during an outbreak.[14] A provider who feels informed, necessary, and competent is more likely to report to work. Numerous venues exist for education, including professional meetings, conferences, and online resources, and these should be relevant to the audience's level of specialization, workplace, and their likely level of involvement in an outbreak. The ability of the laboratory system to actively rule out, refer, and confirm begins at the hospital and with the physician. The infectious disease physician will have a primary role in working with local academic and community partners in coordinating an appropriate sampling protocol and coordinating with the public health agency.[15]

An ongoing communication process must be in place prior to any outbreak incident. The local public health authority must be responsible for optimizing this process. Ideally, clinicians should know and have frequent contact with the public health authority in their locality. Each physician should know how to contact the local health department 24 hours-a-day to report suspect cases and for consultation purposes. In practice, the day-to-day pressures of infectious disease practice is likely to weigh more heavily on the minds of many community physicians than the lower likelihood (perceived or real) that an outbreak of significance will occur in their community on any given day, and most may expect that public health will notify the community of a problem. More complex than assuring that community physicians know their public health contacts is assuring that community infectious disease physicians can themselves be readily contacted for notification. A ready means of notifying infectious disease physicians and primary care providers in diverse settings, and at diverse times, is certainly an area of need. Including a variety of media is more likely to optimize the success of such communications. Vital resources may include television and radio broadcasts, fax, e-mail, and wireless media such as cell phones and personal digital assistants.[14]

SUMMARY

In the light of current global ecology and sociopolitical pressures, "preparedness" against microbial terror coincides with prevention, detection, and treatment of emerging and re-emerging pathogens. Microbial threats will persist beyond real or perceived terrorist pressures, and are more likely to affect public health on a global scale than bioterrorism; diversion of resources from these issues are more likely to disrupt than benefit human health. Preparedness for both infrequent bioterrorist events and chronic daily threats may be pursued synergistically and in a multifaceted approach. Importantly, "preparedness" should not be measured or judged by perceived failures when an outbreak or terrorist event "gets past the safeguards". The latter is inevitable as ecological mechanisms and human ingenuity adapt to public health actions. Rather, public health efforts are necessarily work-in-progress and must evolve as needs arise. This may, at times, require more flexibility than political climates currently allow.

The role of the infectious disease physician, and indeed, the public health system, continues to evolve in unprecedented ways. Infectious disease physicians are a diverse and immeasurable resource to public health preparedness and response to infectious agents. Many infectious disease physicians follow the media, recognize the likelihood that they will be involved as both experts and first-responders to an infectious disease crisis, and are anxious, therefore, to be included in preparedness efforts. Few have been. Infectious disease physicians are typically in leadership roles in their hospitals, related to infection control, and are among the first to be contacted by colleagues and local institutions seeking advice about unusual infections, their opinions regarding infectious disease issues in the news, and bioterrorism preparedness in general. At some level, it is often assumed by other health care workers that infectious disease physicians are "in the loop" with public health planning. It should be ensured that indeed they are.

REFERENCES

1. *JAMA: Acinetobacter baumannii* infections among patients at military medical facilities treating injured U.S. service members, 2002–2004, *MMWR* **53**:1063–1066.
2. Maragakis, L. L., Cosgrove, S. E., Song, X., *et al.*, 2004, An outbreak of multidrug-resistant *Acinetobacter baumannii* associated with pulsatile lavage wound treatment, *JAMA*. **292**:3006–3011.

3. Guarner, J., Johnson, B. J., Paddock, C. D., *et al.*, 2004, Monkeypox transmission and pathogenesis in prairie dogs, *Emerg. Infect. Dis.* **10**:426–431.
4. Guarda, J. A., Asayag, C. R., and Witzig, R., 1999, Malaria re-emergence in the Peruvian Amazon region, *Emerg. Infect. Dis.* **5**:209–215.
5. Federation of American scientists program for monitoring emerging diseases (ProMED), available at: http://www.fas.org/promed/. Accessed March 23, 2005.
6. Bartlett, J. G., March 25, 1999, Statement of John G. Bartlett, MD, President, Infectious diseases society of America, before the subcommittee on public health committee on public health, education, labor and pensions, *United States Senate, on "Bioterrorism: Our Frontline Response, Evaluating U.S. Public Health and Medical Readiness"*, available at: http://www.idsociety.org/Template.cfm?Section=Bioterrorism1&CONTENTID=6944&TEMPLATE=/ContentManagement/ContentDisplay.cfm
7. Duchin, J. M., 2003, Bioterrorism, WebMD, available at: www.webmd.com.
8. Schoch-Spana, M., December 1999, A west Nile virus post-mortem, [bulletin archive]. *Bio-Defense Quarterly.* http://www.upmc-biosecurity.org/pages/publications/archive/quarter1_3.html. Accessed March 23, 2005.
9. Drociuk, D., Gibson, J., and Hodge, J., 2004, Health information privacy and syndromic surveillance systems. (in: Syndromic surveillance: reports from a national conference, 2003). *MMWR* **53** :221–225.
10. Interim planning guide, improving local and state agency response to terrorist incidents involving biological weapons, Monograph on the internet, 2000, Department of Defense, 1–32, available at: http://www.chem-bio.com/resource/2000/bwirp_interim_plan_guide.pdf#search='Department%20of%20Defense%20Interim%20Planning%20Guide%20Terrorist%20Incidents%20Involving%20Biological%20Weapons'
11. Lazarus, R., Kleinman, K., Dashevsky, I., *et al.*, 2002, Use of ambulatory-care encounter records for detection of acute illness clusters, including potential bioterrorism events. *Emerg. Infect. Dis.* **8**:753–760.
12. Reingold, A., 2003, If syndromic surveillance is the answer, what is the question? *Biosecurity Bioterrorism,* **1**:77–81.
13. Cochrane, D. G., 2004, Perspective of an emergency physician group as a data provider for syndromic surveillance. (in Syndromic surveillance: reports from a national conference, 2003), *MMWR.* **53**:209–214.
14. Chan-Yeung, M., 2004, Severe acute respiratory syndrome (SARS) and health-care workers, *Int. J. Occup. Environ. Health.* **10**:421–427.
15. Gebbie, K., Rosenstock, L., and Hernandez, L. M. (eds.), 2003, Who will keep the public healthy? Educating public health professionals for the 21st century [monograph on the internet], National Academies Press, Washington, available at: http://www.nap.edu/openbook/030908542X/html/.
16. Katona, P., 2004, Bioterrorism preparedness: practical considerations for the hospital and physician, *Infect. Med.* **21**:427–432.
17. Perkins, B., Popovic, T., and Yerskey, K., 2002, Public health in the time of bioterrorism, *Emerg. Infect. Dis.* **8**:1015–1018.

4

Modulation of Innate Immunity to Protect Against Biological Weapon Threat

KEN ALIBEK and CATHERINE LOBANOVA

1. BIOTERRORISM AND BIOLOGICAL WEAPONS THREAT

Quelling terrorism presents one of the biggest challenges of the twenty-first century. The potential for infectious diseases to be used as weapons has been recognized for centuries. Throughout history, there have been numerous attempts to kill people using various means to initiate outbreaks and epidemics.(1) During the twentieth century, however, this potential has been realized on a large scale with the development of production methods capable of producing large quantities of pathogenic micro-organisms to infect and disease humans and livestock.(2) For years, numerous countries have carried out research on the possibility of developing and using biological weapons as mass casualty weapons in military conflicts and large-scale wars.

Although the international community attained an understanding of the danger posed by biological weapons, a movement to prohibit the development and use of biological weapons began simultaneously. By the beginning of the 1970s, it resulted in the creation and ratification of a treaty banning the research and manufacture of biological weapons. Many countries, however, continued to work clandestinely in the area of developing and manufacturing biological weapons until the end of the twentieth century, but the threat of using biological weapons in a military setting was obviously diminishing because of the end of the cold war.

KEN ALIBEK and CATHERINE LOBANOVA • National Center for Biodefense, George Mason University.

As this threat diminished, however, a new threat began to emerge—the threat of bioterrorism. Though it is not entirely recognized by the world as a new form of terrorism, we in the United States fully understand that we are on the edge of new incoming threat[3] and we understand the necessity to respond to this challenge in the nearest future.

Biological weapons are mass casualty weapons based on bacteria, viruses, rickettsiae, fungi, or toxins. Biological weapons are unique in their diversity compared to nuclear, chemical, or conventional weapons. Dozens of different biological agents can be used to make a biological weapon and each agent produces a markedly different disease.

To date there have been few terrorist attacks using biological weapons, but terrorist groups are exhibiting increasing interest in unconventional weapons. In addition, advances in biotechnology have made biological weapons progressively less expensive and easier to produce. Although the most advanced and effective versions of biological weapons require sophisticated equipment and scientific expertise, which is only likely to be possessed by a nation or a state-sponsored terrorist group, there is considerable concern that primitive biological weapons can be produced in a small area by someone with minimal equipment and limited training. The general consensus among defense experts has been that a terrorist biological attack is not a matter of "if" but a matter of "when." Clearly, the anthrax bioterrorist attack that followed the September 11, 2001, terrorist attacks has demonstrated that "when" is indeed "now."

The continuously growing threat of bioterrorism makes the issues of understanding the threat, and the types of medical defense needed to address the threat, critical issues. A more comprehensive understanding of the bioterrorism threat will direct us to the appropriate avenues to address this threat and will result in novel prophylactic and therapeutic methods and means for diseases caused by biological weapons.

Biological weapons can be deployed in three ways:

- Contamination of food and water supplies, which are ingested by the victims.
- Release of infected vectors, such as mosquitoes or fleas, which then bite the victims.
- Creation of an aerosol cloud that is inhaled by the victims.

Water contamination is the least effective method for disseminating biological weapons, particularly in countries with effective water treatment systems. The effectiveness of a biological attack on the water supply would be limited both by the presence of disinfecting agents such as chlorine in the water, and also by dilution of the agent in the enormous volume of water present. Food contamination would most likely be used in a terrorist rather than in a military attack, since it is difficult to contaminate enough food to gain a military advantage. The agents that can be disseminated by food or water contamination are limited to those for which the intestinal tract can serve as a portal of infection.

Release of infected vectors is not a particularly efficient method for military purposes, but could be used for terrorist purposes to produce disruption of various vital activities and panic. The choice of agent would be limited to those agents that are naturally disseminated by vectors.

For either military or terrorist purposes, creation of an aerosol cloud—usually accomplished by explosion or spraying—is by far the most efficient and effective mode of deploying biological weapons. Aerosol dispersion is the only method that can effectively be used against large target areas. Practically any biological threat agent can cause a pulmonary infection and such infections are often more severe and more lethal than the naturally occurring form of the infection. Thus, effective biological defense must first and foremost involve protection against an aerosol attack.

In military scenarios, the use of aerosolized biological weapons and the methods used to deploy them are usually designed to provide devastating effects over large areas. For example, in the late 1980s Soviet military planning determined that one medium-range bomber equipped with two 2-ton spray tanks would effectively cover more than 1000 square kilometers and result in extensive casualties across the entire area. However, even a smaller scale military or terrorist attack would have the potential to cause a significant number of casualties, as well as tremendous panic, and a disruption of the economy and other vital activities. For example, the Council of Economic Advisors estimates that the cost of a smallpox attack could approach $177 billion dollars per week.[4, 5]

This enormous destructive potential stems from the peculiarities and uncertainties inherent in a biological attack:

- A biological weapons attack may go undetected until victims begin to fall ill, which will complicate diagnosis, treatment, and containment efforts.
- Once a biological attack has been detected, additional time will likely pass before the causative agent has been conclusively identified, again complicating diagnosis, treatment, and containment efforts.
- It will be difficult to determine the size and perimeter of the contaminated areas, making it difficult to estimate the number of exposed, identify those exposed, and determine where to conduct decontamination operations if necessary.
- The target population is not likely to be vaccinated against the threat agent (for most threat agents, no vaccine is available; if a vaccine is available, the majority of the population will not be vaccinated because these agents generally do not present a current public health threat).
- The few vaccines that exist against biological threat agents will be of very little use once an attack has taken place, since they do not reach full effectiveness until days to weeks after inoculation.

- Treatment options are limited or nonexistent for the majority of biological threat agents.
- The public and military health services are not well equipped in terms of personnel, equipment, or pharmaceuticals to accommodate a widespread epidemic.
- A biological attack will incite panic and result in an influx of patients—many of whom are not ill or were not even exposed—to already overburdened health care facilities.

Thus, the potential results of a biological weapons attack—whether against the military or civilians—are large numbers of diseased and dead, a panicked populace, an overwhelmed health care system, and significant disruption of economic and military activity.

1.1. Current Medical Defense Against Biological Weapons

Medical defense against biological weapons can be divided into three categories:

- Early pre-exposure prophylaxis: protective means administered long before a biological attack; this includes vaccines and means for passive immunization.
- Urgent pre- and post-exposure prophylaxis: preemptive use of protective means before an imminent attack or administered after an attack has taken place, but before the patient develops symptoms.
- Treatment: etiologic, pathogenic, and symptomatic therapy administered after the patient has developed symptoms.

Vaccines are currently the primary means of preexposure prophylaxis. However, the utility of vaccines is limited by:

- Time to reach effectiveness: a vaccine takes weeks to months to reach full effectiveness.
- Specificity: because it works by prompting the body to produce specific antibodies; currently available vaccines protect only against one specific disease.
- Lack of availability: for most biological weapons threat agents, no vaccine has been developed.

Vaccines can serve as effective pre-exposure prophylaxis only when:

- The target population is well defined and can be identified well in advance of an attack.
- The biological threat agents in the enemy's biological weapons arsenal are known.
- Vaccines for those agents have already been developed.
- The biological agents used are not genetically altered strains capable of circumventing a vaccine.

Some military and almost all terrorist scenarios will not meet all of these criteria.

Certain drugs can be used as pre-exposure or post-exposure chemoprophylaxis, either alone or, for certain diseases, in combination with vaccines. Drugs are more rapid-acting than vaccines. However, chemoprophylaxis entails some of the same problems as vaccines. Although antimicrobial drugs are not as specific as vaccines—for example, a single antibiotic can be effective against several types of bacteria—they still can be useful against a very limited number of threat agents. Thus, the agent must first be identified for an effective chemoprophylaxis to be selected. As with vaccines, no effective chemoprophylaxis has been developed for most threat agents. The situation is not much better for the treatment of diseases caused by biological weapons, as effective treatments are available for only a limited number of the possible threat agents.

It is unlikely that the situation with vaccines, chemoprophylaxis, and treatment will improve significantly in the near future. Current vaccines are by their very nature, specific, and thus it will take a long time to start understanding how to develop a "universal" vaccine to protect against a wide variety of pathogens. Antimicrobial drugs usually work by disrupting specific elements of microbial metabolism; these specific elements are common only to certain limited groups of microbes. Developing a truly broad-spectrum antimicrobial drug would entail finding an element (or elements) of microbial anabolic or catabolic metabolism that is common to a number of pathogens but does not exist in human cells (e.g., microbial enzymes).[6] Even though this approach shows promise, it is unlikely to have such products in the near future.

The aims of medical defense against biological weapons are to ensure the survival of the target population and minimize loss of manpower. These must be accomplished in situations where the target population is large and probably also poorly defined; where the attack is not immediately apparent or its scale is unknown; and where the biological agent used in the attack is not immediately defined. Therefore, our biodefense arsenal should also focus on approaches that are *prophylactic, rapid-acting, long-lasting, effective against a broad spectrum of threat agents, effective against aerosol attacks, and relatively easy to deliver to a large population.*

Successful *prophylactic* approaches are likely to have a greater impact than successful treatment approaches for three reasons. First, successful prophylaxis means no illness and no loss of manpower capability, whereas with successful treatment there is still a period of illness and consequent manpower loss. Second, prophylaxis is much less manpower- and resource-intensive than treatment, and in general has a higher probability of success. Third, if the biological agent is contagious, successful prophylaxis will contain an epidemic much more quickly and effectively than successful treatment.

Rapid-acting approaches do not require that the potential target population be identified months or weeks in advance, as is the case if vaccines are employed as prophylactic measures. A rapid-acting approach increases the likelihood of preventing disease symptoms from developing, as it expands the window in which prophylaxis will be successful.

A successful approach must be sufficiently long-lasting so that the frequency of its administration does not present an overwhelming operational burden or interfere with the war-fighting capability of troops.

Approaches that are effective against a broad spectrum of threat agents will eliminate the limitations of specificity that existing chemoprophylaxis and vaccines have. Such approaches can be used as pre-exposure prophylaxis even when the exact threat agent is unknown, or can be used as post-exposure prophylaxis even when the attack agent has not yet been conclusively identified. A broad-spectrum approach will also provide effective prophylaxis against agents for which one does not currently exist.

A successful approach must be effective against an aerosol attack, as this is the most likely mode of attack in a military scenario. Furthermore, in a terrorist scenario, an aerosol attack will likely result in far greater numbers of casualties than attacks delivered by other means.

An approach that is relatively easy to deliver is essential to address the logistics of a limited number of medical personnel administering prophylaxis to an enormous potential target population.

The target effectiveness of any eventual product is protection against 10—20 LD_{50} (ID_{50}) of common biological threat agents.[(7)] This level of effectiveness is predicted to provide protection for 90–99% of potential casualties, a prediction based on both Soviet and American data. In the 1970s and 1980s, Soviet biological weapons scientists analyzed the distribution of LD_{50} among experimentally infected animals in field tests with aerosol biological weapons. They found that in standard military situations, using existing dispersion methods, the great majority (90–99%) of subjects were infected by 1–10 LD_{50} (ID_{50}) of the agent.[(6)] In 1991, an US estimate of the average concentration of anthrax (biological weapon) to which American soldiers in Operation "Desert Storm" could be exposed was 6 LD_{50}.[(8)] Thus, in theory, prophylaxis that protects against 10 LD_{50} of an agent could be expected to protect 90–99% of the potential casualties of a biological attack.

When developing protection against biological weapons it is important to envision which prophylactic preparations would be applicable to which scenarios:

- Potential attack: Intelligence indicates that troops are at high risk of biological attack (similar to the situation encountered by troops deployed in the Persian Gulf War).
- Suspected attack:
  - An unusual explosion or release has been detected that may have been a biological attack.
  - A biological attack is suspected based on an unusual disease outbreak.

- Confirmed attack:
  - A biological attack has been confirmed by biological detection equipment.
  - A biological attack has been confirmed based on an epidemiological evaluation of an unusual disease outbreak.

One possible way to develop this type of protection is to find effective and safe approaches and means to modulate the host's immune system's response to biological threat agents. Innate immunity might be considered a feasible "target" for such modulation due to its peculiarities to respond to any "non-self" agents without having a memory, which is the major requirement for the adaptive immunity.[9] We need to explore possible ways to develop protection based on the innate immunity modulation for both full protection and providing "a golden hour" timetable until specialized medical care is available, or newly formed specific immunity would eliminate the pathogen in the host.

2. IMMUNE SYSTEM

The immune system is composed of many different networks of cellular and soluble components that interact to eliminate pathogens. The immune system is conventionally divided into at least two distinguishable subsystems that predetermine two different types of immune responses: nonspecific and specific. Against biological agents, the first line of defense is nonspecific (innate) immunity, which is usually followed by acquired (specific) immune responses. Innate immunity responds quite rapidly and does not require a previous "memory" to a pathogen to attack and eliminate it. Adaptive (acquired) or specific immunity is a much better described and "used" part of the immune system. Among the conditions required to initiate the elimination of a pathogen is "memory" meaning that in order to react, the immune system should have encountered the pathogen previously, or the host should live long enough during the same infection in order to develop this memory. It is important to mention that the adaptive immunity exists only in vertebrates; all other multicellular organisms have just the innate immune system for battling infectious diseases.[10] Although, in humans and other vertebrates, innate immunity is considered to be the early initial response to a pathogen, it alone is potentially capable of eradicating infection of the host. In other cases, when the innate immune system is not capable of eliminating the pathogen, it may reduce significantly the pathogen load, allowing for a slower spread of the infection, reduced pathology, and a more effective clearing of the infection by the later induced response.[11]

When comparing the innate and adaptive immune systems, several differences can be appreciated.[12] The innate immune system has a rapid response

TABLE I
Cellular and Soluble Components of the Immune System

Type of immune system	Cellular components	Soluble components
Innate immunity	Macrophages Monocytes Neurophils Dendritic cells Basophils Eosinophils NK cells LAK cells Epithelial cells	Cytokines Chemokines Complement proteins Acute phase proteins Long petraxins Defensins Cathelicidins Histatins
Adaptive immunity	T-lymphocytes B-lymphocytes	Specific immunoglobulins
Overlapping (Innate–Adaptive) Immunity	Macrophages Dendritic cells NK cells	Cytokines Chemokines

(hours to a few days) whereas the adaptive system has a delayed response (days to weeks). The innate system reacts to a "pattern" rather than to a specific epitope that is characteristic of adaptive immunity. The innate immune system has no memory response and will be activated to the same degree when encountering a particular infectious organism on repeated occasions. In contrast, the adaptive immune system has memory and can respond more rapidly to a second challenge, and with an enhanced activity against the infectious agent. The most important components of both systems is provided in Table I.

As is seen from Table I, there exist both soluble and cellular components of the innate and adaptive immune systems, which play an essential role within the specific system. For example, neutrophils and defensins have a primary role in innate immunity, whereas B-lymphocytes and specific antibodies are essential players within the adaptive immunity. However, there are many cellular and soluble components that participate in the response of both the innate and adaptive immunities, as well as the many components that bridge and orchestrate nonspecific and specific immune responses. The following examples demonstrate some interconnections within the immune system:

- Macrophages and dendritic cells present antigens to generate T-memory lymphocytes
- T-helper lymphocytes produce both pro-inflammatory and anti-inflammatory cytokines that modulate the activity of innate immunity phagocytes and NK cells

- Formation of membrane attack complex of complement as well as the formation of C3 and C5 complement proteins that play an important role in opsonization and chemotaxis can be initiated via the classical pathway as a result of antigen–antibody complex formation
- Chemokines produced mostly by neutrophils play an important role in the chemotaxis of lymphocytes

Of course, there exist a number of other examples of a mutual interaction and influence between the innate and adaptive immune systems. It shows that the division based on the allocation of specific cellular and soluble components into one or another part of the immune system has a significant degree of convention. Attempts to develop a logical division between these two systems will not succeed because they are highly interconnected.

3. INNATE IMMUNITY

As with the immune system itself, innate immunity consists of two major components: soluble and cellular. However, some scientists divide innate immunity into a three-element system composed of mechanical, chemical and cellular components.[13] The mechanical component is the physical barrier provided by the skin and mucosa, supported by physical functions such as cilial action, motility, desquamation, and mucus secretion. The cellular component includes the cells capable of killing pathogens through various mechanisms upon activation. This component includes macrophages and monocytes, neutrophils and other granulocytes, NK and LAK cells, epithelial, mast, and dendritic cells. The third component of the innate immune system is the chemical component, which includes soluble and cell-associated pattern recognition molecules, antimicrobial proteins and peptides (e.g., acute phase reactants, complement proteins, etc.), and chemokines and cytokines that mediate the immune response and function.

A common feature of the innate immunity is that all mechanisms of this response have evolved to recognize and respond to conserved structures of microorganisms. These recognized features include particular bacterial carbohydrates, lipopolysaccharide (LPS; recognized by Toll-like receptors), or forms of bacterial DNA (unmethylated cytosine-guanosine-rich areas known as CpG sequences) that are not seen in human genes.

Even though this classification is more comprehensive, it cannot be considered a true classification of the innate immunity. Many elements of the first component do not constitute true immunity. Some of them are actually mechanical barriers with other major functions, and their participation in the protection against infectious agents is a sort of extension of their basic functions. The inclusion of cell-associated pattern recognition molecules is hard to justify as a separate component of the innate immunity since these molecules are a part of the cellular component as many other surface molecules that participate in signal transduction. Proceeding from this, it is still logical to divide the innate immune system into the cellular and soluble component subsystems.

4. PULMONARY INNATE IMMUNITY

In biological weapons defense, the role of the respiratory tract's innate immunity cannot be overestimated. As stated above, significant casualties will likely be the result of an aerosol biological weapons attack. The great majority of people diseased as a result of such an attack will be infected via the respiratory system, the main portal of all aerosol infections. The ability of the respiratory tract's innate immunity to effectively respond to the invading pathogens is critical if the host is to reduce the probability of getting diseased. Pulmonary innate immunity has a number of various components and pathways involved in the eliminating of microbial pathogens.

4.1. Major Components of Pulmonary Innate Immunity

Lysozyme is a major constituent of lavage fluid and sputum, and represents an important antimicrobial defense, particularly against Gram-positive bacteria. It is produced by glandular serous cells, surface epithelial cells, and macrophages. It was shown that elimination of pulmonary pseudomonas and streptococcal infections could be dramatically enhanced by this enzyme. Furthermore, clinical study of susceptibility and resistance to acute bronchitis showed a correlation of protection with levels of macrophage-derived lysozyme.[(5)] The mechanism of this enzyme's action is well understood; lysozyme has a direct lytic effect on bacterial peptidogycans.

Lactoferrin is another component of the constitutive defenses in the airways and is produced by serous cells as well as neutrophils. Like many of the other mucous constituents, it is able to kill and agglutinate bacteria that it recognizes on the basis of carbohydrate motifs, as well as stimulating superoxide production by neutrophils. As with lysozyme, the concentration of lactoferrin is markedly increased in the lower respiratory tract in subjects with chronic bronchitis.[(8)]

The *α- and β-defensins* show broad microbicidal activity against Gram-negative and Gram-positive bacteria, fungi, and some viruses. They are not specific to the lungs, and are found in other mucosal membranes including the gut and reproductive tract. They act by inducing permeabilization and are up-regulated in the lungs in response to the inflammatory cytokine, interleukin-1 (IL-1).[(9)] Binding of defensins to complement components also leads to triggering the alternative complement cascade.

Collectins are a large family of proteins that have the ability to recognize and bind to carbohydrates on the surface of pathogens including both bacteria and viruses. This binding results in a number of defensive mechanisms, including activation of the alternative pathway of complement cascade as well as activation of macrophages and lymphocytes. Key members of the collectin family include surfactant proteins A and D (SP-A and SP-D) and mannan-binding lectin (MBL).

Complement is another major component of innate immunity. Complement-mediated cytolytic activity is described as an accessory reaction to antigen–antibody interaction and can also be activated by another pathway that is triggered by susceptible foreign surfaces such as bacteria and yeast cell walls.(14) A third pathway, the lectin pathway, which, although it is probably an ancient pathway, was only recently discovered.(15) Complement proteins and complement receptors play an important role in the respiratory mucosal immune system. Complement cascade can be directly activated in response to an encounter with invading pathogens through any of the above-described pathways.(16) Although there are a large number of complement components that play an important role in the respiratory defense, the third component of complement, C3, has a central role in all three pathways. C3 fragments are deposited on the surface of pathogenic bacteria, "marking" them for macrophage phagocytosis. There are other mechanisms and components involved in the pulmonary defense against respiratory pathogens. For example, the C5a receptor apparently plays an important, nonredundant role in the transport and activation of neutrophils and macrophages.(17) Even though debate still continues as to which complement activation pathway plays the primary role, there is no debate that this cascade is critical for the protection of the respiratory tract.

Immunoglobulin A, the major immunoglobulin found in the healthy respiratory tract, is considered the most important immunoglobulin in the pulmonary mucosal immune system. Like other immunoglobulins, it must be made and secreted by B- lymphocytes. However, unlike the other immunoglobulins found in the lung, specifically IgG and IgE, it is not part of the T-lymphocyte-dependent immune response. It prevents the adherence and absorption of antigens and is capable of neutralizing various pathogens, including viruses, directly within the epithelium. In addition, IgA antibodies are potential participants in inflammatory reactions as they can bind to receptors on a variety of leukocytes and are apparently capable of activating the alternative complement pathway.(18) As IgA-producing plasma cells are short-lived (with a half-life of 5 days), the process of B-cell differentiation into IgA plasma cells is important in maintaining effective mucosal immunity. Cytokines IL-4 and IL-5 have been found to regulate the two major steps in differentiation of IgA B cells: first, isotype switch differentiation of surface IgM-bearing B cells into surface IgA-bearing B cells and, second, terminal differentiation of IgA B cells into IgA-producing plasma cells.(19) In addition, the human airway epithelium constitutively produces IL-2, tumor growth factor (TGF)-β, IL-6, and IL-10 factors, which are essential for B-cell clonal proliferation, IgA isotype switch, and differentiation into IgA-producing plasma cells.[20] The IgA response is an important component of rapid, local lung immune responses to infection by viruses and bacteria that use the respiratory tract and the portal and the primary site of infection.

4.2. Interferons and Other Cytokines

Another rapid response to infection (especially to viral infection) comes from the type I interferons (IFNs), IFN- α, and IFN-β. This is part of the direct, autonomous response of cells in the respiratory tract, including epithelial cells and macrophages, to viral infection. As a consequence of type I interferon release, a number of factors are released that have the ability to interfere with intracellular replication. These interferons, together with some other cytokines including IFN-γ, IL-3, M-CSF, GM-CSF, and TNF-α, play an important role in the pulmonary defense by priming alveolar macrophages for selected functions. Primed macrophages respond to secondary stimuli to become fully activated, that is, becoming capable of killing intracellular parasites, tumor cell lysis and secretion of mediators of inflammation including TNF-α, prostaglandin E2, IL-1 and IL-6, which in turn activate other mechanisms of pulmonary defense against various pathogens entering the respiratory tract. Cytokines also increase the microbiostatic and killing capacities of neutrophils against bacteria and fungi. IFN-γ and GM-CSF independently amplify neutrophil antibody-dependent cytotoxicity. Anti-inflammatory cytokines IL-4 and IL-10 inhibit the production of IL-8 and the release of TNF-α and IL-1, thus blocking neutrophil activation.[21] Various cytokines are basic regulators of neutrophil functions. Many of them, including hematopoietic growth factors and pyrogens, have shown to be potent neutrophil priming agents. The pyrogenic cytokines IL-1, TNF, and IL-6 all prime various pathways that contribute to the activation of NADPH (β-nicotinamide-adenine dinucleotide phosphate, reduced) oxidase. Proinflammatory cytokine IL8, also known as neutrophil-activating factor, is a potent chemoattractant; it synergizes with IFN-γ, TNF-α, GM-CSF and granulocyte colony stimulating factor (G-CSF) to amplify various neutrophil cytotoxic functions. Neutrophils also synthesize and secrete small amounts of some cytokines, including IL-1, IL-6, IL-8, TNF-α and GM-CSF; they may act in an autocrine or paracrine manner.

A summary of the functions of the major cytokines in the pulmonary immune system is provided in Table II.

4.3. Known Pathways of Pathogen Elimination by the Pulmonary Innate Immunity

There are numerous known pathways along which the various components of the pulmonary mucosal immune system interact. One nonspecific pathway

TABLE II
Major Cytokines Activities in Pulmonary Host Defense

Cytokine	Roles in pulmonary host defense
IFN-α/β	• induce antiviral state • induce cell growth • induce class I MHC antigens • activate monocytes/macrophages • activate natural killers • activate cytotoxic T cells • modulate immunoglobulin synthesis in B cells
IFN-γ	• induces antiviral state • induces cell growth • induces class I and II MHC antigens • activates monocytes/macrophages • activates natural killers • activates cytotoxic T cells • modulates immunoglobulin synthesis in B cells • induces F_c receptors in monocytes • inhibits the growth of nonviral intracellular pathogens
IL-1	• activates monocytes/macrophages • enhances activation of T cells
IL-2	• induces T cell proliferation and differentiation • co-stimulates B cell proliferation and differentiation • augments natural killers • induces adhesion molecules
IL-8 (and other chemokines)	• promotes neutrophil chemotaxis and activation (neutrophil activating factor) • activates T cells
IL-12	• induces natural killer activity and Th1 responses • induces IFN-γ production • enhances cell mediated cytotoxicity
IL-15	• induces T cell proliferation and differentiation (similar to T cell activities of IL2)
GM-CSF	• activates macrophages • activates granulocytes • activates eosinophils
TNF-α	• induces monocytes/macrophage activation • induces signaling pathways that lead to proliferation

for macrophage elimination of intracellular microorganisms and viruses (virus-infected cells) proceeds as follows:

- Antigen (Ag) is taken up by macrophages and presented (complexed with class II MHC) to CD4+ cells, which were previously exposed to the Ag

- CD4+ cells (after engaging Ag-MHC) release IFN-γ and IL-2, macrophage and lymphocyte chemotactic factors
- IFN-γ induces endothelial cells to increase expression of adhesion molecules that leads to intensive egress of macrophages and lymphocytes through the endothelial barrier
- IFN-γ and IL-2 act as proliferation and activation signals leading to the expansion of T-cell memory clones and newly arrived T cells
- T cells (activated by IFN-γ and IL-2) start secreting MIF (migration inhibiting factor) that prevent macrophages from leaving
- Simultaneously T cells start secreting IFN-γ and GM-CSF that lead to activation of macrophages; the activated macrophages are able to eliminate intracellular bacteria and virus-infected cells
- Activated macrophages secrete IL-1 and TNF-α to potentiate the secretion of IFN-γ and GM-CSF.

Other specific and nonspecific pathways for the elimination of viral pathogens include:

- IL-12 pathway: IL-12 activates natural killer cells, which produce IFN-γ that leads to induction of antiviral state
- Cytotoxicity pathway: IFN-α/β activate natural killer cells, inducing the natural killers' cytotoxic activity
- Antibody-dependent cellular cytotoxicity pathway: once virus-specific antibodies are introduced, natural killer cells, using their immunoglobulin receptors, lyse virus-infected cells
- IL-8 (and other chemokines) pathway: many chemokines rapidly and profoundly activate neutrophils that results in the neutrophils' adherence to opsonized microorganisms, whereas MIP-1-α activates inflammatory response of natural killers

There are a number of other mechanisms for eliminating intracellular microorganisms (specifically viruses). For example, the antibody effector system reviles its antiviral activity through the complement-mediated cytotoxicity and/or antibody-dependent cell-mediated cytotoxicity of virus-infected cells, as well as through neutralization of viruses and opsonization of virus particles. The CD4+ subset effector system also exhibits antiviral activity by aiding antiviral antibody production.

Effective pulmonary host defense against bacterial and viral infections requires cytokine-dependent activation and recruitment of phagocytic cells and adequate lymphocyte cell response. Initiation, maintenance, and resolution of the inflammatory response in the setting of infection are also dependent on the expression of cytokines. Many studies have demonstrated the importance of cytokine regulation in host immune response, including in pulmonary (mucosal) host defense, and several studies have indicated that immunologic manipulation of cytokine expression has considerable potential for the treatment of serious lung infections.

Many of the pathways described above lead to the release of mediators that have the effect of increasing migration of neutrophils (which constitute the first line of defense against infectious agents or "non-self" substances that penetrate the body's physical barriers) to the lung. Within a few hours of experimental infection or LPS injection, there is massive neutrophil recruitment until these cells constitute 60–80% of bronchial alveolar lavage (BAL) cells. This can occur through the stimulation of factors including IL-8, complement activation, or the release of chemokines. Activated neutrophils have an enormous capacity to phagocytize and neutralize bacteria, and can also secrete various factors including defensins, TNF-α, interleukin-1β (IL-1β) and IL-6 that leads to the additional neutrophil recruitment and the enhanced alveolar macrophage response, both of these contributing to the accelerated clearance of the infection.[18]

There is another group of phagocytic cells that play a critical role in lung defense. These cells are alveolar macrophages whose role in non-specific pulmonary defense cannot be overestimated. They play a pivotal role in host lung defense mechanisms. Alveolar macrophages are highly versatile and specialized cell types with an impressive repertoire of functions that are expressed according to the activation status. Activated macrophages engulf bacterial pathogens, take them into the regional lymph nodes, and eliminate them either in route or in the lymph nodes.

Viruses attempting to infect cells in the lung face potential attack from one more cell-type of the rapid, innate response, that is, natural killer (NK) cells. NK cells derive from the same hematopoietic lineage as T-lymphocytes, but, unlike those cells, do not have to mature in the thymus and do not express re-arranged antigen receptors. NK cells use what is now appreciated to be a large number of families of cellular receptors to survey the body, looking for cells that, for reasons of viral infection or transformation, have altered expression of human leukocyte antigen (HLA) class I tissue antigens. If the NK cells fail to receive a cellular signal that normal HLA class I has been recognized, they enter a program of activation leading to lysis of the infected cell and release of interferon-γ (IFN-γ). This may in turn lead to recruitment of other cells. For example, in experimental RSV infection, there is an extremely rapid antiviral NK cell IFN-γ response that precedes and leads to recruitment of virus-specific, cytotoxic T-lymphocytes.[19] Local release of IL-12 is an important early event leading to stimulation of NK cells for such rapid anti-viral responses in the lung.[20]

5. MODULATION OF IMMUNITY FOR PROTECTION AGAINST INFECTION

The immune system can be stimulated nonspecifically in an antigen-independent manner by modulators of immune function. A diverse array of recombinant, synthetic, and natural immunomodulatory preparations are being investigated extensively in clinical and preclinical studies.

Immunotherapeutics may act by neutralizing or direct lysis of pathogens, by interfering with microbial gene expression and growth, by enhancing and activating immune cells or a combination thereof.[22]

There are a certain number of publications on the use of immunotherapy for the treatment of bacterial infection, but probably it would be fair to say that none of them can be comparative with the therapeutic effect of traditional antibacterial therapy. More progress has been made in the use of immunomodulators for the treatment of certain viral infections including influenza, HIV, and hepatitis.[23–25] Currently available antifungal chemotherapy often fails to eradicate fungal infections and application of immunotherapy to enhance impaired host immune responses has attracted attention with some positive results. There is increasing evidence that CSFs can alter the course of established fungal infections[26] and IFN-γ can be useful as an adjunctive therapy for treatment of certain fungal infections.[27] The use of immunotherapeutics in oncology has also been extensively studied.

Even more promising results are seen in regard to immuno-prophylactics. However, changing the fine balance of immune mechanisms used to fight a particular pathogen may lead to an altered inflammatory response and in turn to inflammatory tissue damage.

Immunomodulators can be classified into the following main categories: whole microbes—attenuated strains, whole heat-killed bacteria; microbial products—bacterial homogenates, isolated fractions; compounds of natural origin—calf thymic hormones, glucans, plant fractions; synthetic compounds—isoprinosine, muramyl peptides; and compounds of endogenous origin—cytokines.[28] Some antibiotics possess immunomodulatory features as well.

Probiotics, especially lactic acid bacteria, are immunomodulatory.[29] They are mostly used to prevent antibiotic-associated diarrhea and to treat other diarrheal illnesses caused by bacteria, but there is some evidence that they may have an effect on some respiratory infections as well. This concept has been proven correct for bacterial extracts, which have been used as immunomodulators to prevent recurrent infections of the respiratory tract. It is known that activated lymphocytes from the gut are capable of conferring protective immunity by disseminating into the intestinal tract and other mucosal tissues located in respiratory and urogenital organs. Thus, stimulation of the gut-associated lymphoid tissue can lead to the induction of a generalized response by the whole mucosal-associated lymphoid tissue. Functional tests on alveolar macrophages have shown increases in motility and in the production of superoxide anion and chemiluminescence.[27]

Therapy with thymus extract preparations has been studied and confirms the effectiveness of treatment with thymomodulin in patients with recurrent respiratory infections, even though the immunological background of clinical improvement remains to be elucidated.[30] Methyl inosine monophosphate (MIMP), a novel thymomimetic purine immunomodulator, was proven to be capable of enhancing a wide variety of immune responses. For example, following influenza challenge infection, mice that were administered MIMP demonstrated complete survival and reduction of viral load.[31]

An immunomodulator, glucan phosphate, induces protection in murine polymicrobial sepsis through stimulation of the PI3K pathway and may have some effectiveness for preventing or treating sepsis and/or septic shock.[32] Lentinan, one of the glucanes that is isolated from an edible Japanese mushroom and currently licensed for antitumor therapy in Japan, also demonstrates certain protection against influenza and some other viruses.[27]

Regardless of a large number of publications, which show promising results on the use the above-described and other exogenous mediators of innate immune response, the most promising results, however, in our opinion, should be expected from the use of immunomodulators of endogenous origin because they are natural elements of the immune system's function and response.

The concept of using endogenous immunomodulators for protection and treatment is not new. Several publications over the past two decades have confirmed the high efficacy of cytokines to protect against and treat viral and bacterial diseases.[33–35]

Over the last twenty years interferons have received considerable attention as potential preparations to be used for the treatment of a number of acute and chronic disorders. For example, IFN-γ therapy has been licensed for the treatment of chronic granulomatosis disease and is considered to be a promising therapeutic anticancer and antibacterial agent.[36]

However, this approach is not currently used to protect against biological weapon threat agents. Due to the high probability of a smallpox bioterrorist attack and the severe consequences thereof, we studied how effectively cytokines protected against orthopoxvirus infections.[37, 38] It was shown that some cytokines, namely IL-15 and especially IFN-α and IFN-γ, could be very good prospects for antiorthopoxviral defense.

Additional confirmation that endogenous mediators are good prospects is that a number of cytokines are FDA-approved for treatment or prevention of certain diseases. These include several types of interferon-α (used for the treatment of hairy cell leukemia, Kaposi's sarcoma, condyloma acuminatum, and chronic hepatitis B and C), interferon-β-1b (used for the treatment of multiple sclerosis), interferon-γ (for treatment of condyloma acuminatum), interleukin-2 (used for the treatment of metastatic renal cell carcinoma), epoetin alfa (used in conjunction with zidovudine for the treatment of HIV infection), granulocyte colony-stimulating factor (used to prevent infections after chemotherapy) and granulocyte-macrophage colony-stimulating factor (used to prevent infections in myeloid reconstitution after autologous bone marrow transplantation and in the case of bone marrow engraftment failure or delay).

In addition, extensive study of leukocytic and recombinant interferons (and other cytokines) as well as interferon inducers indicates that these products have therapeutic and prophylactic effects against many other infectious diseases, including those that could be used in biological weapons. Human studies have indicated the utility of interferons in the treatment of acute pneumonia[39] and respiratory viral infections,[40] hemorrhagic fever renal virus infection,[41] tick-borne encephalitis,[42] and Japanese encephalitis[43]; of interferon-γ and GM-CSF in human papilloma virus infection[44]; of IL-2 in tuberculosis infection[45]; and of GM-CSF against fungi and DNA-containing viruses.[46] In addition, interferons have shown promise in the prophylaxis of postsurgical wound infections and postpartum infections,[47] and colony-stimulating factor in the prevention of infection in myelosuppressed patients.[48] Numerous animal studies indicate that various cytokines are also effective in the prophylaxis or treatment of other infectious agents including Caraparu virus,[49] Rift Valley fever virus,[50, 51] herpes simplex virus,[52] Banzi virus,[52] Semliki Forest virus,[52, 52] Hantaan virus,[53] Venezuelan equine encephalitis virus,[54] Marburg and Ebola viruses,[55] yellow fever virus,[56] dengue virus,[57] smallpox virus,[58] and various *Yersiniae*[59] and *Candida albicans.*[60]

Cytokine therapy can be accompanied by adverse effects, which are sometimes severe. These adverse effects are generally associated with relatively high doses of a single cytokine. Generally, for the best-studied cytokines—the various interferons and interleukin-2—lower doses are without such side effects.[61–64] In addition, localized administration techniques—specifically inhalational administration—have demonstrated a lower rate of adverse effects (see below). Furthermore, a recent unpublished study has indicated that up to 60% of the molecules of recombinant cytokine preparations have incorrect conformations, increasing the likelihood of adverse effects. In transmucosal administration, such as inhalational administration, the conformation of the cytokine preparations is reconfigured into the correct (more active and less toxic) form as they pass through the epithelial and endothelial cells, thereby decreasing the incidence of adverse effects in comparison with intravenous or subcutaneous administration.[65]

Inhalational administration of cytokines has been shown effective in a number of clinical cases. Interleukin-2 (IL-2) has been administered by inhalation in a number of trials to treat pulmonary metastases of renal cell carcinoma in humans, often in combination with subcutaneously administered interferon-α. Response to the treatment was generally good and toxicity was low.[66–68] Another study of IL-2 compared quality of life for metastatic renal cell carcinoma patients treated with inhalational and intravenous IL-2. Inhalational IL-2 treatment stabilized patient quality of life for a mean of 13.4 months. In contrast, patients receiving intravenous IL-2 had a marked decrease in quality of life, and 3 of 10 patients had to drop out of the study due to treatment-related toxicity.[69]

This research confirms the probability of using cytokines as well as other endogenous and exogenous modulators of innate immune response for prophylaxis and possible therapy of infectious diseases including those caused by biological weapons.

6. MODULATION OF INNATE IMMUNITY TO PROTECT AGAINST BIOLOGICAL WEAPON THREAT—A SUMMARY

Currently, in the field of biodefense the major efforts are focused on the development of vaccines and therapeutic preparations. However, these efforts are insufficient and another approach should be considered for protecting troops and civilian population against biological weapons or bioterrorism, especially in the event of a large-scale biological attack using viral pathogens.

It is important to understand that in the field of biodefense medical protective measures could be divided into three major types:

- Early preexposure prophylaxis: protective means administered long before a biological attack; this includes vaccines and means for passive immunization
- Urgent pre- and post-exposure prophylaxis: preemptive use of protective means before an imminent attack or administered after an attack has taken place but before the patient develops symptoms
- Treatment: etiologic, pathogenetic, and symptomatic therapy administered after the patient has developed symptoms.

The inclusion of urgent prophylaxis is actually one of the major differences between protection against biological weapons and naturally occurring infections. In the field of protection against naturally occurring infections, attention is not usually paid to urgent prophylactic measures, nor are many preventive preparations for both pre- and post-exposure prophylaxis developed. In the case of a large-scale BW attack, however, the situation would be completely different since large numbers of exposed people (or people suspected of having been exposed) could appear. This situation predetermines a necessity to develop different approaches and preparations that should be prophylactic, rapid-acting, long-lasting, effective against a broad spectrum of threat agents, effective against aerosol attack, and relatively easy to deliver to a large population.

Understanding that the number of biological agents, which could be used as weapons of terror, is large and quite diversified, it is impossible at this point of life science development to devise a preparation that would have a broad-spectrum therapeutic effect, especially against viral biological agents. However, one of the possible ways to develop such prophylactic preparations is through increasing the nonspecific immune system's response to biological agents.

Particular attention should be paid to innate immunity, which does not require that the host have encountered the pathogen before to be activated (as is the case with adaptive immunity), and which would serve as the most suitable target for accent.

Currently, preclinical and clinical testing is underway on a diverse array of recombinant, synthetic, and natural immunomodulatory preparations. The most promising results are likely to occur with the use of immunomodulators of endogenous origin because they are natural elements of the immune system function and response. Several cytokines have been already approved by the FDA for use in the treatment and/or prevention of certain diseases, thus supporting the concept of using endogenous mediators. Recombinant versions of most known cytokine{XE "Cytokines"} mediators of the immune response have been developed, produced, and used experimentally for determining their role in immune processes.

Recent research has demonstrated that cytokines can be used to treat and prevent many diseases of viral and bacterial origin. Many research data have indicated that certain cytokines may either up-regulate or down-regulate both nonspecific and specific immune responses against a broad spectrum of viruses and bacteria. Preventive preparations based on the innate immunity modulation could protect against biological agents by themselves or by providing "a golden hour" timetable necessary for the specific immunity formation to completely eliminate the pathogen. Proceeding from this understanding, it is extremely important to explore possible ways to research and develop broad-spectrum protective preparations based on the innate immunity modulation.

REFERENCES

1. Geissler, E., and Van Courtland Moon, J. E., 1999, Biological and toxin weapons: development and use from the middle ages to 1945, in: *Sipri Chemical and Biological Warfare Studies,* Vol. 18, Oxford University Press, Oxford.
2. Alibek, K., and Hendelman, S., 1999, *Biohazard: The Chilling True Story of The Largest Covert Biological Weapons Program in the World—Told from Inside by the Man Who Ran It,* Random House, New York.
3. *Biological Threats and Terrorism: Assessing the Science and Response Capabilities,* 2002, Institute of Medicine, National Academy Press, Washington, DC.
4. Broad, W. J., 2002, White House Debate on Smallpox Shows Plan for Wide Vaccination, *New York Times.* **1**:20.
5. Broad, W. J., 2002, White House debate on smallpox shows plan for wide vaccination, *New York Times.* **1**:20.
6. Dunn, B. M., ed., *Proteases of Infectious Agents,* 1999, Academic Press, New York, p. 220.
7. Alibek, K., and Patrick, W., Top officials of the USSR/Russian and the USA biological weapons programs, Personal communication.
8. Patrick, W., Biological weapons consultant to the U.S. Government, Personal communication.
9. Beutler, B., 2004, Innate immunity: an overview, *Mol. Immunol.* **40**:845–859.
10. Medzhitov, R., and Janeway, C. A., Jr., 1998, An ancient system of host defense, *Curr. Opin. Immunol.* **10**:12–15.
11. Lee, S. H., Webb, J. R., and Vidal, S. M., 2002, Innate immunity to cytomegalovirus: the Cmv1 locus and its role in natural killer cell function, *Microbes Infect.* **4**:1491–1053.
12. Levy, J. A., Scott, I., and Mackewicz, C., 2003, Protection from HIV/AIDS: the importance of innate immunity. *Clin. Immunol.* **108**:167–174.
13. Basset,C., Holton, J., O'Mahony, R., *et al.,* 2003, Innate immunity and pathogen-host interaction., *Vaccine.* **21**:S12–S23.
14. Morgan, B. P., and Harris, C. L., 1999, *Complement Regulatory Proteins,* Academic Press, New York.

15. Matsushita, M., and Fujita, T., 1992, Cleavage of the third component of complement (c3) by mannose-binding protein in association with a novel C1s-like serine protease. *J. Exp. Med.* **176**:1497–1502.
16. Kadioglu, A., and Andrew, P. W., 2004, The innate immune response to pneumococcal lung infection: the untold story. *Trends Immun.* **25**:143–149.
17. Hopken, U. F., *et al.*, 1996, The C5a chemoattractant receptor mediates mucosal defence to infection. *Nature* **383**:25–26.
18. Lamm, M. E., 1997, Interaction of antigens and antibodies at mucosal surfaces, *Annu. Rev. Microbiol.* **51**:311–340.
19. Strober, W., 1990, Regulation of IgA B-cell development in the mucosal immune system, *J. Clin. Immunol.* **10**:56S–63S.
20. Salvi, S., and Holgate, S. T., 1999, Could the airway epithelium play an important role in mucosal immunoglobulin{XE "Immunoglobulins"} A production? *Clin. Exp. Allergy.* **29**:1597–1605.
21. Štvrtinová, V., Jakubovský, J., and Hulín I., 1995, Inflammation and fever, *Pathophysiology: Principles of Diseases,* Academic Press, New York.
22. Hengel, H., and Masihi, K. N., 2003, Combinatorial immunotherapies for infectious diseases, *Int. Immuno. Pharmaco.* **3**:1159–1167.
23. Garaci, E., Rocchi, G., Perroni, L., *et al.*, 1994, Combination treatment with zidovudine, thymosin alpha 1 and interferon-alpha in human immunodeficiency virus infection, *Int. J. Clin. Lab. Res.* **24**:23–28.
24. Sneller, M. C., 1996, Consensus symposium on combined antiviral therapy; overview of interferon and IL-2 combinations for the treatment of HIV infection, *Antiviral Res.* **29**:105–109.
25. Baker, D. E., 2003, Pegylated interferon plus ribavirin for the treatment of chronic hepatitis C. *Rev. Gastroenterol Disord.* **3**:93–109.
26. Nemunaitis, J., Shannon-Dorcy, K., and Appelbaum, F. R., 1993, Long-term follow-up of patients with invasive fungal disease who received adjunctive therapy with recombinant human macrophage colony-stimulating factor, *Blood.* **82**:1422–1427.
27. Phillips, P., Forbes, J. C., and Speert, D. P., 1991, Disseminated infection with *Pseudollescheria boydii* in a patient with chronic granulomatous disease: response to gamma-interferon plus antifungal therapy, *Pediatr. Infect. Dis. J.* **10**:536–539.
28. Masihi, K. N., 2000, Immunomodulatory agents for prophylaxis and therapy of infections,. *Int. J. Antimicrob. Agents,* **14**:181–191. Review.
29. Cano, P. G., and Perdigon, G., 2003, Probiotics induce resistance to enteropathogens in a re-nourished mouse model. *J. Dairy Res.* **70**:433–40.
30. Galli, L., de Martino, M., Azzari, C., *et al.*, 1990, Preventive effect of thymomodulin in recurrent respiratory infections in children, *Pediatr. Med. Chir.* **12**:229–32.
31. Masihi, K. N., and Haden, J. W., 2002 May, Protection by methyl inosine monophosphate (MIMP) against aerosol influenza virus infection in mice. *Int. Immunopharmacol.* **2**(6):835–841.
32. Williams, D. L., Li, C., Ha, T., *et al.*, 2004 Jan, Modulation of the phosphoinositide 3-kinase pathway alters innate resistance to polymicrobial sepsis. *J. Immunol.* **172**(1):449–56.
33. Hubel, K., Dale, D. C., and Liles, W. C., 2002, Therapeutic use of cytokines to modulate phagocyte function for the treatment of infectious diseases: current status of granulocyte colony-stimulating factor, granulocyte-macrophage colony-stimulating factor, macrophage colony-stimulating factor, and interferon-gamma, *J. Infect. Dis.* **185**:1490–1501. Review.
34. Liles, W. C., 2001, Immunomodulatory approaches to augment phagocyte-mediated host defense for treatment of infectious diseases, *Semin. Respir. Infect.* **16**:11–17. Review.
35. Roilides, E., and Pizzo, P. A., 1993, Perspectives on the use of cytokines in the management of infectious complications of cancer, *Clin. Infect. Dis.***17**:S385–S389. Review.
36. Sechler, J. M., Malech, H. L., White, C. J., *et al.*, 1988, Recombinant human interferon-gamma reconstitutes defective phagocyte function in patients with chronic granulomatous disease of childhood. *Proc. Natl. Acad. Sci.* **85**:4874–4878.

37. Liu, G., Zhai,Q., Schaffner, D. J., et al., 2004, Prevention of lethal respiratory vaccinia infections in mice with interferon-alpha and interferon-gamma, *FEMS Immunol Med Microbiol.* **40**:201–206.
38. Liu, G., Zhai ,Q., Schaffner, D., *et al.*, 2004, IL-15 induces IFN-beta and iNOS gene expression, and antiviral activity of murine macrophage RAW 264.7 cells, *Immunol. Lett.* **91**:171–178.
39. Bezrukov, K. Yu., *et al.*, 1991, Use of concentrated interferon{XE "Interferons"} in acute pneumonia in young children, *Current Problems in the Chemotherapy of Bacterial Infections: Theses of Reports of the All-Union Conference,* Medbioekonomika, Moscow.
40. Bezrukov, K. Yu., *et al.*, 1991, A basis for the use of interferon{XE "Interferons"} and vitamin E in acute respiratory infections in children, in *Current Problems in the Chemotherapy of Bacterial Infections: Theses of Reports of the All-Union Conference,* Medbioekonomika, Moscow.
41. Ivanov, K. S., Koshil, O. I., Petrushenko, Iu. M., Kapustin, V. M., 1992 Dec., The use of reaferon in the combined therapy of patients with hemorrhagic fever with renal syndrome. *Voem. Med. Zh.* (12):46–7.
42. Malinovskaia, V. V., *et al.*, 1995, The interferon{XE "Interferons"} system in acute tick-born encephalitis and effect on the dynamics of clinical laboratory indicators using different methods of interferon therapy, *Vopr Virusol.* **40**:234–238.
43. Harinasuta, C., *et al.*, 1985, The effect of interferon-alpha A on two cases of Japanese encephalitis in Thailand, *Southeast Asian J. Trop. Med. Public Health* **16**:332–336.
44. Gaspari, A. A., *et al.*, 1997 , Successful treatment of a generalized human papillomavirus infecction with GM-CSF{XE "Granulocyte/macrophage colony stimulating factor"} and interferon-gamma immunotherapy in a patient with a primary immunodeficiency and cyclic neutropenia, *Arch. Dermatol.***133**:491–496.
45. Johnson, B. J., *et al.* 1997, rhuIL-2 adjunctive therapy in multidrug resistant tuberculosis: a comparison of two treatment regimens and placebo, *Tuber. Lung Dis.* **78**:195–203.
46. Barbaro, G., 1997, Effect of recombinant human GM-CSF{XE "Granulocyte/macrophage colony stimulating factor"} on HIV-related leukopenia: a randomized, controlled clinical study, *AIDS* **11**:1453—1461.
47. Kuznetsov, V. P., *et al.*, 1991, Combined preparations of interferon{XE "Interferons"} and cytokines{XE "Cytokines"} in bacterial infections. N. F. Gamaleya Scientific Research Institute of Epidemiology and Microbiology, Moscow. Current Problems in the Chemotherapy of Bacterial Infections: Theses of Reports of the All-Union Conference, October 22–24, Medbioekonomika, Moscow.
48. Viero, P., *et al.*, 1997, G-CSF after or during intensive remission induction therapy for adult acute lymphoblastic leukemia: effects, role of patient pretreatment characteristics, and costs, *Leuk. Lymphoma,* **26**:153–161.
49. Brinton, M. A., *et al.*, 1993, Characterization of murine Caraparu Bunyavirus liver infection and immunomodulator-mediated antiviral protection, *Antiviral Res.* **20**:155–171.
50. Morrill, J. C., *et al.*, 1991, Recombinant human interferon-gamma modulates Rift Valley fever virus infection in the rhesus monkey, *J. Interferon* Res. **11**:297–304.
51. Kende, M., 1985, Prophylactic and therapeutic efficacy of poly (I, C)–LC against Rift Valley fever virus infection in mice, *J. Biol. Response Mod.* **4**:503–511.
52. Pinto, A. J., *et al.*, 1990, Comparative therapeutic efficacy of recombinant interferon-alpha, -beta, and -gamma against alphatogavirus, bunyavirus, and herpesvirus infections, *J. Interferon* **10**:293–298.
53. Tamura, M., *et al.*, I987, Effects of human and murine interferons{XE "Interferons"} against hemorrhagic fever with renal syndrome virus (Hantaan virus), *Antiviral Res.* **8**:171–178.
54. Bulychev, L. E., *et al.*, 1995, The efficacy of the therapeutic and prophylactic actions of immunomodulators in experimental actions of immunomodulators in experimental infection due to the causative agent of Venezuelan equine encephalomyelitis,{XE "Venezuelan equine encephalomyelitis (VEE)"} *Antibiot. Khemoter* **40**:28–31.

55. Sergeev, A. N., *et al.*, 1995, The efficacy of the emergency prophylactic and therapeutic actions of immunomodulators in experimental filovirus infection, *Antibiot. Khemoter* **40**:24–27.
56. Arroyo, J. I., *et al.*, 1988, Effect of human gamma-interferon on yellow fever virus infection, *Am. J. Trop. Hyg.* **38**:647–650.
57. Hotta, H., *et al.*, 1984, Effect of interferons{XE "Interferons"} on dengue virus multiplication in cultured monocytes{XE "Monocytes"}/macrophages,{XE "Macrophages"} *Biken. J.* **27**:189–193.
58. Loginova, S. Ia., *et al.*, 1997, Study of the effectiveness of recombinant alpha-2-interferon and its inducers in primates with orthopoxvirus infection, *Vopr Virusol.* **42**:186–188.
59. Nakajima, R., and Brubaker, R. R., 1993, Association between virulence of *Y. pestis* and suppression of gamma-interferon and tumor necrosis factor{XE "Tumor necrosis factor"}-alpha, *Infect. Immun.* **61**:23–31.
60. Kuhlberg, B. J., *et al.*, 1998, Recombinant murine granulocyte colony stimulating factor protects against acute disseminated Candida albicans infection in nonneutropenic mice, *J. Inf. Dis.* **177**:175–181.
61. Davey, R. T., Jr. *et al.*, 1997, Subcutaneous administration of interleukin-2 in human immunodeficiency virus type 1-infected person, *J. Infect. Dis.***175**:781–789.
62. Tilg, H., *et al.*, 1993, Pilot study of natural human interleukin-2 in patients with chronic hepatitis B. Immunomodulatory and antiviral effects, *J. Hepatol.* **19**:259–267.
63. Jacobsen, E. L., *et al.*, 1996, Rational interleukin{XE "Interleukins"} 2 therapy for HIV positive individuals: daily low doses enhance immune function without toxicity, *Proc. Natl. Acad. Sci.* **93**:10405–10410.
64. Quinnan, G. V., Jr., 1998, Cytokines{XE "Cytokines"} in the treatment and prevention of infectious diseases, in: *Infectious Diseases* (2nd ed., Gorbach, S. L., *et al.*, eds.) W. B. Saunders Company, Philadelphia, pp. 92–101.
65. Professor Vladimir Zavialov, Director, Lyubuchany Institute of Engineering Immunology, Lyubuchany, Russia, Personal communication.
66. Huland, E., *et al.*, 1999, Treatment of pulmonary metastatic renal-cell carcinoma in 116 patients using inhaled interleukin-2 (IL-2)., *Anticancer Res.* **19**:2679–2683.
67. Huland, E., *et al.*, 1994, Inhaled interleukin-2 in combination with low-dose systemic interleukin-2 and interferon alpha in patients with pulmonary metastatic renal-cell carcinoma: effectiveness and toxicity of mainly local treatment, *J. Cancer Res. Clin. Oncol.* **120**: 221–228.
68. Nakamoto, T., *et al.*, 1997, Inhalation of interleukin-2 combined with subcutaneous administration of interferon for the treatment of pulmonary metastases from renal cell carcinoma, *Int. J. Urol.* **4**:343–348.
69. Heinzer, H., *et al.*, 1999, Subjective and objective prospective, long-term analysis of quality of life during inhaled interleukin-2 immunotherapy. *J. Clin. Oncol.* **17**:3612–3620.

5

Smallpox: Pathogenesis and Host Immune Responses Relevant to Vaccine and Therapeutic Strategies

MICHELE A. KUTZLER, KENNETH E. UGEN, and DAVID B. WEINER

1. INTRODUCTION

Recently, smallpox has been a disease of only historical interest since the certification of its eradication by the World Health Organization on May 8, 1980.[1] However, there is a growing awareness and apprehension regarding possible bioterrorist threats with these concerns escalating since the tragedy on September 11, 2001. Accordingly, there is an increased interest in understanding smallpox-induced pathogenesis as well as in the development of new vaccines and therapeutics. This chapter will discuss the history of smallpox infection and its eradication. Discovery of methods for protection against naturally occurring smallpox infection will also be discussed, as well as clinical and epidemiological features of infection, virus structure and pathogenesis as well as host defense and the immune response. An improved understanding of the disease is leading to new methods of prophylaxis and therapy that are discussed in this chapter. In addition, current vaccination strategies will also be reviewed since the development and licensure of novel smallpox vaccines

MICHELE A. KUTZLER and DAVID B. WEINER • Department of Pathology and Laboratory Medicine, University of Pennsylvania, School of Medicine, PA 19104, USA. KENNETH E. UGEN • Department of Medical Microbiology and Immunology and Center for Molecular Delivery, University of South Florida College of Medicine, Tampa, FL 33612, USA.

that may be safely used to immunize those with exposure and/or risk factors is currently of high priority.

2. HISTORY OF SMALLPOX INFECTION AND ITS ERADICATION

Once prevalent throughout the world as an endemic infection, wherever concentrations of population were sufficient to sustain transmission, smallpox usually found its major reservoir in children because there is no animal reservoir for smallpox. Therefore, the virus had to spread continually from human to human to survive with epidemics occurring when travelers carried the agent to outlying populations that lacked immunity. Smallpox was first described in South Africa by de Korte in 1904 and in the United States by Chapin in 1913, and subsequently became prevalent throughout the United States, parts of South America, Europe as well as some areas of Eastern and Southern Africa.[2] The first evidence that smallpox emerged as a pathogen was some time after the first human agricultural settlements, about 10, 000 B.C., while the first scientific evidence for smallpox was identified in the mummified remains of the 18th Egyptian dynasty and Ramses V.[3] Written descriptions of smallpox typically did not appear until the fourth century A.D. in China and the tenth century in southwestern Asia. However, earlier descriptions, although rare, did appear such as the one by Thucydides in 430 B.C. in Athens.[4] There is no Greek or Latin word for smallpox, although the name *variola,* derived from the Latin *varius,* meaning pimple was first used during the sixth century by Bishop Marius of Switzerland. By the tenth century, the word *poc* or *pocca,* a bag or pouch, was used to describe smallpox and the prefix *small* was used to distinguish variola, the "small pox", from syphilis, the "great pox".[5]

The first immunization procedure was termed "variolation", in which material from pustules or scabs from infected persons were deliberately inoculated into the skin, a method first carried out in India sometime before A.D. 1000.[6] This method resulted in an infection that was usually less severe than an infection acquired naturally by inhalation of droplets. Importantly, the method of variolation was brought to England by Lady Mary Wortley Montague. She had been disfigured by smallpox in 1715 and while in Istanbul with her ambassador husband she became aware of the practice of variolation. She, in fact, had her son and daughter "variolated" which led to some acceptance of this method, which then spread throughout England. Although a small percentage of individuals "purposely" infected by the variolation method did not survive, the mortality was considerably lower than in those naturally exposed and infected with smallpox. In 1796, Edward Jenner discovered that infection with a more benign poxvirus caused by cowpox virus, prevented subsequent smallpox infection. He called the material *vaccine,* from the Latin *vacca,* meaning cow. The process of *vaccination* then began to be employed widely in many countries of Europe, and within a decade, it had been transported to countries throughout the world. Several reasons existed for opposition to the use of Jenner's vaccine strategy including being able to find cows infected with cowpox, or in some cases, there was a significant opposition among religious leaders who opposed the principle of infecting humans with an animal-derived serum as being unnatural. Moreover, confidence in the procedure was challenged when some individuals who had been previously "vaccinated" acquired smallpox infection. With the flank of the calf offering an adequate and safer supply, the numbers of vaccinations in Europe increased and subsequently the incidence of smallpox in more industrialized countries diminished more rapidly. However, in less developed areas, smallpox infection continued until the middle of the twentieth century due to the development of a freeze-dried vaccine.[7] With such a valuable vaccine available, a global eradication plan was initiated by the Pan American Sanitary Organization in 1950, followed by a plan in 1958 by the Union of Soviet Socialist Republics proposed to the World Health Assembly that a global smallpox eradication plan should be undertaken. However, it was not until 1966, when the World Health Organization provided more funding that a more intensified program was initiated. Interestingly, in 1967, an estimated 10 to 15 million smallpox cases occurred in 31 countries in which the disease was endemic. Therefore, the campaign carried out massive vaccinations in each country reaching at least 80% of the population, and also developed a system to contain cases and outbreaks. A total of 3, 234 cases of smallpox were reported from Eastern Africa to the World Health Organization[8] in the period January 1–December 6, 1977. The last reported indigenous known case of smallpox occurred in Somalia on October 26, 1977 in the Merca District. The source of this case was a known outbreak in the nearby district of Kurtuware and all 211 contacts were traced, revaccinated, and kept under surveillance. The last known case of smallpox in Ethiopia occurred on August 9, 1976, in El Kere Region, while in Kenya, the last case was on February 5, 1977, in the Mandera District (1977). In 1980, the World Health Organization declared smallpox globally eradicated.[2,9,10]

3. CLINICAL AND EPIDEMIOLOGICAL FEATURES

Smallpox is a viral disease unique to humans, which is spread from person to person by inhalation of air droplets or aerosols. Twelve to 14 days after infection, an average patient has a 2- to 5-day period of high fever, malaise, and prostration with headache and backache followed by the development of maculopapular rash over the face which then spreads to the extremities. These clinical symptoms are listed in Table I as compiled and summarized by the Centers for Disease Control and Prevention (CDC). The rash appears on the mucosa of the mouth and pharynx, the face, and the forearms and spreads to the trunk and legs, and becomes vesicular and then pustular, characterized by round, tense, and deeply embedded in the dermis when crusts begin to

TABLE I
Clinical Symptoms of Smallpox as Described by the CDC

Clinical Stage/Sympton	Duration	Contagious	Description
Incubation Period	7 to 17 days	NO	Following initial exposure, sympton free
Initial Symptoms	2 to 4 days	YES	Fever, malaise, head and body aches, vomiting "prodromal phase"
Early Rash	4 days	YES	Rash emerges as small red spots on tongue and mouth Rash develops into sores that break open Rash then appears on skin, starting on face and spreading to arms and legs, then to hands and feet Within 24 hours, rash spreads to all parts of the body As rash appears, fever falls and person may feel better By the third day, the rash becomes raised bumps By the fourth day, the bumps fill in with thick, opaque fluid and often have a depression in the center. Fever will rise at this time and remain high until scabs form over the bumps
Pustular Rash	5 days	YES	The bumps become pustules, sharply raised, round and firm
Pustules and Scabs	5 days	YES	The pustules begin to form crust and then scab
Resolving Scabs	6 days	YES	Scabs fall off, leaving marks on skin that eventually become pitted scars
Scabs Resolved	—	NO	Scabs have fallen off and person is no longer contagious

form by the 8th or 9th day. Eventually scabs form, which separate and leave pigment-free skin, and pitted scars. In 5 to 10% of smallpox patients, more rapidly progressive, malignant disease develops. Late in the 1st week or during the 2nd week of illness, death occurs due to the effects of overwhelming viremia.[11,12] On occasion, a severe and fatal hemorrhagic form occurs with extensive bleeding into the skin and gastrointestinal tract. Importantly, vaccination before exposure or within 2 to 3 days of exposure offers complete protection against the disease, while vaccination as late as 4 to 5 days post-exposure may protect against death. Although the epidemiology of smallpox infection including morbidity and mortality has been described, the molecular mechanisms of smallpox-induced death are unclear.

Variola virus spreads most readily in the dry and cool winter months but can be transmitted in any climate and in any part of the world. The age distribution of cases depends primarily on the degree of smallpox susceptibility in the population. In most areas, cases predominated among children because adults were protected by immunity induced by vaccination or previous smallpox infection.[13] In rural areas that had not been previously vaccinated or had smallpox infections, the distribution would be similar to the age distribution of the population.[13] The triumph of the global smallpox eradication has led to an irony in that the ensuing worldwide cessation of vaccination has rendered many of today's population susceptible to infection, which has resulted in smallpox becoming a significant bioterror weapon.

4. VIRUS STRUCTURE AND CLASSIFICATION

Poxviruses are the largest and most complex viruses that infect humans and belong to the genus *Orthopoxvirus*, family Poxviridae, which includes the agents vaccinia, monkeypox, cowpox, camelpox, and ectromelia.[14] Poxviruses are made of a single molecule of double-stranded DNA and have the ability to replicate in the cytoplasm rather than the nucleus of susceptible cells.[15] The linear genome contains approximately 200 genes ranging in size from 130-kb to 260-kb with those in the central region encoding proteins involved in virus uncoating, genome replication or virion structure. The DNA polymerase has conserved the sequences of these genes, and the flanking regions contain genes encoding proteins that modify the intra- and extracellular environment so that virus can replicate and spread. The virions are large and brick-shaped and range in size from 160 nm to 300 nm. Poxvirus replication occurs in the cytoplasmic inclusions, where infecting virions are partly uncoated by cellular enzymes and fully uncoated by viral enzymes released from the virion core. The replication cycle can be divided into functions controlled by early (pre-replicative) gene products and late (post-replicative) gene products. Most virions remain within cells and lack the outer envelope found on naturally released virions. In addition, each infected cell produces two different kinds of virions.[16] The majority of intracellular mature virions remain within necrotic cells and are shed in skin debris or saliva droplets, where they serve as sources of infection. The second type of virion makes up a small percentage that acquire an additional membrane and are transported to the cell surface where they become extracellular enveloped virions that are responsible for cell to cell spread and may participate in systemic dissemination. The membrane antigens may be the targets for humoral immunity and neutralizing antibody responses; however, the core antigens are not expressed on the viral membrane and therefore are only the targets of cellular immunity.

5. PATHOGENESIS, HOST DEFENSE, AND THE IMMUNE RESPONSE

Naturally occurring smallpox infection is initiated by the implantation of variola virus on the oropharyngeal or respiratory mucosa. Replication at the point of entry is followed by infection of mononuclear phagocytic cells in regional lymph nodes, possibly with further spread through the bloodstream to similar cells in the liver, spleen, and other tissues. Virions in droplets expressed from nasal and oropharyngeal secretions are far more infectious than virions bound in the scab itself. After migration and multiplication in regional lymph nodes, an asymptomatic viremia develops around the 3rd or 4th day, followed by multiplication of virus in the spleen, bone marrow, and lymph nodes. A secondary viremia begins around the 8th day, accompanied by fever and toxemia. This means that the incubation period has ended when the release of inflammatory mediators from infected cells caused fever and other symptoms, and the spread of virus (either within infected monocytes or as free virions) to capillaries in the skin and mucosal membranes initiates the rash. The virus, contained in leukocytes, localizes in small blood vessels of the dermis beneath the oral and pharyngeal mucosa and subsequently infects adjacent cells. In the skin, this process results in the characteristic maculopapular lesions, and later vesicular and pustular lesions. After reaching the skin, disease severity is determined by the ability of host responses to limit viral replication during the incubation period, as reflected by the level of secondary viremia. Secondly, once viral dissemination has occurred, many features of severe illness are the result of host inflammatory responses including the release of chemokine, cytokines, and other mediators into the bloodstream, causing vascular dysfunction, coagulopathy, and multiorgan failure resembling septic shock.[17,18]

Both the humoral and cellular immune responses play important roles in host defense against smallpox infection. Specifically, the humoral response to smallpox infection results in the production of short-lived IgM and persistent IgG, and it may be elicited by inactive virions or viral antigen by non-enveloped or enveloped virus. There are three classes of antibodies important in the host immune response to infection.[16,19–22] The first are antibodies elicited against both non-enveloped virus and enveloped virus resulting in the neutralization of viral infection. The second types of antibodies are those that combine with complement to lyse virus-infected cells. The third are antibodies that combine with circulating antigen to produce immune complexes resulting in some of the toxic symptoms seen in the host during smallpox infection. These classes of antibodies are listed in Table II. Hemagglutinin-inhibiting and neutralizing antibodies could be detected beginning about the 6th day of illness, or about

TABLE II
Classes of Antibodies Important to Host Immune Responses to Smallpox

Elicited against enveloped and non-enveloped virus resulting in neutralization of virus infection
Those combining with complement to lyse virally infected cells
Those which combine with circulating antigen to produce immune complexes resulting in some of the toxic symptoms noted in the host during smallpox infection

18 days after infection, while complement-fixing antibodies could be identified approximately 2 days later.[23,24]

Moreover, in addition to B cell responses, the cellular response generated following infection also elicits CD4+ T helper and cytotoxic CD8+ T cells, as well as natural killer cells to combat infection.[19,25] The direct effects of T cells on virus-infected cells as well as secreted products including IFN-gamma play a role in host immune responses to smallpox infection. Therefore, it is likely that both humoral and cell-mediated immunity are important in protecting against smallpox. Patients with genetic defects in either B cell or T cell immunity are at increased risk of complications following smallpox vaccination.[26] Moreover, various animal models have also shown that adoptive transfer of either neutralizing antibodies[27–29] or virus-specific T cells[30,31] can provide full protection against vaccinia infection.

6. FEATURES OF SMALLPOX MAKING IT A LIKELY BIOTERROR AGENT

Of the potential biological weapons, smallpox poses by far the greatest threat, albeit because of its clinical and epidemiological properties. Smallpox poses a serious threat because a large segment of the population that has not been previously vaccinated is now susceptible, due to the continued vaccination program being halted several decades ago coincident with the eradication of the disease. It is currently debated by virologists and immunologists as to what percentage of the population vaccinated 30 years ago is protected in terms of morbidity and mortality. It is expected that the case fatality rate after infection with smallpox in the non-vaccinated/non-protected individuals is 30%. Moreover, virus, in an aerosol form, can survive for 24 hours or more and is highly infectious at low doses.[32] Other features of smallpox that make it a likely bioweapon candidate is that it can be produced in large quantities, is stable for storage and transportation, and is spread from person to person.[33]

7. HISTORY AND POTENTIAL OF SMALLPOX AS A BIOWEAPON

Smallpox was first used as a biological weapon during the French and Indian Wars (1754–1767) by British forces in North America. Soldiers distributed blankets that had been used by smallpox patients with the intent of initiating outbreaks among American Indians. Epidemics occurred, killing more than 50% of affected tribes. However, following the global campaign to eradicate smallpox globally in 1977, the World Health Organization required that all countries cease vaccination (1980). The WHO committee later recommended that all laboratories destroy their stocks of variola virus in June 1999.[34] The deliberate reintroduction of smallpox as an epidemic disease would be an international crime of unprecedented proportions, but it is now regarded as a possibility because the last 4 years have been marked by escalating concerns in the United States about the threat of biological weapons. This is not unconceivable, as Dr. Kenneth Alibek, a former first deputy chief of research and production for the Russian biological weapons program, has reported that smallpox virus had been mounted in intercontinental ballistic missiles and in bombs for strategic use.[35] Former Soviet scientists successfully weaponized many agents and created missile delivery systems for smallpox, while active research was undertaken to engineer more virulent strains. Moreover, with the collapse of the Soviet Union, microbe stocks, including the smallpox virus and other technologies, have possibly found their way into the hands of unknown individuals, increasing the risk of transfer of these materials to terrorists. At least 17 nations are believed to have had offensive biological weapons programs, and scientists with this type of expertise are believed to have been actively recruited by Libya, Iran, Syria, Iraq, and North Korea.[11,12,36,37]

With increasing awareness has come a growing attempt to defend against the possibility of biological warfare and terrorism. One of the best defenses will continue to be vaccines and other treatment options, and this requires the development of new and improved vaccines and treatment against smallpox.

8. SMALLPOX VACCINES AND ANTIVIRAL THERAPIES

The events of September 11, 2001, coincident with the use of anthrax as a bioweapon, underscored the potential for use of biological agents as weapons. This concern prompted the Bush administration to make recommendations for the use of a smallpox vaccine in a pre-event vaccination program. This has prompted revisiting the safety concerns for the existing smallpox vaccine, (Dryvax), as well as the need for developing an efficacious but safe vaccine against smallpox. To that end, eradication of naturally occurring smallpox has not eliminated the need for an improved smallpox vaccine. The current threat posed by potential bioterrorist attacks has brought forth the need for new vaccination strategies due to the large numbers of individuals living in North America who are elderly, women of child-bearing age, non-vaccinia-virus (VACV) vaccinated, and immuno-compromised (HIV infection, transplantation recipients, intravenous drug users as well as other individuals on immunosuppressive therapies). Although the traditional live VACV vaccine for smallpox has proven to be effective in conferring protection where only 1 out of 10, 000 individuals vaccinated experienced significant adverse effects during the 1960s, in today's society, where there are a large number of susceptible individuals, the potential complications arising from adverse events following vaccination are likely unacceptable. It is imperative that instead of abandoning the current live VACV vaccine, leaving us vulnerable to terrorist attacks, a new strategy that improves the safety of the current vaccine, yet increases its potency, is developed and implemented.

8.1. Smallpox Vaccine Strategies and Related Issues

The currently available smallpox vaccine is a lyophilized preparation of live vaccinia virus (VACV), generated from a New York City calf lymph strain (NYCCL), obtained from infecting cows by scarification, with subsequent lesional removal by scraping.[38] This vaccine is one of the oldest and most successful vaccines ever developed. However, this live virus vaccine also had reports of several adverse complications.[39–53] Dermatologic and central nervous system disorders were the most frequently recognized adverse events, including vaccinia necrosum, a complication with case fatality rates of 75 to 100% that occurred almost exclusively in persons with cellular immunodeficiency.[49] Generalized vaccinia was reported, resulting in rare blood-borne dissemination of virus in normal persons. More rare diseases such as pericarditis,[54] arthritis,[55] and malignant tumors at vaccination scars[56] have been described in case reports. Eczema vaccinatum was associated with case-fatality rates of up to 10% overall and 30 to 40% in children less than 2 years old.[57] Moreover, approximately seven to nine deaths per year were attributed to vaccination. Infants were identified as the highest-risk population, with death resulting from postvaccinal encephalitis.[43,44] Thus, the adverse events associated with the current smallpox vaccine are well documented, and new strategies must be developed to prevent further complications.

An important concern, as indicated earlier, is that there are a significant number of immunocompromised (HIV infected individuals) and elderly populations, as well as pregnant women, intravenous drug users, transplant recipients, and individuals on immunosuppressive drugs, who are at significant risk of developing adverse reactions after smallpox vaccination. These risks groups are listed in Table III. In North America, the fact that an overwhelming number of people, in theory, could be hospitalized due to serious complications, is of major concern, and many people could even die. Currently, several options are

TABLE III
Groups at Risk of Adverse Reactions to Vaccinia-Based Smallpox Vaccine

Pregnant women
Therapeutically induced immunosuppresssion such as in those receiving cancer chemotherapy or anti-organ rejection drugs
Persons with HIV infection
Persons with history of eczema
Intravenous drug users
Potentially the young or elderly due to immune incompetence

available. Firstly, we can go forth and use the currently stockpiled vaccine, and risk a significantly higher rate of complications than occurred in the 1960s. Alternatively, we could design and manufacture a novel efficacious and safe vaccine, disregarding the current one. However, this obviously leaves the country very vulnerable, and without protection, until such a vaccine is developed. A third strategy would be to continue with the current vaccine while pursuing studies to improve on its safety profile, while not interfering with its potent immune responses.

Several vaccine-related issues need to be addressed to ensure public safety. These include the need for a modern alternative to the live animal-produced stock and to determine immune correlates relevant for the twenty-first century in order to test new, safer vaccine candidates. Hammarlund *et al.*[58] provide evidence that vaccine-induced immunity persists for many years, perhaps lifelong. This is an interesting finding since millions worldwide, and about 90% of individuals in the United States over the age of 35, were vaccinated before the end of the mass vaccination campaigns. Although the status of their immunity against smallpox has been under debate,[38,59–61] Hammarlund *et al.*[58] measured T cell immunity against vaccinia virus in 306 vaccinees, up to 75 years after their last vaccination. Interestingly, within the first 7 years after vaccination, CD4+ and CD8+ T cell responses remained high and then declined slowly over decades, with the decline in CD4+ T cells occurring more slowly. Yet even between 41 and 75 years after vaccination, most vaccinees showed some CD4+ and some CD8+ T cell immunity. Conversely, the humoral immune response in these cohorts showed that most maintained stable antibody responses for up to 75 years after vaccination, suggesting lifelong immunity. Studying the usefulness of additional vaccination for people later in life and the expansion of their T cell responses is important. The persistent immune responses observed by Hammarlund *et al.*[58] suggest that side- effects of vaccination, such as eczema vaccinatum, should occur infrequently in revaccinated individuals because most side-effects of vaccinia are observed upon primary series of immunization. Other implications from this study are that many people in the United States over 40 years of age are likely have some immunity to smallpox, aiding in "herd immunity", therefore, the focus should be on the young population and the immunocompromised for development of new vaccines as they are unprotected.

Weltzin *et al.*[62] developed a new tissue culture method for producing smallpox vaccine that bypasses the methodology requiring scraping the hides of cows infected with vaccinia virus, resulting in replacement stocks for those without immunity. This is important because the current vaccine stocks would probably not fulfill the demands of unvaccinated individuals in the United States.[38,59–61] Weltzin *et al.*, adapted the existing DryVax vaccine, which is derived from the crossprotective vaccinia virus, to a human cell line for production in tissue culture. In a small clinical study in humans, 100% subjects vaccinated with the new vaccine candidate (ACAM-100) versus 97%

DryVax- vaccinated subjects exhibited the hallmark of vaccine take, a significant cutaneous reaction at the site of inoculation/scarification. The vaccines had similar safety profiles with each participant experiencing at least one mild to moderate adverse event. Moreover, DryVax induced higher antibody titers, while ACAM-100 vaccination seemed to result in stronger CD4+ T cell responses. Taken together, the two aforementioned studies provide new strategies toward the goal of development of next-generation vaccines.

Modified vaccinia virus ankara (MVA) was generated by more than 500 passages of vaccinia virus in chick embryo fibroblasts, during which it acquired multiple deletions and mutations and lost the capacity to replicate efficiently in human and most other mammalian cells.[63–66] MVA is being considered as a replacement for the present smallpox vaccine for those with a high risk of adverse complications because immune responses elicited by one or two doses of MVA should approach, although not necessarily equal, those of the licensed smallpox vaccine, or for more general use as a pre-vaccine since MVA should reduce the reaction to a subsequent smallpox vaccination without blocking the resulting immune response. Earl *et al.*[67] compare the highly attenuated MVA with the licensed DryVax vaccine in a monkey model, since licensing includes comparative immunogenecity and protection studies in non-human primates. After two doses of MVA or one dose of MVA followed by DryVax, antibody binding and neutralizing titers as well as T cell responses were equivalent or higher than those induced by DryVax alone. After the challenge with monkeypox virus, non-immunized animals developed more than 500 pustular skin lesions and became gravely ill or died, whereas vaccinated animals were healthy and asymptomatic. These findings of similar humoral and cellular immune responses to MVA and DryVax in non-human primates and substantial protection against a monkeypox virus challenge are important steps in the evaluation of MVA as a replacement vaccine for those with increased risk of severe side-effects from the standard live vaccine, or as a pre-vaccine. Perhaps one approach would be to vaccinate with MVA before a smallpox threat, with the thought that standard vaccine or MVA would be used as a boost. However, further experiments would need to be carried out to determine the longevity of protection, and the consequences of delayed boosting and dosage effects as well as, most importantly, other approaches to be used in the case of immunocompromised individuals. An important study by Wyatt *et al.*[68] examines the safety of MVA in immune-deficient mouse models and shows that overlapping immune responses protect immune-competent and immune-deficient mice against a lethal intranasal challenge with a pathogenic vaccinia virus.

Although vaccinia virus is highly immunogenic and is known to confer long-lasting protective immunity to smallpox, the adverse events associated with current vaccine strategies pose a significant obstacle to successful vaccination campaigns. Another novel strategy for improved safe vaccines against smallpox includes the use of DNA vaccines. DNA vaccines induce antigen-specific immune responses following the direct injection of non-replicating plasmids into a host target tissue.[69] Once injected, plasmids drive the synthesis of specific foreign proteins within the inoculated host and mimics natural infection. The host provides post-translational modifications to antigen that faithfully reproduce native conformations. These host-synthesized viral proteins then become the subject of immune surveillance via both the MHC class-I and class-II pathways. These processes lead to elicitation of protective immunity against an infectious agent, or pathogen, primarily by activating both the humoral and cellular arms of the immune system.[70–73] Moreover, DNA vaccines can be constructed to function with many safety features as well as the specificity of a subunit vaccine, there is little risk of reversion to a disease-causing form, and there is no risk for secondary infection as the material injected is non-replicating, and non-infectious. In addition to their added safety, DNA vaccines are highly flexible; encoding genes for immunologic inhibition, or cross-reactivity (autoimmunity) can be altered or deleted altogether. DNA vaccines possess greater stability, and can be easily manufactured on a large scale. The unique features of nucleic immunization make it well suited as an immunization/immune therapy strategy especially when safety in immunocompromised individuals is a concern.

Gene gun-delivered-DNA vaccine approach used to test several vaccinia genes and gene combinations for immunogenicity and protective efficacy in mice resulted in 100% protection of those mice challenged with a lethal dose of vaccinia.[74,75] The authors then moved on to a study of a DNA vaccine comprised of four vaccinia virus genes (L1R, A27L, A33R, and B5R) administered by gene gun in rhesus macaques and were able to first demonstrate significant immunogenicity of the plasmid in this animal model, and later demonstrated protection from severe disease following challenge with monkeypox virus.[75,76] The authors selected these four immunogens due to their role as targets of neutralizing or otherwise protective antibody responses.[27–29,77–79] Animals vaccinated with a single gene L1R, which encodes a target of neutralizing antibodies, developed severe disease but survived with clinical symptoms of the monkeys challenged with a four-fold lower dose of virus. These data support the notion that a subunit (gene- or protein-based) poxvirus vaccine has the potential to mimic the protection afforded by live vaccinia administered by scarification. Such a vaccine would contribute greatly to vaccination strategies aimed at reducing the health hazards of the present smallpox vaccine.

8.2. Smallpox Antiviral Therapies

Vaccination against variola has clearly been responsible for the elimination of naturally occurring smallpox infections in the world. However, the concern that bioterrorists may use smallpox as a bioweapon has stimulated the interest to characterize and develop antiviral agents against this poxvirus as an alternative or adjunct to vaccination. In addition, effective antiviral agents are important for the treatment of the potentially serious and life-threatening complications that can occur from smallpox vaccination. Currently, the only available treatment of infectious complications resulting from vaccination against smallpox is vaccinia immune globulin (VIG) that is generated from hyperimmune sera from vaccinees.[80,26,81,82,47] Novel antiviral agents against variola will therefore be useful as both a therapy against infection as well as an alternative/adjunct to VIG for the treatment of vaccine-induced complications.

Another drug utilized for the treatment of vaccine-induced complications was methisazone,[83] which although toxic was reported to accelerate the resolution of eczema vaccinatum and was beneficial against progressive vaccinia.[83,84] However, the lack of properly controlled studies using this drug made it difficult to access the specific efficacy of this drug. As such, methisazone is no longer in use. Several other antivirals have been used to treat vaccinia infection in animals, but many of them have proven to have too much systemic toxicity for human use. These antivirals include ribavirin,[85,86] cidofovir,[86–89], 5-iodo-2′-deoxyuridine,[90,91] 2-amino-7-[(1,3-dihydroxy-2-propoxy)methyl]purine[88] adenine arabinoside[92,93] and trifluorothymidine.[94,95] Of these ribavirin, cidofovir, and trifluorthymidine have had some clinical use for the treatment of vaccinia vaccine-induced conditions as well as having utility against other disorders. Other agents with anti-vaccinia activity have included various nucleoside analogues[96] as well as interferon.[97,98] Although some of these pharmacologics have been demonstrated to have anti-vaccinia activity, the accepted standard for the treatment of vaccine-induced complications has been the VIG preparation. However, it is clear that a more efficient and standardized antibody/antisera preparation is needed particularly if widespread vaccination/re-vaccination is required to be implemented in the future. Generation of a cocktail of human or humanized monoclonal antibodies against vaccinia would potentially be useful as an alternative or replacement for the VIG preparation.

In addition to the development of antiviral and immune-based approaches to treat smallpox vaccine-induced adverse events, it is likewise important to develop new prophylactics and therapeutics for smallpox infection itself. Safer vaccines are needed, which have a lower incidence of induction of adverse events that are associated with the current vaccinia-based vaccine preparation. Such novel vaccines would utilize non-live attenuated preparations such as DNA vaccines. In terms of the development of novel antivirals against variola it is important to have a comprehensive knowledge of the cell and molecular biology of poxviruses. Novel prophylactic/therapeutic targets would include variola enzymes[99] as well as viral.[100] Byrd and colleagues have recently generated a structural model of the vaccinia virus 17L proteinase using a homology-based bioinformatics approach and a large library (excess of 50, 000) of chemical compounds some of which have shown some antiviral activity.[100] To date, however, the only drug accepted to possess potential directly against smallpox has been cidofovir ((S)-1-(3-hydroxy-2-phosphonylmethoxypropyl) cytosine = HPMPC). This drug has been previously established to possess antiviral activity against cytomegalovirus (CMV) and is approved for clinical use for the treatment of CMV retinitis in AIDS patients[101]. In addition, cicofovir has been shown to have biological activity against other herpes viruses including human herpes viruses types 6, 7, and 8 as well as against varicella zoster virus and some polyoma, papilloma, and adenoviruses.(86) Notably, and importantly, it has been demonstrated that cidofovir could successfully be used as a preventative and therapy against lethal vaccinia infection in severe combined immune deficient (SCID) mice.(88,102,103) In addition, it has been shown to have efficacy against cowpox infections in mice and monkeys.(104–106) Lastly, cidofovir has been demonstrated to show efficacy against poxvirus infection in humans, i.e., molluscum contagiosum and orf (sheep pox).(107–110) These observations establish cidofovir to currently have the greatest potential as an antiviral agent against variola infection. It is anticipated that for clinical use against variola in humans cidofovir could be used in cases where infected individuals are unable to obtain a dose of the vaccine within 4 days after the initial contact with the dsisease.

9. CONCLUSIONS

The presence of resurgent smallpox infection is always a concern, especially given the enormous efforts that have been made to eradicate what has been characterized as one of the most devastating of all diseases. Unfortunately, smallpox as a bioterrror agent is a legitimate threat, with safety issues with the current vaccine stocks being of major concern. Moreover, the manufacturing process used to create the smallpox vaccine previously used is not suitable for today's vaccine production standards. MVA is a likely first next-generation approach for novel smallpox vaccines; however, there are major issues as to what the correlate of protection is and what should be in the boost injection for these vaccines. Importantly, immune-deficient individuals would continue to be a high-risk group for these live attenuated-based vaccines. Due to safety and manufacturing issues, DNA vaccine strategies and other recombinant strategies are likely important tools for the approaching development of novel vaccines for smallpox, particularly in the immunocompromised. Most critical to the development of smallpox strategies is the development of quantitative cellular immunological assays and determination of baseline immune responses to facilitate vaccine development and possible use as surrogate correlates. In addition, the discovery of new sources of non-immune/vaccine-based therapies outside of vaccines is important and these studies are currently underway.

REFERENCES

1. World Health Organization on smallpox, 1980, *J. Trop. Med. Hyg.* **83**:47–48.
2. Fenner, F., 1982, A successful eradication campaign. Global eradication of smallpox, *Rev. Infect. Dis.* **4**:916–930.

3. Hopkins, D. R., 1985, Smallpox entombed, *Lancet.* **1**:175.
4. Hopkins, D. R., and Berce, Y. M., 1985, The spread of Jenner's vaccine: social mobilization in the early nineteenth century, *Assignment Child.* **69–72**:225–230.
5. Cooray, M. P., 1965, Epidemics in the course of history, *Ceylon Med. J.* **10**:88–96.
6. Dixon, M. F., 1970, Smallpox vaccination, *Br. Med. J.* **2**:539.
7. Collier, L. H., 1955, The development of a stable smallpox vaccine, *J. Hyg. (London)* **53**:76–101.
8. WHO at the crossroads. Report of the Director-General to the thirtieth world health assembly on the work of WHO in 1976, 1977, *WHO Chron.* **31**:207–238.
9. Fenner, F., 1980, The global eradication of smallpox, *Med. J. Aust.* **1**:455.
10. Fenner, F., 1993, Smallpox: emergence, global spread, and eradication, *Hist. Philos. Life Sci.* **15**:397–420.
11. Henderson, D. A., 1999, The looming threat of bioterrorism, *Science.* **283**:1279–1282.
12. Henderson, D. A., 1999, Smallpox: clinical and epidemiologic features, *Emerg. Infect. Dis.* **5**:537–539.
13. Henderson, D. A., Inglesby T. V., Bartlett, J. G. *et al.*, 1999, Smallpox as a biological weapon: medical and public health management. Working Group on Civilian Biodefense, *JAMA.* **281**:2127–2137.
14. Fenner, F., and Burnet F. M., 1957, A short description of the poxvirus group (vaccinia and related viruses), *Virology.* **4**:305–314.
15. Joklik, W. K., 1962, Some properties of poxvirus deoxyriboncleic acid, *J. Mol. Biol.* **5**:2165–2174.
16. Boulter, E. A., and Appleyard G., 1973, Differences between extracellular and intracellular forms of poxvirus and their implications, *Prog. Med. Virol.* **16**:86–108.
17. Mitra, A. C., Chatterjee, S. N., Sarkar, J. K. *et al.*, 1966, Viraemia in haemorrhagic and other forms of smallpox, *J. Indian Med. Assoc.* **47**:112–114.
18. Hotchkiss, R. S., and Karl, I. E., 2003, The pathophysiology and treatment of sepsis, *N. Engl. J. Med.* **348**:138–150.
19. Sissons, J. G., and Oldstone, M. B., 1980, Antibody-mediated destruction of virus-infected cells, *Adv. Immunol.* **29**:209–260.
20. Sissons, J. G., and Oldstone, M. B., 1980 Killing of virus-infected cells by cytotoxic lymphocytes, *J. Infect. Dis.* **142**:114–119.
21. Sissons, J. G., and Oldstone, M. B., 1980, Killing of virus-infected cells: the role of antiviral antibody and complement in limiting virus infection, *J. Infect. Dis.* **142**:442–448.
22. Sissons, J. G., Oldstone, M. B., Schreiber, R. D. *et al.*, 1980, Antibody-independent activation of the alternative complement pathway by measles virus-infected cells, *Proc. Natl. Acad. Sci. U S A.* **77**:559–562.
23. Downie, A. W., Saint Vincent L., Goldstein, L. *et al.*, 1969, Antibody response in non-haemorrhagic smallpox patients, *J. Hyg. (London).* **67**:609–618.
24. Downie, A. W., L. Saint Vincent L., Rao, A. R. *et al.*, 1969, Antibody response following smallpox vaccination and revaccination, *J. Hyg. (London).* **67**:603–608.
25. Fenner, F., 1979, Portraits of viruses: the poxviruses, *Intervirology* **11**:137–157.
26. Kempe, C. H., 1960, Studies smallpox and complications of smallpox vaccination, *Pediatrics.* **26**:176–89.
27. Czerny, C. P., and Mahnel H., 1990, Structural and functional analysis of orthopoxvirus epitopes with neutralizing monoclonal antibodies, *J. Gen. Virol.* **71**:2341–2352.
28. Galmiche, M. C., Goenaga J., Witteck, R. *et al.*, 1999, Neutralizing and protective antibodies directed against vaccinia virus envelope antigens, *Virology.* **254**:71–80.
29. Ramirez, J. C., Tapia, E., Esteban, M. *et al.*, 2002, Administration to mice of a monoclonal antibody that neutralizes the intracellular mature virus form of vaccinia virus limits virus replication efficiently under prophylactic and therapeutic conditions, *J. Gen. Virol.* **83**:1059–1067.

30. Zinkernagel, R. M. and Althage, A., 1977, Antiviral protection by virus-immune cytotoxic T cells: infected target cells are lysed before infectious virus progeny is assembled, *J. Exp. Med.* **145**:644–651.
31. Derby, M., Alexander-Miller M., Tse, R. *et al.*, 2001, High-avidity CTL exploit two complementary mechanisms to provide better protection against viral infection than low-avidity CTL, *J. Immunol.* **166**:1690–1697.
32. Wehrle, P. F., Posch, J., Richter, K. H. *et al.*, 1970, An airborne outbreak of smallpox in a German hospital and its significance with respect to other recent outbreaks in Europe, *Bull. World Health Organ.* **43**:669–679.
33. LeDuc, J. W., and Becher, J., 1999, Current status of smallpox vaccine, *Emerg. Infect. Dis.* **5**:593–594.
34. Breman, J. G., and Henderson D. A., 1998, Poxvirus dilemmas—monkeypox, smallpox, and biologic terrorism, *N. Engl. J. Med.* **339**:556–559.
35. Alibek, K., 2004, Smallpox: a disease and a weapon, *Int. J. Infect. Dis.* **8**:S3–S8.
36. Henderson, D. A., 1998, Bioterrorism as a public health threat, *Emerg. Infect. Dis.* **4**:488–492.
37. Henderson, D. A., 2002, Countering the posteradication threat of smallpox and polio, *Clin. Infect. Dis.* **34**:79–83.
38. Rosenthal, S. R., Merchlinsky, M., Kleppinger, C. *et al.*, 2001, Developing new smallpox vaccines, *Emerg. Infect. Dis.* **7**:920–926.
39. Dolgopol, V. B., . Greenberg M., Aronoff, R. *et al.*, 1955, Encephalitis following smallpox vaccination, *AMA. Arch. Neurol. Psychiatry.* **73**:216–223.
40. Neff, J. M., Lane, J. M., Pert, J. H. *et al.*, 1967, Complications of smallpox vaccination. I. National survey in the United States, 1963, *N. Engl. J. Med.* **276**:125–132.
41. Neff, J. M., Levine, R. H., Lane, J. M. *et al.*, 1967, Complications of smallpox vaccination United States 1963. II. Results obtained by four statewide surveys *Pediatrics.* **39**:916–923.
42. Lane, J. M., Ruben, F. L., Neff J. M. *et al.*, 1969, Complications of smallpox vaccination, 1968, *N. Engl. J. Med.* **281**:1201–1208.
43. Lane, J. M., F. L. Ruben, Neff, J. M. *et al.*, 1970, Complications of smallpox vaccination, 1968: results of ten statewide surveys, *J. Infect. Dis.* **122**:303–9.
44. Lane, J. M., . Ruben, F. L., Abrutyn, E. *et al.*, 1970, Deaths attributable to smallpox vaccination, 1959 to 1966, and 1968, *JAMA.* **212**:441–444.
45. Mellin, H., Neff, J. M., Garber, H. *et al.*, 1970, Complications of smallpox vaccination, Maryland 1968, *John Hopkins Med. J.* **126**:160–168.
46. Neff, J. M., and Drachman R. H., 1972, Complications of smallpox vaccination, 1968 surveillance in a Comprehensive Care Clinic, *Pediatrics.* **50**:481–483.
47. Goldstein, J. A., Neff, J. M., Lane, J. M. *et al.*, 1975, Smallpox vaccination reactions, prophylaxis, and therapy of complications, *Pediatrics* **55**:342–347.
48. Neff, J. M., Lane, J. M., Fulginiti, V. A. *et al.*, 2002, Contact vaccinia–transmission of vaccinia from smallpox vaccination, *JAMA* **288**:1901–1905.
49. Fulginiti, V. A., Papier A., Lane, J. M. *et al.*, 2003, Smallpox vaccination: a review, part I. Background, vaccination technique, normal vaccination and revaccination, and expected normal reactions, *Clin. Infect. Dis.* **37**:241–250.
50. Fulginiti, V. A., Papier A., Lane, J. M. *et al.* 2003, Smallpox vaccination: a review, part II. Adverse events, *Clin. Infect. Dis.* **37**:251–271.
51. Neff, J. M., Lane, J. M., Fulginiti, V. A. *et al.*, 2003, Smallpox and smallpox vaccination, *N. Engl. J. Med.* **348**:1920–1925; author reply 1920–1925.
52. Poland, G. A., and Neff, J. M., 2003, Smallpox vaccine: problems and prospects, *Immunol. Allergy Clin. North Am.* **23**:731–743.
53. Greenberg, R. N., Schosser, R. H., Plummer, E. A. *et al.*, 2004, Urticaria, exanthems, and other benign dermatologic reactions to smallpox vaccination in adults, *Clin. Infect. Dis.* **38**:958–965.
54. Cangemi, V. F., 1958, Acute pericarditis after smallpox vaccination, *N. Engl. J. Med.* **258**:1257–1259.
55. Holtzman, C. M., 1969, Postvaccination arthritis, *N. Engl. J. Med.* **280**:111–112.

56. Marmelzat, W. L., 1968, Malignant tumors in smallpox vaccination scars: a report of 24 cases, *Arch. Dermatol.* **97**:400–406.
57. Copeman, P. W., and Wallace H. J., 1964, Eczema Vaccinatum, *Br. Med. J.* **5414**:906–908.
58. Hammarlund, E., Lewis, M. W., Hansen, S. G. *et al.*, 2003, Duration of antiviral immunity after smallpox vaccination, *Nat. Med.* **9**:1131–1137.
59. Mack, T. M., Noble, J., Jr., Thomas, D. B. *et al.*, 1972, A prospective study of serum antibody and protection against smallpox, *Am. J. Trop. Med. Hyg.* **21**:214–218.
60. Kaplan, E. H., Craft, D. L., Wein, L. M. *et al.*, 2002, Emergency response to a smallpox attack: the case for mass vaccination, *Proc. Natl. Acad. Sci. U S A.* **99**:10935–10940.
61. Frey, S. E., Newman F. K., Yan, L. *et al.*, 2003, Response to smallpox vaccine in persons immunized in the distant past, *JAMA.* **289**:3295–3299.
62. Weltzin, R., Liu, J., Pugachev, K. V. *et al.*, 2003, Clonal vaccinia virus grown in cell culture as a new smallpox vaccine, *Nat. Med.* **9**:1125–1130.
63. Meyer, H., Sutter, G., Mayr, A. *et al.*, 1991, Mapping of deletions in the genome of the highly attenuated vaccinia virus MVA and their influence on virulence, *J. Gen. Virol.* **72**:1031–1038.
64. Carroll, M. W. and Moss B., 1997, Host range and cytopathogenicity of the highly attenuated MVA strain of vaccinia virus: propagation and generation of recombinant viruses in a nonhuman mammalian cell line, *Virology* **238**:198–211.
65. Blanchard, T. J., Alcami, A., Andrea, P. *et al.*, 1998 "Modified vaccinia virus Ankara undergoes limited replication in human cells and lacks several immunomodulatory proteins: implications for use as a human vaccine, *J. Gen. Virol.* **79**:1159–1167.
66. Drexler, I., Heller K., Wahren, B. *et al.*, 1998, Highly attenuated modified vaccinia virus Ankara replicates in baby hamster kidney cells, a potential host for virus propagation, but not in various human transformed and primary cells, *J. Gen Virol.* **79**:347–352.
67. Earl, P. L., Americo J. L, Wyatt, L. S. *et al.*, 2004, Immunogenicity of a highly attenuated MVA smallpox vaccine and protection against monkeypox, *Nature.* **428**:182–185.
68. Wyatt, L. S., Earl, P. L., Eller, L. A. *et al.*, 2004, Highly attenuated smallpox vaccine protects mice with and without immune deficiencies against pathogenic vaccinia virus challenge, *Proc. Natl. Acad. Sci. U S A.* **101**:4590–4595.
69. Chattergoon, M. A., Robinson T. M., Boyer, J. D. *et al.*, 1998, Specific immune induction following DNA-based immunization through in vivo transfection and activation of macrophages/antigen-presenting cells, *J. Immunol.* **160**:5707–5718.
70. Cohen, A. D., Boyer J. D., Weiner, D. B. *et al.*, 1998, Modulating the immune response to genetic immunization, *FASEB. J.* **12**:1611–1626.
71. Koprowski, H., and Weiner, D. B., 1998, DNA vaccination/genetic vaccination, *Curr. Top. Microbiol. Immunol.* **226**:V–XIII.
72. Shedlock, D. J. and Weiner D. B., 2000, DNA vaccination: antigen presentation and the induction of immunity, *J. Leukoc. Biol.* **68**:793–806.
73. Kim, J. J., Yang, J. S., Manson, K. H. *et al.*, 2001, Modulation of antigen-specific cellular immune responses to DNA vaccination in rhesus macaques through the use of IL-2, IFN-gamma, or IL-4 gene adjuvants, *Vaccine.* **19**:2496–2505.
74. Hooper, J. W., Custer, D. M., Schmaljohn, C. S. *et al.*, 2000, DNA vaccination with vaccinia virus L1R and A33R genes protects mice against a lethal poxvirus challenge, *Virology* **266**:329–339.
75. Hooper, J. W., Custer, D. M., Thompson, E. *et al.*, 2003, Four-gene-combination DNA vaccine protects mice against a lethal vaccinia virus challenge and elicits appropriate antibody responses in nonhuman primates, *Virology.* **306**:181–195.
76. Hooper, J. W., Thompson, E., Wilhelmsen, C. *et al.*, 2004, Smallpox DNA vaccine protects nonhuman primates against lethal monkeypox, *J. Virol.* **78**:4433–4443.
77. Czerny, C. P., Johann S., Holze, L. *et al.*, 1994, Epitope detection in the envelope of intracellular naked orthopox viruses and identification of encoding genes, *Virology* **200**:764–777.
78. Wolffe, E. J., Vijaya, S., Moss, B. *et al.*, 1995, A myristylated membrane protein encoded by the vaccinia virus L1R open reading frame is the target of potent neutralizing monoclonal antibodies, *Virology* **211**:53–63.

79. Law, M., and Smith, G. L., 2001, Antibody neutralization of the extracellular enveloped form of vaccinia virus, *Virology*. **280**:132–142.
80. Kempe, C. H., Berge, T. O., England, B. *et al.*, 1956, Hyperimmune vaccinal gamma globulin; source, evaluation, and use in prophylaxis and therapy, *Pediatrics* **18**:177–188.
81. Sussman, S. and Grossman, M., 1965, Complications of smallpox vaccination. Effects of vaccinia immune globulin therapy, *J. Pediatr.* **67**:1168–1173.
82. Sharp, J. C., and Fletcher, W. B., 1973, Experience of anti-vaccinia immunoglobulin in the United Kingdom, *Lancet* **1**:656–659.
83. Bauer, D. J., 1965 Chemoprophylaxis of smallpox and treatment of vaccinia gangrenosa with 1-methylisatin 3-thiosemicarbazone, *Antimicrobial Agents Chemother.* **5**:544–547.
84. McLean, D. M., 1977, Methisazone therapy in pediatric vaccinia complications, *Ann. N Y Acad. Sci.* **284**:118–121.
85. Baker, R. O., Bray, M., Huggins, J. W. *et al.*, 2003, Potential antiviral therapeutics for smallpox, monkeypox and other orthopoxvirus infections, *Antiviral Res.* **57**:13–23.
86. De Clercq, E., 2001, Vaccinia virus inhibitors as a paradigm for the chemotherapy of poxvirus infections, *Clin. Microbiol. Rev.* **14**:382–397.
87. Neyts, J., and De Clercq, E., 1993, Efficacy of (S)-1-(3-hydroxy-2-phosphonylmethoxypropyl) cytosine for the treatment of lethal vaccinia virus infections in severe combined immune deficiency (SCID) mice, *J. Med. Virol.* **41**:242–246.
88. Smee, D. F., Bailey, K. W., Sidwell, R. W. *et al.*, 2002, Treatment of lethal cowpox virus respiratory infections in mice with 2-amino-7-[(1,3-dihydroxy-2-propoxy)methyl]purine and its orally active diacetate ester prodrug, *Antiviral. Res.* **54**:113–120.
89. Hanlon, C. A., Niezgoda M., Shankar, V. *et al.*, 1997, A recombinant vaccinia-rabies virus in the immunocompromised host: oral innocuity, progressive parenteral infection, and therapeutics, *Vaccine* **15**:140–148.
90. Kaufman, H. E., Nesburn, A. B., Maloney, E. D. *et al.*, 1962, Cure of vaccinia infection by 5-iodo-2′-deoxyuridine, *Virology* **18**:567–569.
91. Neyts, J., Verbeken, E., DeClerq, E. *et al.*, 2002, Effect of 5-iodo-2′-deoxyuridine on vaccinia virus (orthopoxvirus) infections in mice, *Antimicrob. Agents Chemother.* **46**:2842–2847.
92. Hyndiuk, R. A., Okumoto, M., Damiano, R. A. *et al.*, 1976, Treatment of vaccinial keratitis with vidarabine, *Arch. Ophthalmol.* **94**:1363–1634.
93. Worthington, M., and Conliffe M., 1977, Treatment of fatal disseminated vaccinia virus infection in immunosuppressed mice, *J. Gen. Virol.* **36**:329–333.
94. Hyndiuk, R. A., Seideman, S., Leibsohn, J. M. *et al.*, 1976, Treatment of Vaccinial keratitis with trifluorothymidine, *Arch. Ophthalmol.* **94**:1785–1786.
95. Lee, S. F., Buller, R., Chansue, E. *et al.*, 1994, Vaccinia keratouveitis manifesting as a masquerade syndrome, *Am. J. Ophthalmol.* **117**:480–487.
96. Snoeck, R., Holy, A., DeWolf-Peters, C. *et al.*, 2002, Antivaccinia activities of acyclic nucleoside phosphonate derivatives in epithelial cells and organotypic cultures, *Antimicrob. Agents Chemother.* **46**:3356–3361.
97. Jones, B. R., Galbraith, J. E., Al-Hussaini, M. K. *et al.*, 1962, Vaccinial keratitis treated with interferon, *Lancet* **1**:875–879.
98. De Clercq, E., and De Somer P., 1968, Protective effect of interferon and polyacrylic acid in newborn mice infected with a lethal dose of vesicular stomatitis virus, *Life Sci.* **7**:925–933.
99. Harrison, S. C., Alberts, B., Ehrenfeld, E. *et al.*, 2004, Discovery of antivirals against smallpox, *Proc. Natl. Acad. Sci .U S A* **101**:11178–11192.
100. Byrd, C. M., Bolken T. C., Mjalli, A. M. *et al.*, 2004, New class of orthopoxvirus antiviral drugs that block viral maturation, *J. Virol.* **78**:12147–12156.
101. Safrin, S., Cherrington, J., Jaffe, H. S. *et al.*, 1997, Clinical uses of cidofovir, *Rev. Med. Virol.* **7**:145–156.
102. Smee, D. F., Bailey, K. W., Sidwell, R. W. *et al.*, 2001, Treatment of lethal vaccinia virus respiratory infections in mice with cidofovir, *Antivir. Chem. Chemother.* **12**:71–76.

103. Smee, D. F., Bailey, K. W., Wong, M. H. *et al.*, 2001, Effects of cidofovir on the pathogenesis of a lethal vaccinia virus respiratory infection in mice, *Antiviral Res.* **52**:55–62.
104. Bray, M., Martinez M., Smee, D. F. *et al.*, 2000, Cidofovir protects mice against lethal aerosol or intranasal cowpox virus challenge, *J. Infect. Dis.* **181**:10–19.
105. Smee, D. F., Bailey, K. W., Wong, M. *et al.*, 2000, Intranasal treatment of cowpox virus respiratory infections in mice with cidofovir, *Antiviral Res.* **47**:171–177.
106. Smee, D. F., Bailey, K. W., Sidwell, R. W. *et al.*, 2000, Treatment of cowpox virus respiratory infections in mice with ribavirin as a single agent or followed sequentially by cidofovir, *Antivir. Chem. Chemother.* **11**:303–309.
107. Davies, E. G., Thrasher A., Lacey, K. *et al.*, 1999, Topical cidofovir for severe molluscum contagiosum, *Lancet* **353**:2042.
108. Ibarra, V., Blanco, J. R., Oteo, J. A. *et al.*, 2000, Efficacy of cidofovir in the treatment of recalcitrant molluscum contagiosum in an AIDS patient, *Acta. Derm. Venereol.* **80**:315–316.
109. Toro, J. R., Wood, L. V., Patel, N. K. *et al.*, 2000, Topical cidofovir: a novel treatment for recalcitrant molluscum contagiosum in children infected with human immunodeficiency virus 1, *Arch. Dermatol.* **136**:983–985.
110. Geerinck, K., Lukito, G., Snoeck, R. *et al.*, 2001, A case of human orf in an immunocompromised patient treated successfully with cidofovir cream, *J. Med. Virol.* **64**:543–549.

6

Bacillus anthracis: Agent of Bioterror and Disease

CHRISTOPHER K. COTE, DONALD J. CHABOT, ANGELO SCORPIO, THOMAS E. BLANK, WILLIAM A. DAY, SUSAN L. WELKOS, and JOEL A. BOZUE

1. INTRODUCTION

Anthrax is an ancient disease described over three thousand years ago by many cultures, including the Greeks, Egyptians, Romans, and Hindus. More recently, in the nineteenth century, anthrax played a central role in the development of the germ theory of disease. In addition, the disease was integral to the development of Koch's postulates[(1)] as well as the pioneering vaccine work of Greenfield,[(2)] and Pasteur.[(3)]

Anthrax was also the first disease for which different clinical manifestations were ascribed to infection with a single agent. Those diseases, which include cutaneous anthrax, gastrointestinal anthrax, and inhalational anthrax, all result from infection with the Gram-positive, spore-forming bacterium *Bacillus anthracis* (Fig. 1). *B. anthracis* infections begin with germination of environmentally resistant spores within host tissues to produce vegetative, aerobic bacilli that replicate to high numbers and eventually kill the host.

CHRISTOPHER K. COTE, SUSAN L. WELKOS, JOEL A. BOZUE • Bacteriology Division, United States Army Medical Research Institute of Infectious Diseases, Fort Detrick, Frederick, MD, 21702. DONALD J. CHABOT, ANGELO SCORPIO, THOMAS E. BLANK, WILLIAM A. DAY • Headquarters, United States Army Medical Research Institute of Infectious Diseases, Fort Detrick, Frederick, MD, 21702.

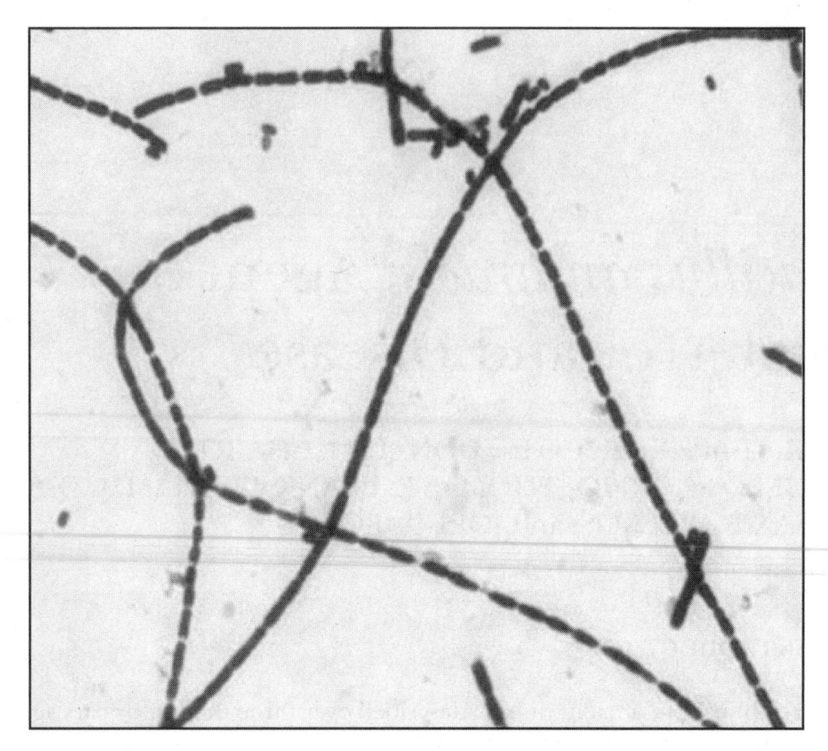

FIGURE 1. Gram stain of *Bacillus anthracis* vegetative cells.

In the twentieth century, the idea of using *B. anthracis* as a biological weapon was realized by several nations. Research programs were implemented for both offensive and defensive purposes. Most recently in the twenty-first century, *B. anthracis* was shown to be an effective agent of bioterrorism. The anthrax letters of 2001 clearly illustrated the potential damage that could be caused by intentionally dispersing even small volumes of *B. anthracis* spores. The effects of the anthrax letters remain evident to date.

This chapter will briefly describe some of the characteristics associated with *B. anthracis* being used as an agent of bioterror. We will address clinical presentation, virulence attributes, and aspects of pathogenesis that eventually lead to a fulminant infection and development of anthrax.

2. CLINICAL PRESENTATION OF ANTHRAX

Anthrax infection occurs naturally in three forms: cutaneous, gastrointestinal, and inhalational. Cutaneous infection, the most common form, usually results from handling livestock products contaminated with *B. anthracis* spores. Entry of the spores at the site of a skin abrasion may result in the formation of a pruritic papule 1 to 12 days after exposure. The papule then develops into a small painless black eschar, which can appear similar to an insect bite. An infected individual may also present with malaise, headache, and fever.[4] If not treated, the localized infection may disseminate and become systemic. Treating cutaneous anthrax with antibiotics, such as ciprofloxacin and doxycycline, results in a mortality rate of less than 1%. Failure to administer antibiotics, however, may result in a case fatality rate of greater than 20%.[5] Intravenous therapy followed by oral therapy is the recommended treatment for cutaneous anthrax with signs of systemic involvement or lesions on the head and neck. As recommended by the Centers for Disease Control (CDC), treatment should be given for 7–10 days for naturally acquired cutaneous infections or 60 days in the case of cutaneous anthrax due to a bioterrorist attack.[6] Gastrointestinal anthrax is considered to be a relatively rare form of the disease and is usually caused by ingesting undercooked, contaminated meat. After an incubation period of 1–7 days, pharyngeal lesions develop at the base of the tongue with concurrent fever and lymphadenopathy. Anthrax bacilli surviving to the lower intestine may cause inflammation, resulting in abdominal pain and bloody diarrhea.[5] The relatively mild early symptoms of gastrointestinal anthrax make diagnosis difficult, resulting in a higher mortality rate (26–60%) than for cutaneous anthrax.[7,8] In July 2000, an investigation was conducted into human consumption of meat contaminated with *B. anthracis*. After eating contaminated hamburger, there were two reports of gastrointestinal illness, including diarrhea, abdominal pain, and fever.[9] Both recovered without treatment and were placed on a prophylactic antibiotic regimen of ciprofloxacin.[9]

Inhalational anthrax, as with gastrointestinal anthrax, has an insidious clinical progression and can rapidly lead to systemic infection before effective antibiotic treatment is administered. After an incubation period of 2 to 60 days, the initial nonspecific symptoms include malaise, myalgia, fatigue, and low-grade fever but can quickly proceed to septic shock and respiratory failure if untreated. Inhalational anthrax is initiated when spores are deposited in the lungs and then subsequently engulfed by alveolar macrophages. Spores are then transported by the lymphatic system to the mediastinal space where germination may occur, causing in many cases a lymphadenopathy and mediastinal widening that is readily apparent in a chest radiograph. The vegetative bacilli synthesize an extracellular antiphagocytic capsule, begin to multiply rapidly, and secrete the two anthrax exotoxins, causing hemorrhagic lymphadenitis. Bacteremia quickly develops with dissemination to multiple organs. Resistance of encapsulated anthrax bacilli to phagocytosis and serum killing facilitate massive growth of the organisms in the blood and lymphatic system, producing a fulminant infection with symptoms of fever, dyspnea, diaphoresis, and shock. At this stage of the disease, systemic infection can lead to death within hours.[10]

Rapid progression of disease from inhalational infection has a mortality rate of 80–90%.

3. *B. anthracis* AND BIOTERROR

The October 2001 anthrax cases in the United States were caused by intentionally contaminating letters with *B. anthracis* spores. A relatively large area of the United States, including the District of Columbia, Florida, Connecticut, New Jersey, and New York, was affected.[11] A total of 22 cases of anthrax were reported from contact with the letters containing *B. anthracis* spores, 11 cutaneous and 11 inhalational. Epidemiological data indicated that the median incubation period for the inhalational anthrax patients was 4 days, and initial symptoms presented as fever, fatigue, cough, dyspnea, nausea, and vomiting. Six of the first 11 patients were in the initial phase of infection when they were treated with antibiotics treatment while five exhibited late-stage symptoms at the time of treatment. The six patients who were treated at the early stage of the disease survived while the five patients treated after the infection had progressed died. The initial phase of the infection for all patients was difficult to diagnose because of a lack of specific symptoms. However, all of the patients had abnormal chest X-rays, including mediastinal widening and pleural effusions. Additionally, all of the patients cultured prior to receiving antibiotic therapy were positive for *B. anthracis*, indicating that the bacilli reach the blood stream early in infection. Using combination antimicrobial therapy proved valuable, as six of 11 patients survived, more than the previously reported inhalational anthrax survival rate of 15%.[11]

The anthrax scare of 2001 brought to public consciousness the real potential of *B. anthracis* as a potent biowarfare agent. Although recognized and examined as a biological weapons threat for decades,[12] this latest incident has spurred a flurry of government, academic, and private institutional research aimed at combating *B. anthracis* as an agent of bioterror. Although cutaneous anthrax is the most common naturally occurring form of the disease, the ability of *B. anthracis* spores to resist environmental stress and the ease of spore dispersal makes aerosol delivery a far more potent biological weapon. Furthermore, the nonspecific symptomology associated with inhalational anthrax adds to the danger of *B. anthracis* as an agent of bioterror. One of the first documented incidents of mass inhalational anthrax was in 1979, after an accidental environmental release of anthrax spores from a military research facility in Sverdlovsk in the former Soviet Union.[13] The affected, exposed individuals were confined to a narrow zone of the city, directly downwind of the facility and extending approximately 4 kilometers. Of the 77 documented patients exposed to anthrax spores, approximately 80% succumbed to infection, indicating the lethality of the organism, even when dispersed over a relatively large area. The only documented cases of deliberate spore dispersal, other than the anthrax letters of 2001, are of a Japanese cult, Aum Shinrikyo, which released aerosols of *B. anthracis* spores and botulinum toxin in Tokyo on several occasions.[14] In this case, the strain of *B. anthracis* used was an attenuated animal vaccine strain and the attacks failed to produce illness.[15]

4. EVOLUTION INTO A PATHOGEN

Two studies support the idea that *B. anthracis* is a single clonal lineage of *Bacillus cereus*. Electrophoretic characteristics of broadly conserved proteins, sequence comparisons of 16S ribosomal RNA genes, and the fact that the sequences of the chromosomes of the organisms are highly similar, has led some investigators to suggest that *B. cereus* and *B. anthracis* could represent a single species.[16,17] This observation is striking when considering the very different virulence potentials of the two organisms.

B. cereus is considered an opportunistic pathogen, associated with gastrointestinal illness and periodontitis. *B. cereus* is most often identified as a commensal organism present in insect and mammalian digestive systems.[154] In contrast, *B. anthracis* replicates within mammalian host tissues and can cause systemic disease resulting in host death. Transformation of the commensal ancestor into the pathogenic *B. anthracis* likely involved horizontal transfer of whole blocks of genes that encode traits required for growth and survival within host tissues.[18]

B. anthracis expresses two toxins and an anti-phagocytic capsule that are not normally produced by *B. cereus*. Genes encoding these factors are present on two large plasmids designated pXO1 (toxin plasmid) and pXO2 (capsule plasmid), which are not part of the ancestral *B. cereus* genome. Recently, sequences of each plasmid, the *B. anthracis* chromosome, and the *B. cereus* chromosome were completed. Analysis of the pXO1 sequence data revealed that the DNA is a chimeric molecule comprised of regions acquired from other genomes.[19,20] The region encoding the anthrax toxins and regulatory elements is, however, only present in the *B. anthracis* pXO1 plasmid and was acquired after horizontal transfer of a 45-kb block of genes termed a pathogenicity island. Similarly, several regions and genes of the capsule plasmid are also present in the genomes of other organisms.[19–21] Perhaps, not surprisingly, additional significant genome modifications were required to accommodate the significant change in growth environments.

B. anthracis lives in an environment that presents very different selective pressures from those that shaped the ancestral organism. Several observations suggest that *B. anthracis* has become highly adapted to a pathogenic lifestyle and has traded fitness in the ancestral niche for fitness in the new environment, mammalian tissue and blood. *B. anthracis* spores have not been clearly shown to germinate in soil, nor have *B. anthracis* bacilli been demonstrated to grow in soil.[22] These characteristics are in sharp contrast to the closely related nonpathogenic ancestor *B. cereus* that thrives in soil. Consistent with these observations, several traits expressed by the nonpathogenic ancestor *B. cereus* are not expressed by *B. anthracis*. These include, under the usual assay conditions, hemolysin, lipase, and lecithinase activity, as well as germination in response to inosine as the sole germinant. Lack of expression of each of the enzymatic activities in *B. anthracis* has been traced to a nonsense mutation in the gene encoding the global transcriptional regulator PlcR.[23] PlcR is an autoregulated trans activator that serves as a pleiotropic regulator of multiple extracellular virulence factors in *B. cereus* and *B. thuringiensis*.

Additional adaptations to a pathogenic lifestyle may include modifications of germinant receptors that trigger spore outgrowth. If *B. anthracis* traded fitness in soil for fitness in animal tissues, then eliminating factors that direct germination in response to signals in soil would be beneficial, as they would ensure that the *B. anthracis* spore does not germinate in an environment in which the organism is unfit. Consistent with this prediction, germination signals that trigger spore outgrowth differ between *B. anthracis* and *B. cereus.*[24] Collectively, these observations suggest that the evolution induced by horizontal transfer enabled what was once a soil and commensal organism to survive within mammalian blood and tissues.

5. SPORE STRUCTURE AND FUNCTION

5.1. Sporulation and Germination

When faced with a nutrient-limiting environment, species of *Bacillus* are able to encase themselves in a resilient, multilayered shell, referred to as a spore. Once converted to this form, *Bacillus* spores are resistant to a variety of extreme environmental conditions, such as temperature, desiccation, and ultraviolet light, making *B. anthracis* spores particularly suited for use as a bioterror agent. The dormant spore is the infectious form of *B. anthracis.*[25] Once in the spore form, *B. anthracis* may remain dormant for decades, awaiting a nutritionally adequate environment in which to germinate. Spore germination is crucial for the initial infection and subsequent development of disease. The dormant spores must contain machinery that will recognize and sense sufficient levels of germinants to initiate transformation from an ungerminated spore to a vegetative, replicating bacillus.

Sporulation has been investigated in detail in *B. subtilis*, and those studies are currently being applied to *B. anthracis*. As described by Moir and Smith, bacterial sporulation is divided into six stages, 0, II, III, IV, V, and VI based on electron microscopy.[25] The initiation stage or commitment to sporulation, stage 0, is followed by stage II, which is characterized by a septum forming towards one end of the bacterial cell. The larger region, in the now compartmentalized cell, continues to develop and eventually engulfs the smaller compartment, forming the forespore within the mother cell in stage III. An additional membrane is added to the forespore, and during stage IV, peptidoglycan is deposited between the inner and outer membranes. In addition, calcium ions are incorporated into the inner cell compartment and dipicolinate is accumulated within the forespore. Dipicolinate is unique to bacterial spores and is not found in vegetative bacteria. During stage V, coat proteins are synthesized by the mother cell and assembled on the forespore. During stage VI, the forespore gains its key characteristic traits; heat resistance and refractility, as detected by phase contrast microscopy. Finally, the mother cell is lysed and the mature spore is released.

Sporulation has been hypothesized to be induced by a complex signal pathway that is mediated by at least five sensor kinases responsible for the phosphorelay.(26–28) This phosphorelay ultimately phosphorylates the Spo0A protein, which is both a transcriptional repressor of genes expressed in the vegetative bacillus and a transcriptional activator of genes involved in sporulation.(29) It is the levels of the phosphorylated form of the Spo0A protein that governs when the organism will undergo sporulation.(30)

Germination is equally important when describing the initial infection of *B. anthracis* and the subsequent pathogenesis of anthrax. The machinery required to detect an environment favorable for germination must be present and responsive to the presence of appropriate germinants. A host of germination (*ger*) genes have been described, and the roles their encoded proteins play in response to different germinants have been examined in several *Bacillus* species: the *gerA* operon in *B. subtilis*,(31–33) the *gerI* operon in *B. cereus*,(34) and the *gerX*, *gerS*, and *gerH* operons in *B. anthracis*.(24,35–37) The *gerX* operon in *B. anthracis* is located on the toxin plasmid pX01 and is situated between the *pagA* gene and the *atxA* gene.(35,36) Guidi-Rontani *et al.* illustrated that a *gerX* null mutant strain of *B. anthracis* did not germinate efficiently *in vivo* or *in vitro*, and that the strain was less virulent in a mouse model of infection.(35) Ireland and Hanna described the chromosomally encoded *gerS* operon in *B. anthracis*, and they reported that the proteins encoded by the *gerS* locus most likely act as receptors for germinants containing aromatic ring structures.(24) Weiner *et al.* identified the *gerH* operon in *B. anthracis* and established that its gene products are important for intiating two germination pathways.(37) The major pathway affected was an amino acid-inosine pathway, while the alanine-aromatic amino acid pathway was affected to a lesser extent.(37) Thus, multiple germination operons may be required to encode appropriate germinant sensor machinery and to promote the efficient germination of *B. anthracis* spores.

The transformation of a dormant spore into a vegetative, dividing bacillus can be considered in three stages (1) activation, (2) germination, and (3) outgrowth.(38) Upon commitment to germinate, the ungerminated spores (bright/refractile as detected by phase contrast microscopy) become slightly engorged and dark (dark/nonrefractile as detected by phase contrast microscopy) (Fig. 2). This conversion from ungerminated to germinated spores

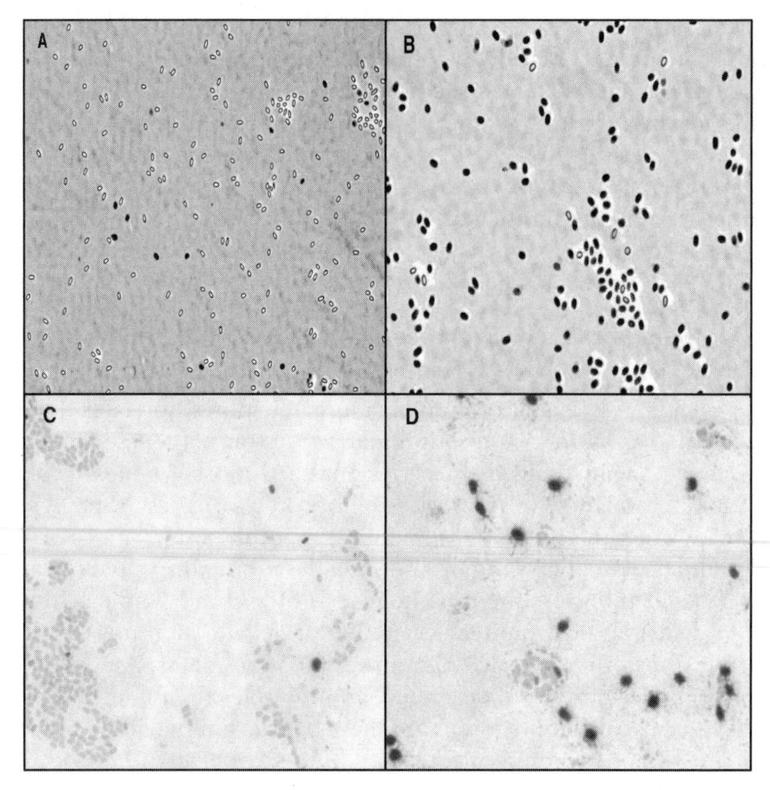

FIGURE 2. The extent of germination of dormant spores *in vitro* can be detected microscopically as a decrease in refractility under phase contrast and an increase in staining by dyes.[39] Phase micrographs of spore suspensions containing mainly ungerminated (UG), refractile spores (A) or germinated, nonrefractile spores (B). Malachite green spore stains of suspensions with mostly UG green-stained (light) spores and rare germinated ones stained purple (dark) with the Wright/Giemsa counterstain (C), or approximately equal numbers of UG and germinated organisms (D). Reprinted with permission from Elsevier.[39]

may also be detected by staining the samples with a suspension of malachite green dye (Fig. 2).[39] As the germinated spores begin to outgrow to vegetative cells, noticeable enlargement and elongation occurs, and the cells begin to divide rapidly.[40]

Most *B. anthracis* strains tested germinate *in vitro* at relatively the same rate, approximately 90% of the spores are germinated within 2 hours regardless of the virulence of that strain.[41] Some factors that may affect germination rates include temperature, spore concentration, and germinants available to the spores. *B. anthracis* germination is induced at temperatures ranging from 22 to 37°C. It has been hypothesized that optimal germination at temperatures less then 37°C may play a role in virulence, particularly establishing a cutaneous infection.[41] Spore concentration is also important for efficient spore germination *in vitro*. Large numbers of spores concentrated in a small volume of medium germinate slower then a less concentrated sample.[41] It has been hypothesized that this phenomenon may be a direct response to the amount of germinant available in the medium to the spores. Finally, the specific germinant and the amounts of germinant available to the spores are crucial for inducing germination.

There are reported differences associated with optimal germinant components. Sera collected from different animals affect the germination rates of spores. For example, spores germinate more rapidly when exposed to bovine sera or sheep sera as compared to spores exposed to mouse, guinea pig, or horse sera.[24,41] Complex media such as brain heart infusion broth can induce rapid germination of *B. anthracis* spores. In addition, studies have identified single components that are efficient in stimulating spore germination. L-alanine is a major germinant and can induce germination by itself; other amino acids and nucleosides, such as inosine, function as co-germinants.[24,37,42] Significant advances have been made in understanding the process of germination and sporulation of *Bacillus* species, yet numerous gaps remain in aspects of these processes, especially as they relate to the pathogenic species, such as *B. anthracis* (for example, location and identification of the germinant receptor).

5.2. Spore Coat Proteins and Exosporium

Spore formation has been studied much more extensively in *B. subtilis* than in *B. anthracis*. However, more attention has been focused on understanding the spore coat of *B. anthracis* recently, as the coat may serve as an important diagnostic tool and an antigen for vaccine development. There are many similarities in structure of the spore between *B. subtilis* and *B. anthracis*; both species have a core, cortex, and coat. The main body of the spore, known as the core, houses the chromosome of the bacterium. The cortex, composed of a membrane and a thick layer of peptidoglycan, surrounds the core. The cortex is then surrounded by the spore coat.

Many of the proteins composing the spore coat of *B. anthracis* and *B. subtilis* are homologous, as determined by proteomic analysis.[43] Furthermore, genomic comparisons between the two species showed that coat proteins involved with the assembly of the spore coat, such as SpoIVA and CotE, are similar between the two organisms, suggesting that the construction of the coat occurs by the same mechanism.[44] The SpoIVA and CotE proteins of *B. anthracis*, to date, have not yet been characterized as well as they have been in *B. subtilis*. The SpoIVA protein in *B. subtilis* creates scaffolding around the forespore for protein assembly to occur and to provide proper targeting for the CotE protein.[45–47] The CotE protein also surrounds the forespore and serves

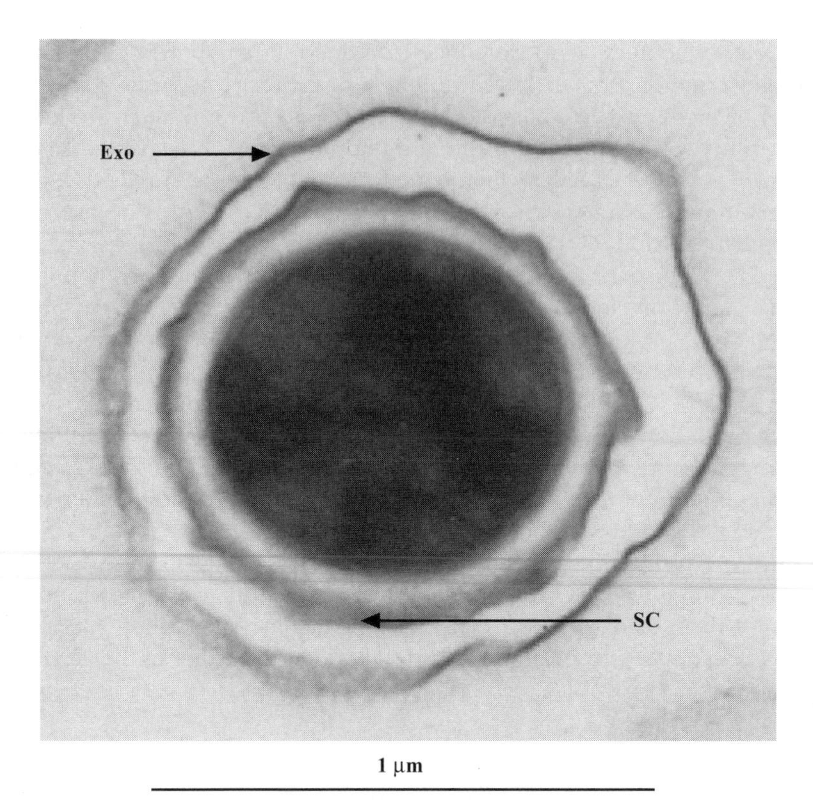

FIGURE 3. Transmission electron micrograph of an ungerminated *B. anthracis* (Ames strain) spore. The spore coat (SC) and the exosporium (Exo) are clearly visible. The exosporium consists of a membrane and a hair-like fringe.

as the site for outer-coat proteins assembly.[46–48] Mutations in the respective genes were constructed in *B. subtilis.* In a *spoIVA* mutant strain, the spore coat is made; however, the coat misassembles by swirling within the mother cell cytoplasm. If the *cotE* gene is inactivated, assembly for the outer coat layer is disrupted.[48]

Although there are similarities in the structure between the spores of *B. subtilis* and *B. anthracis,* important differences do exist. One major structural difference between the two species is the presence of an exosporium (Fig. 3). The exosporium is a loose-fitting layer that envelops the spore of *B. anthracis* but is not present in *B. subtilis.* High-resolution electron microscopy has shown that the exosporium consists of a paracrystalline basal layer and hair-like fibers.[49] Further analysis measured the exosporium membrane to be 100 Å and the hair-like fringe to be about 620 Å. Exosporia are also present on *B. cereus* and

B. thuringiensis. The exosporium membrane of *B. cereus* consists of proteins, polysaccharides, and lipids.[50] As *B. cereus*, *B. anthracis*, and *B. thuringiensis* are genetically very closely related,[16] the exosporium composition of the three species will most likely be similar.

Protein components of the exosporium of *B. thuringiensis*, *B. cereus*, and *B. anthracis* have been identified. The first exosporium protein characterized was from *B. thuringiensis*.[51] In this study, 72-kDa and 205-kDa glycoproteins were identified as different forms of a component of either the exosporium or the spore coat. Recently, Todd *et al.* determined the N-terminal sequence of this purified protein, which corresponded in size and sequence to a protein purified from the exosporium of *B. cereus*.[52]

In further studies with the *B. cereus* exosporium, Todd *et al.*[52] identified 10 components of purified exosporium from the ATCC 10876 strain of *B. cereus*. All of these proteins were also encoded by *B. anthracis*.[52] Two proteins, ExsB and ExsC, have some differences between the species. The *B. cereus* ExsB has 88% homology with the *B. anthracis* protein. However, between residues 17 to 34 of the mature *B. cereus* protein, only four of the 18 corresponding residues of the *B. anthracis* protein are identical. In addition, the *B. cereus* ExsC protein has only 66% identity with the protein encoded on the chromosome of *B. anthracis*. The *exsC* gene of *B. anthracis* has an inefficient ATA initiation codon and may not be expressed.

Sylvestre *et al.* described the first exosporium protein of *B. anthracis*, BclA.[53] This protein has many similarities to mammalian glycoproteins, as the central portion contains a region of GXX motifs, referred to as a collagen-like region (CLR). The CLR is highly polymorphic among strains of *B. anthracis*, with the number of contiguous GXX triplets, varying from 17 to 91. The length of the CLR correlates with the length of the filaments of the exosporium.[54] The *bclA* gene was deleted in the Sterne 7702 strain of *B. anthracis*.[53] When spores were constructed from this mutant, the hair-like fibers normally present on the exosporium were absent. However, the exosporium layer itself remained intact. Further testing of the mutant showed no difference as compared to spores of wild-type *B. anthracis* when assayed under several extreme conditions. In addition, the LD_{50} of the mutant strain was similar to that of the parental strain in a mouse subcutaneous model of anthrax infection.[53]

A study by Steichen *et al.* identified five of the major proteins from the exosporium of spores from the Sterne strain of *B. anthracis*.[53] One of these proteins was BclA.[53] Two other proteins identified in the exosporium were BxpA and BxpB. These proteins did not show any significant homology to any other current open reading frames available in nucleotide databases, although homologs were identified in the unfinished *B. cereus* database. As these proteins are absent in *B. subtilis* that lacks an exosporium, it is believed that these proteins are necessary to form an exosporium.[55]

6. VIRULENCE FACTORS

6.1. Anthrax Toxins

B. anthracis produces two toxins, lethal toxin and edema toxin. Both toxins are classic examples of an AB-type toxin, in that toxin component A is transported to the cytosol of an intoxicated cell by toxin component B.[(56)] The two anthrax toxins are composed of protective antigen (PA) coupled to either the lethal factor (LF) or the edema factor (EF) to produce either lethal toxin (LT) or edema toxin (ET), respectively. These three components, PA, LF, and EF are encoded on the pXO1 virulence plasmid and are synthesized and released from the bacterium as discrete monomeric units before being assembled to form either toxin.

Protective antigen (PA): PA is the toxin component common to both the anthrax LT and the anthrax ET. PA plays the role of transporter molecule to either LF or EF. It is secreted from *B. anthracis* as an 83-kDa (735 amino acids), monomeric protein before being converted to the active PA-63.[(57)] The crystal structure of PA-83 has four protein domains.[(58)] Domain I binds two calcium ions and contains the cleavage site responsible for converting PA-83 to PA-63. Domains II and III are important for oligomerizing PA-63 and exhibit characteristics implicated with membrane insertion. Domain IV binds to the surface receptor on the eukaryotic cell being intoxicated.[(59–61)]

Once outside the bacterial cell, monomeric PA-83 binds to the anthrax toxin receptor. PA-83 then undergoes a cleavage event, at which point the active PA molecule becomes 63 kDa. The PA molecule is cleaved by furin or another related endogenous protease and is then able to form a heptameric ring/pore structure that is responsible for transport into the host cell cytosol.[(62,63)] The heptamer of PA molecules may then accept three molecules of either EF and/or LF.[(64,65)] LF and EF can only stably bind to PA-63 dimers or higher order oligomers such as the heptameric PA-63 ring.[(65–67)] This holotoxin is then endocytosed into the cell. Upon trafficking into the endosome, it is hypothesized that the lower pH triggers the PA heptamer to form a pore within the endosomal membrane, ultimately releasing the EF and/or LF payload into the host cytosol. The structure and function of this PA heptamer has been extensively studied and point mutations have been made to characterize key residues in this process.[(68,69)] In addition to this cell-dependent model, it has been shown that PA can complex with LF in a cell-free environment.[(70)] Ezzell and Abshire illustrated that PA is predominantly found as a 63-kDa protein in the blood of infected animals.[(70)] These authors also reported that this cleavage of the 83-kDa PA protein was being catalyzed by a calcium-dependent, heat-labile serum protease that was found in numerous animal species ranging from rodents to primates.[(70)]

Lethal factor (LF): LF is a 90-kDa zinc-dependent metalloproteinase.[(71–73)] The crystal structure of LF has been determined and has illustrated the presence of four distinct domains.[74] Domain I of LF is responsible for binding to the PA heptamer, domains II and III are responsible for substrate recognition, and domain IV contains the catalytic site.[74] Pannifer *et al.* observed through crystallography and subsequent analyses that the LF has evolved into an enzyme with "high and unusual specificity".[74] For example, LT can enter most cell types, but only certain cells are susceptible to the effects of the toxin.[75,76] It has shown by several groups that macrophages appear to be the primary targets of the lethal toxin effector molecule.[77–78]

Once inside the cell, LF has specificity for mitogen-activated protein kinase kinases (MAPKKs), thus preventing the phosphorylation and subsequent activation of mitogen-activated protein kinases (MAPKs).[80–82] Park *et al.* demonstrated that LF cleaves the MAPKKs at a point between their amino terminal end and the catalytic domain.[83] The inactivation of these kinases selectively induces apoptosis of activated macrophages.[83] This cellular lethality associated with LF has been described as closely resembling necrosis.[79] The exact association between the MAPKK specificity and cellular lethality is still under investigation. Popov *et al.* observed that sublytic doses of lethal toxin *in vitro* induce several characteristics indicating apoptosis in RAW264.7 cells, including change in membrane permeability, loss of mitochondrial function, and fragmentation of DNA.[76] When LF is introduced into the cytosol of macrophages in the absence of PA, LF retains cytolytic potential.[84] These data suggest that the toxic effects associated with LT can be attributed to LF.[84]

The effect of LT on cytokine production is somewhat unclear. In one instance, it was reported that macrophages exposed to LT do not express proinflammatory cytokines.[85] This is in direct contrast with another report that cultured macrophages express high levels of tumor necrosis factor (TNF) and interleukin-1 (IL-1).[79] In the latter report, the authors hypothesize that the high levels of host cytokines induced in macrophages by the LT, mainly IL-1, could be responsible for the systemic shock and subsequent death associated with the anthrax infection.[79] A very recent report describes the inhibitory effect of LT on the activation of interferon-regulatory factor 3.[86] By inhibiting the activation of this regulator, LT also subsequently inhibited cytokine production,[86] lending new support to the work reported by Erwin *et al.*[85]

In addition, LF can affect several other cellular processes. Treating macrophages with LF was associated with a decrease in glycogen synthase kinase-3-beta, a pleuripotent kinase.[87] Tucker *et al.* demonstrated that zebra fish exposed to LT have a delayed pigmentation process, experience cardiac hypertrophy, and have a decrease in cellular tubulin production (2003). Webster *et al.* demonstrated that LF can repress glucocorticoid receptor transactivation even at very low concentrations.[88] Repression of the progesterone receptor B and the estrogen receptor alpha has also been observed. It is hypothesized that by repressing the glucocorticoid and other nuclear receptors, LF renders the infected host more susceptible to infection by removing normally protective anti-inflammatory effects associated with certain hormones.[88] These recent revelations of LF actions on cellular processes suggest that more work needs to be done to obtain a more complete understanding of the effects of LF in an infection.

Edema factor (EF): EF is an 89-kDa, calmodulin-dependent adenylate cyclase that is encoded by the *cya* gene located on pXO1.[89] As an adenylate cyclase, EF increases intracellular levels of cyclic AMP, causing edema.[90] EF has been shown to quickly increase the levels of cyclic AMP in Chinese hamster ovary (CHO) cells by approximately 200-fold.[91] In addition, the rise of intracellular levels of cyclic AMP in neutrophils inhibits phagocytosis.[92] Kumar *et al.* proposed that by raising intracellular levels of cyclic AMP, EF interferes with normal cell signaling pathways thus inhibiting the immune system and subsequently promoting the infection.[93]

EF binds poorly to apo-calmodulin, but more tightly to calmodulin loaded with Ca^{2+}, illustrating that EF requires calmodulin as a Ca^{2+} sensor.[94] The effects of EF have been examined in several cell types including CHO cells, RAW264.7 cells, human neutrophils, and human lymphocytes.[93] Kumar observed that the time course and intensity of the EF effects may vary, but upon entry into the cytosol, there is a calcium influx that precedes the increase in cyclic AMP.[93] Agents that interfere with the calcium influx also inhibit the actions of EF.[93]

Like LF, EF requires PA to enter the host cell to intoxicate and subsequently cause damage. It has been shown that the amino acid residues responsible for binding EF and LF to the PA heptamer are identical.[95] This stretch of amino acids (Val Tyr Tyr Glu Ile Gly Lys) is located in both EF (residues 136–142) and LF (residues 147–153). When this region is mutated in EF, binding to PA is blocked, but adenylate cyclase activity is retained.[95]

6.2. Anthrax Toxin Receptor-Mediated Internalization

A cellular receptor for anthrax toxin was identified as being a 368-amino acid protein with extensive identity (the first 364 amino acids) to a protein known to be upregulated in the colorectal endothelium, TEM8.[96] TEM8 encodes a 27-amino acid signal peptide, a 293-amino acid extracellular domain, and a 23-amino acid region encoding a putative transmembrane region. Additionally, the anthrax toxin receptor (ATR) protein contains a short 25-amino acid cytoplasmic tail that is responsible for the divergence between the TEM8 protein and the ATR protein.[96] Amino acid residues 44 through 216 encode an extracellular von Willebrand factor type A (VWA/I) domain. It is hypothesized that this motif is involved in ATR/protein interactions, as it has been shown to be important for binding integrins to proteins. It is also of key importance to mention the short cytoplasmic tail contains an acidic cluster (AC) motif (EESSEE) that is similar to that seen in furin. It is proposed that the AC motif causes the receptor to localize in the vicinity of or into regions occupied by the protease that is required to cleave PA-83, creating PA-63 which can then heptamerize and enter the cell by endocytosis.[96]

In addition, Abrami *et al.*[97] recently showed that the ATR is associated with raft-like lipid domains composed of cholesterol and glycosphingolipids (2003). The ATR proteins are localized to nonionic, detergent-resistant rafts resulting in heptameric clustering. Upon raft localization and binding of EF and/or LF, the ATR/PA heptamer complex is rapidly internalized into the eukaryotic cell. This internalization is by clathrin-dependent machinery and is compared to the same process observed with B-cell receptors.[97]

For the anthrax toxins to become active, many events must be timed precisely. The PA must bind to the ATR, undergo a cleavage event converting from PA-83 to PA-63, and heptamerize. Then the LF and/or EF must bind to the complex to be translocated into the cell where they can be active. These events must occur outside the cell. Thus, the endocytotic rate associated with the lipid rafts containing the ATR proteins is critical to toxin function.[97]

More recently, Scobie and co-workers have identified a second potential receptor for anthrax toxin. The human capillary morphogenesis protein 2 (CMG2) was shown to contain a VWA/I domain that has 60% amino acid identity with the VWA/I domain from the TEM8/ATR protein.[98] It was established that PA will directly bind CMG2 VWA/I domain and adding soluble CMG2 VWA/I domain protected CHO-K1 cells from intoxication by anthrax LT.[98] It was also noted that both TEM8/ATR and CMG2/ATR are found in a host of different human tissue types, suggesting that these receptors would be relevant to anthrax pathogenesis.

6.3. Toxin Gene Regulation

Early studies demonstrated that while PA is a required component of either anthrax toxin, *B. anthracis* does not express toxin under all environmental conditions. Because PA is an essential component of any effective anthrax vaccine, empirical studies were undertaken to determine conditions that maximize PA production.

Studies demonstrated that PA production is maximal during growth at 37°C in defined medium containing minimal nutrients and bicarbonate.[99,100] This is hypothesized to be the case as several nutrients are less available in blood (several amino acids and iron) and bicarbonate is the buffering system of the blood. Thus, Ristroph and Ivins proposed that *B. anthracis* evolved to exploit unique environmental cues present in host tissues to induce toxin expression.[99] Several *B. anthracis* encoded factors that influence anthrax toxin expression are identified. Two of these proteins include AtxA and PagR, which are encoded by genes within the pXO1pathogenicity island. AtxA is a constitutively expressed transactivator essential for toxin expression and virulence.[101–103] Analysis of the AtxA protein sequence suggests the presence of several helix-turn-helix motifs located in the N-terminal region of the protein that may mediate DNA binding. Similar domains are known in a related *Streptococcus pyogenes* transcription factor essential for virulence named Mga.(104) Several studies indicated that AtxA is a master regulator that controls or influences expression of many *B. anthracis* factors, including chromosomally encoded elements.(105,106) AtxA interferes with ancestral sporulation pathways(23) and is required for optimal *B. anthracis* growth on minimal medium.(105) An additional pXO1 locus that influences toxin expression is a DNA-binding protein, termed PagR, which represses not only toxin expression but also chromosomally encoded S-layer protein expression.(107–109) Several chromosomally encoded loci have also been identified that influence toxin expression. One such protein is the transition state regulator, AbrB, which represses toxin expression during log phase growth *in vitro* by repressing an alternative sigma factor (SigH) which is likely required for RNA polymerase interaction with the toxin gene promoter.(111)

6.4. Capsule: Chemistry and Composition

The other essential virulence determinant associated with *B. anthracis* is the presence of a large capsule surrounding the vegetative bacilli. By phase and immunofluorescence microscopy, capsule first appears on germinating spores as blebs that then coalesce on bacilli as they mature.(110) In electron micrographs, capsule appears as wavy fibers on the outermost surface of bacilli.(112) Though some of the capsule appears to be released from *B. anthracis,* much is tightly bound to the bacillus and protects it from antibodies and other molecules either as a function of its charge or size and density(112,113) (Fig. 4). Like many other bacterial capsules, it consists of a repeating polyanion, but unlike most other capsules, it is a protein, albeit a very unusual one. It is believed to be entirely poly-γ-D-glutamate,(114–117) unlike the capsular material of other *Bacillus* species, which are poly-γ-D, L-glutamate.(118) It is not known if there is secondary or tertiary structure to the capsule. However, one study showed that antiserum directed against capsule reacts more strongly with a branched polyglutamate peptide than linear peptides.(119)

Capsule synthesis is dependent on each of three consecutive genes encoded on pXO2, a 965-kbp plasmid.(101,120) The genes, *capB, capC, and capA,* encode proteins of 44, 16, and 46 kDa, respectively,(121) which are believed to form a membrane complex, similar to that described for *B. licheniformis.*(123,124) All three proteins were identified in a membrane fraction when expressed in *Escherichia coli* minicells.(121) CapC is believed to be an integral membrane protein, as it is very hydrophobic, and is trypsin resistant when expressed in *E. coli* spheroplasts. CapB and CapA each have a hydrophobic stretch of 20 amino acids that likely anchor them in the membrane, while the other portions are likely in the cytoplasm. Urushibata *et al.* suggested that CapB synthesizes capsule on the inner side of the membrane and that CapC transports it across the

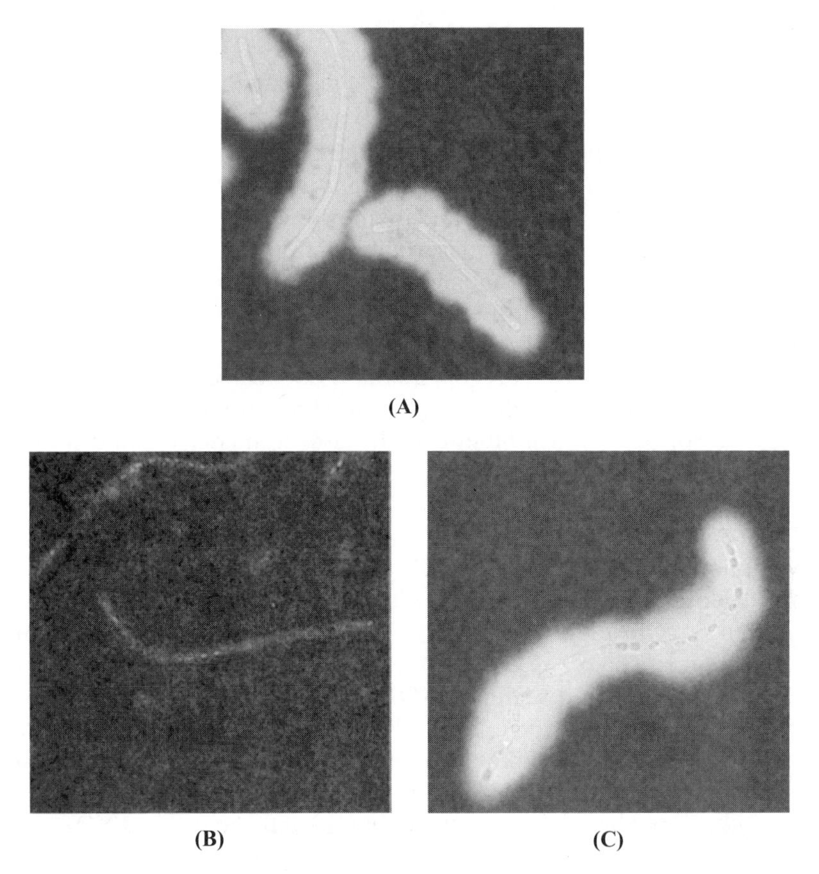

FIGURE 4. India Ink staining of *B. anthracis* strains grown on NBY/bicarbonate agar at 37°C with 20% CO_2 for 48 hours. This stain clearly shows the development of a robust capsule surrounding the bacilli in strains that are $pXO2^+$ (panels A and C). (A) Ames ($pXO1^+$, $pXO2^+$), (B) ΔANR ($pXO1^-$, $pXO2^-$), and (C) ΔAmes ($pXO1^-$, $pXO2^+$).

membrane.[125] In addition to the three genes required for synthesis of the capsule, another gene has been identified on the capsule operon. *DepA,* also known as *capD,* is believed to be responsible for capsule degradation. DepA is a secreted protein and theoretically might also enhance virulence by hydrolyzing host γ-glutamyl compounds.[129] DepA is reported to be an endopeptidase[113,129]; however, it has high homology with γ-glutamyltranspeptidases, which hydrolyze glutathione to glutamate and ammonia.[124] Transcription of *capA–D* can occur independently, as each gene has its own promoter, or transcripts encoding the entire operon can be produced. Northern blots with probes from any of the *cap* genes hybridized to a 6–7-kb transcript from bacteria grown with carbon dioxide/bicarbonate, but transcript was not detected in the absence of carbon dioxide and bicarbonate.[129,113]

6.5. Capsule Gene Regulation

Regulation of capsule production in *B. anthracis* by carbon dioxide and bicarbonate has been known for some time.(114,116,128) The physiological basis of carbon dioxide/bicarbonate regulation is thought to be positive co-regulation in combination with AtxA regulation,(129) and it is believed that this regulatory mechanism results in capsule and toxin production *in vivo*.(117) The equilibrium among carbon dioxide, bicarbonate, and carbonate depends upon pH. Regulation of capsule production by pH may also occur, as one report shows more than 10 times more capsule on *B. anthracis* grown at pH 8.5 than at pH 6.5.(130) Strains that constitutively express capsule have also been described, but mechanisms of deregulation have not been determined.(59,130,131) Perhaps they result from inactivation of an unidentified inhibitor of capsule production. Such an inhibitor might also be made by some strains of *B. subtilis* that have similar genes for capsule synthesis but do not produce capsule.

There are two functional promoters upstream of the capsule operon.(129) Transcription of the operon is positively regulated *in trans* by two proteins that are 25% identical and appear to act on the same 70-bp sequence upstream of the operon.(132) One of the regulators is AcpA, and is encoded on pXO2, more than 11-kbp upstream of the capsule operon.(134) The other regulator, AtxA, is encoded between *cya* and *pagA* on pXO1 and regulates transcription of many genes, including *cya*, *pagA*, and *lef*.(106) Both of these transcription regulators require carbon dioxide/bicarbonate to upregulate *cap* transcription, and *acpA* transcription also requires carbon dioxide/bicarbonate. The significance of *acpA* was illustrated by finding that the mutation of the gene in a $pXO1^-$ strain eliminated transcription of the capsule operon and subsequent expression of capsule, even in the presence of carbon dioxide and bicarbonate.(134)

The possibility of a regulator of capsule synthesis being encoded on pXO1 was suspected because $pXO1^-$ strains generally have reduced capsule expression.(133) Complementation with plasmids containing *atxA* restored capsule production.(135,132) In addition, an *acpA*$^-$ mutant could be complemented by *atxA*.(132) Recently, Drysdale *et al.* reported the discovery of another positive regulator of capsule expression in *B. anthracis*. The new protein was named AcpB, and is 62% similar to the previously described regulator AcpA. Reports suggest that a main regulator of capsule synthesis, AtxA, exerts its effects on the *capBCAD* operon indirectly by positively regulating *acpA* and *acpB*.(136)

6.6. Capsule Function

A number of reports claim that poly-γ-D-glutamate is non-immunogenic (137–139) No peptide-specific antibody was detected to L-glu_{60}D-ala_{40} or D-glu_{60}L-ala_{40} injected into animals and humans.(140) This indicates the weak immunogenicity of D-amino acids in general and poly-D-glutamate in particular and is consistent with the T-independent nature of bacterial polyanionic capsules.[141,142] Conversion of capsule to a T-dependent antigen by complexing with a carrier protein was suggested by the work of Goodman and colleagues, which showed enhanced immunogenicity when *B. anthracis* capsule was noncovalently linked to methylated bovine serum albumin simply by allowing electrostatic interaction at pH 4.[138] More recently, several reports showed dramatically enhanced immunogenicity of capsules of *Bacillus* species after chemically cross-linking to carrier proteins.[139,143] In addition, bacterial components may enhance antibody response to poly-D-γ-glutamate.[138] There are reports of strong anti-capsule responses to boiled *B. anthracis*, though these are only reported after a large number of injections with large doses of bacteria and are not consistent.[119,121] Antibody to capsule has also been detected in human sera; a retrospective study of a 1982 epidemic of cutaneous and oropharyngeal anthrax in Thailand showed that while only 72% of the patients were positive by ELISA for antibody to PA, 95–100% had positive anti-capsule titers.[145] It is not known if the anti-capsule antibody formed in response to natural infection is opsonic or protective. However, the finding that Hartley guinea pigs were not protected after vaccination with live vegetative pXO1$^-$, pXO2$^+$ strains, while they were fully protected by vaccination with pXO1$^+$ strains,[144] suggests that the antibody response to capsule on live bacteria may not be protective.

Capsule appears to be an indispensable virulence factor. Considering the growth conditions required for capsule production, it is remarkable that crude capsule preparations were first shown to be antiphagocytic in 1907 and capsule was associated with virulence in 1915.[146,147] The mechanism(s) of phagocytosis resistance has not been determined, but the phenomenon was dramatically demonstrated *in vitro* by the reduction of phagocytosis of strains NP and Sterne by guinea pig neutrophils after adding exogenous capsule.[148] Exogenous capsule also enhanced virulence of the Pasteur II strain in guinea pigs. Other reports described capsule-mediated phagocytosis inhibition in neutrophils and macrophages.[122,143] It is tempting to speculate that the gamma peptide linkages or the D-configuration unique to *B. anthracis* capsule might be required for this activity, but a report shows poly-α-L-glutamate can also inhibit phagocytosis.[117]

Complement-binding to capsule, perhaps in conjunction with S-layer proteins, has been reported.[13] Capsule-induced inhibition of anthracidal activity of normal horse serum and guinea pig leukocyte extracts has also been reported.[117] Capsule might also camouflage bacilli from the immune system by binding host proteins, such as lysozyme, protamine, and albumin.[137] Inability to degrade polymers of D-amino acids within lysosomes is yet another possible virulence mechanism.[141,149,150]

A recent report attributed virulence of a laboratory strain to antiphagocytic and complement binding properties of released capsule fragments.[113] Makino and colleagues attributed release of the capsule fragment to the activity of DepA, encoded by the fourth gene of the capsule operon.[113] Their results suggest capsule fragments might act as sinks for complement and perhaps other mediators of innate immunity.

6.7. Accessory Virulence Factors

As described above, the well-defined factors that contribute to the virulence of *B. anthracis* are the LT, ET, and capsule. Outside of these, few bacterial products contribute to the progression of anthrax. The possibility that such accessory factors exist is raised by four types of findings. (1) Mutations in specific genes or pathways reduce the virulence of *B. anthracis* strains. (2) Analysis of the recently sequenced *B. anthracis* Ames strain genome[17,127,151] revealed genes encoding factors predicted to contribute to virulence. (3) Variations in virulence and ability to overcome vaccine-induced immunity in certain animal models have been demonstrated between strains. (4) Attenuated strains exist that retain the genes for the known virulence factors.

B. anthracis is a member of the *B. cereus* group of bacteria, which includes *B. thuringiensis*, an insect pathogen, and *B. cereus*, generally considered an opportunistic pathogen of mammals. The genomes of these bacilli are quite similar.[17,152,153] A niche for *B. cereus* group strains as commensal inhabitants of invertebrate and even mammalian guts has been proposed.[154,155] Thus, *B. cereus* group chromosomes presumably contain a number of genes that permit colonization of and persistence in host environments. Furthermore, a number of virulence genes that are capable of damaging host cells and tissues have been identified in *B. cereus* and *B. thuringiensis*. Many of these genes, encoding enterotoxins, phospholipases, hemolysins, and proteases, are also present in the *B. anthracis* chromosome.[17,153] Although they would seem to be obvious candidates for *B. anthracis* virulence genes, there is currently little evidence of their expression *in vivo* or contribution to anthrax. In *B. cereus* and *B. thuringiensis*, many of these genes are regulated by PlcR, a transcriptional activator that is truncated and presumably inactive in *B. anthracis*.[156] The *in vitro* expression of an active phosphatidylcholine-specific phospholipase C (cereolysin A), sphingomyelinase (cerolysin B), and cholesterol-dependent cytolysin (cereolysin O) by *B. anthracis* lacking full-length PlcR was recently documented.[158,159] This suggests that orthologs of *B. cereus* PlcR-regulated genes may be expressed in *B. anthracis* under certain conditions in the absence of full-length PlcR. Heterologous expression of the *plcR* gene in *B. anthracis* induces protease, lecithinase, and hemolytic activities *in vitro* with variable success.[109,159] However, the critical observation that the virulence of a recombinant $plcR^{+}$ strain of *B. anthracis* is not increased in mice indicates that PlcR-regulated factors may not perceptibly contribute to the pathogenicity of *B. anthracis*.[23] This may be at odds with an earlier report that expression of cereolysin A and B in a recombinant, virulent *B. anthracis* creates a vaccine-resistant strain that can be overcome only by vaccination with an attenuated strain expressing these factors.[159]

Regulatory factors contribute indirectly to the virulence of *B. anthracis* by controlling the expression of known and perhaps unknown virulence factors. A mutation in the *atxA* gene, whose product activates the expression of toxin, greatly reduces the virulence of *B. anthracis* in mice.(102) In bacteria, multiple sigma factors associate with RNA polymerase and modify the expression of distinct sets of genes. In *B. anthracis*, only sigma B (σ^B) has yet been studied; this regulator responds to the stresses of stationary phase and heat shock.(160) A σ^B mutant was partially attenuated in mice, suggesting this factor plays a supporting role *in vivo*. The genes comprising the *B. anthracis* σ^B-regulon are unknown, with the exception of a putative bacterioferritin.(160)

Many other bacterial products likely contribute to the growth of *B. anthracis* in the host environment, while not being strictly required for bacterial viability. Such factors contribute indirectly to virulence. These may include systems involved in acquisition of iron(162) and other nutrients obtained from the host environment, as well as the synthesis of required nutrients that the host does not adequately provide. As an example, transposon mutants of *B. anthracis* deficient in the synthesis of aromatic compounds are attenuated in guinea pigs and mice.(163) A *gerX* mutant displays reduced virulence in mice, apparently by decreasing or delaying spore germination.(35,36) It is anticipated that many additional factors that contribute to germination, colonization, and persistence in the host will be identified in the future.

Significant variations in virulence between strains of *B. anthracis* have been reported.(164,166) This is intriguing, because this bacterium is thought to exhibit very little genetic variation.(165) Variations in ability to overcome vaccine-induced immunity between strains in guinea pig infections have also been noted, but such variation has not been noted in rabbits and nonhuman primates.(167–172) Furthermore, several *B. anthracis* strains, including several that have been used as large-animal vaccines, appear to carry the genes specifying the major virulence factors, yet paradoxically are significantly reduced in their ability to cause disease in animals.(173–176) It is important to point out that toxin activity in such strains has not been assayed and could be deficient. Transduction studies suggest that certain strains are attenuated in a manner that is not dependent on the pXO1 or pXO2 plasmids.(59,166) The specific defects leading to variations in virulence, vaccine resistance, and attenuation of various strains await identification.

7. *B. anthracis* AND MACROPHAGE INTERACTIONS

Early studies on the pathogenesis of anthrax were performed to determine the optimal host, route, and host cell for infection to occur. Normal epithelial surfaces were shown to be relatively resistant to spores of *B. anthracis*. Young *et al.* challenged guinea pigs by varying routes including oral, intraocular, intrarectal, and intra-vaginal.(177) The animals were also able to resist the challenge with as many as 1×10^8 spores. In addition, cutaneous infection in guinea pigs was not possible unless the epithelium was injured. These results indicated that *B. anthracis* spores lack the ability to invade intact epithelial tissue. However, if *B. anthracis* spores were inhaled and exposed to the pulmonary epithelial surface, a fatal infection ensued.(177) Despite the sensitivity of lung tissue to challenge with *B. anthracis,* no lesions were apparent in the lung tissue. Young *et al.* concluded that the lung serves only as a point of invasion for *B. anthracis* and that the organisms are carried to some other part of the body to multiply.

Joan Ross provided histological evidence that alveolar macrophages were the site of spore germination and multiplication.(178) For these studies, spores were introduced into the trachea or bronchi of guinea pigs and then a time course of infection was followed. Immediately after challenge, most spores were found phagocytized by alveolar macrophages or next to cells lining the alveolar wall. Within 35 minutes, most of the spores were phagocytized and more phagocytic cells became free within the alveoli. Extracellular *B. anthracis* spores were not apparent.(178) Histological sections at this time point showed the bacteria to counterstain with methylene blue, after acid-fast staining. This color change indicated that the *B. anthracis* cells germinated. One hour postinfection, the number of blue staining bacteria increased. By 2 hours, the macrophages containing the spores passed into the lymphatic vessels. Furthermore, no free spores were seen in the lymphatic vessels or lymph glands.(178)

The cytological features of anthrax within macrophages were described by Shafa *et al.*(179) For these studies, alveolar macrophages were obtained from rabbits and incubated with *B. anthracis* Sterne strain spores. The spores were quickly engulfed by macrophages. The spores then germinated and multiplied intracellularly and the macrophage eventually disrupted.

Since these seminal studies, molecular approaches have been initiated to provide further insight into the interaction between anthrax spores and macrophages. However, the fate of anthrax spores within macrophages continues to be debated. Guidi-Rontani *et al.* demonstrated that *B. anthracis* Sterne strain (7702) spores were able to germinate within BALB/c mouse alveolar macrophages as they are fusing with lysosomes. Using strains carrying *lacZ* transcriptional fusions to *atxA*, *lef*, *pagA*, and *cya*, it was shown that β-galactosidase activity was present in a phagolysosome of RAW264.7 macrophages, indicating that the germination of the spore was associated with phagolysosomal fusion and the expression of these virulence genes. It is postulated that *B. anthracis* evolved to respond to sudden changes within the macrophage brought on by fusion with lysosomes to express its virulence factors.(36)

As the *B. anthracis* spores germinate within the phagolysosome of a macrophage, numerous antimicrobial factors must be overcome. Dixon *et al.* performed studies to follow the fate of the *B. anthracis* Sterne strain spores within RAW264.7 macrophages. From these studies, *B. anthracis* spores germinated within the phagosome of the macrophage. As the vegetative bacilli were vulnerable to the hostile antimicrobial environment of the phagolysosome, the bacteria escaped the phagosome, replicated in the cytoplasm, and were eventually released from the macrophage. The studies were extended to examine mutant strains of *B. anthracis* to define the function of various virulence genes in macrophage intracellular survival.[180] When using strain RP-31, which is deficient in LT but retains a functional *atxA* gene, germinated vegetative cells were present within the macrophages by 30 minutes after infection. By 2 hours, bacilli were escaping the phagolysosome. After 3 hours, replicating vegetative cells were seen within the cytoplasm, leading to eventual macrophage permeability.[180] These results suggested that LT is not required for phagolysosomal escape or release from the macrophage. However, as described above, virulence genes present on pXO1, including *pagA* and *lef*, are expressed as the spores germinate within the phagosome.[36] When using strain UT60 that is able to produce LT but contains a mutated *atxA* gene, or ΔSterne, which is $pXO1^{-}$, the spores were able to germinate, but the bacilli accumulated in the macrophage and were not released. These results provided evidence that for *B. anthracis* to escape from the macrophage into the extracellular environment, a functional AtxA regulator was needed.

Other studies showed that when macrophages were challenged with *B. anthracis* spores, the bacteria were hindered in their ability to replicate intracellularly. The survival of the Sterne strain 7702 was compared to Sterne mutant strains carrying mutations in *pagA*, *lef*, or *cya* in Swiss murine peritoneal macrophages and RAW264.7 cells.[181] No differences in uptake between the Sterne strain and its derived mutants were observed in RAW264.7 or Swiss murine peritoneal macrophages. However, 3 hours after uptake of the spores by macrophages, spores of mutant *B. anthracis* strains lacking LF and EF were less viable. Further microscopic analysis of the spores in the primary peritoneal macrophages demonstrated that the spores devoid of the toxins did germinate but were killed before producing the S-layer component of the vegetative cell wall. In contrast, Sterne spores were able to germinate within the phagosome of the macrophage and survive, but not multiply.[181] The macrophages harboring *B. anthracis* Sterne strain 7702 had lost membrane integrity and it appears that this cytotoxicity is LF mediated. Culture medium from the infected macrophages was still toxic to uninfected macrophages.[181] However, if the medium had monoclonal antibodies to either PA or LF added to it, the uninfected macrophages were protected. In contrast, the data of Dixon *et al.* showed that macrophages incubated with a LF-deficient strain of *B. anthracis*, RP-31, still experienced membrane permeability, as measured by ^{51}Cr release.[180]

Macrophages have also been shown to be sporicidal *in vitro*. Peritoneal macrophages isolated from A/J and CBA/J mice were challenged with Sterne spores. After 60 minutes of incubation, macrophages from both strains of mice had similar phagocytic rates. The survival of *B. anthracis* was followed in the macrophages for 24 hours; however, extensive killing of the bacteria was observed.[182] Results similar to these were observed in later studies with the fully virulent Ames strain of *B. anthracis* and with culture medium that supported phagocytosis but not extracellular spore germination, Dulbecco's minimal essential medium with 10% horse serum.[183] Welkos *et al.* compared the survival of spores from either the fully virulent strain of Ames or the vaccine Sterne strain phagocytized by RAW264.7 cells or primary peritoneal macrophages from C3H/HeN mice.[183] For both strains, germination of the spores was observed within the macrophages and then the number of intracellular organisms decreased over 24 hours. One difference noted for the interaction with macrophages between Ames and Sterne spores was that Sterne spores were much more adherent to macrophages and difficult to remove by washing. When macrophages were treated with inhibitors of endosomal acidification and then challenged with Ames spores, the spores germinated and were then able to replicate intracellularly and escape from the macrophage. Taken together, these results suggest that loss in spore viability after phagocytosis may be associated with the antimicrobial environment of the macrophage phagosomem, and that macrophages may be beneficial to the host *in vivo*.

The contribution of macrophages to the pathogenesis of anthrax infection has also been an area of debate. To determine the role of macrophages in intoxication, Hanna *et al.* depleted macrophages by silica injections in 6-month-old BALB/c mice.[79] This procedure eliminated macrophages from the blood, peritoneum, liver, and spleen. The silica-treated, macrophage-depleted mice were resistant to a lethal dose of LT. The mice could be re-sensitized to LT by providing cultured RAW264.7 cells to the macrophage-depleted mice. In contrast, if supplied other cultured cells , Vero, CHO-K1, or IC-21, a toxin-insensitive murine peritoneal macrophage cell-line, the mice were still resistant to LT.[79] These results provide evidence that macrophages serve as mediators of the lethal action of LT *in vivo*.

In response to LT, macrophages may release cytokines. These cytokines might be responsible for host death by systemic shock. When RAW 264.7 cells were exposed to varying amounts of LT, the macrophages produced both IL-1 and TNF in a dose-response manner.[79] *In vitro* studies were then expanded to *in vivo* murine studies. Mice were depleted of IL-1 with antiserum to IL-1, leading to partial protection against challenge with LT. If mice were provided both anti-IL-1 and anti-TNF, the animals were fully protected against a LT challenge.

Other *in vitro* data indicate a different effect of LT on macrophages.[85] Exposure of the LT-sensitive macrophage cell lines, RAW 264.7 or J774 cells, or LT-resistant murine macrophage cell line, IC-21, to lipopolysaccharide (LPS) from *E. coli*, resulted in the expression of significant levels of TNF-α and IL-1β. In contrast, when the cell lines were first exposed to LT, none of the macrophage cell cultures expressed either TNF-α or IL-1β in response to LPS. The levels of LT tested ranged from 10^{-6} to 10^{2} μg/ml.[85] The authors suggested that the LT may impair the host response to the anthrax infection, thereby hindering the inflammatory response. Recently these reports have been supported by new observations suggesting LT can inhibit the activation of the IFN-regulatory factor 3 and also subsequent cytokine production.[86] These contradictory reports concerning the role of LT in cytokine induction remain unresolved. Clearly, the nature of the early interactions between *B. anthracis* and macrophages in the pathogenesis of disease are complex and will require greater attention.

8. VACCINE AND THERAPEUTIC APPROACHES

Current vaccines used for prophylactic therapy against anthrax infection consist of either cell-free culture filtrates or live attenuated vaccines. The licensed vaccine against anthrax in the United States, Biothrax® (formerly AVA, Anthrax Vaccine Adsorbed, BioPort, Lansing, MI), has been in use since 1970.(184) It consists primarily of aluminum hydroxide-adsorbed supernatant material from fermentor cultures of *B. anthracis,* V770-NP1-R.(185,186) PA is the primary toxin component detected in the medium used. It is recommended for individuals at risk of exposure to *B. anthracis* spores, such as wool workers, laboratory staff, and military personnel. The vaccine is administered subcutaneously in 0.5-ml doses at 0, 2, and 4 weeks and then at 6, 12, and 18 months followed by yearly boosters.(187) It is highly effective in rhesus macaques and rabbits(172) and although there is no human clinical protection data, a vaccine similar to Biothrax® protected humans in a field study in a setting of industrial exposure to high concentrations of spores.(188) A live attenuated vaccine strain of *B. anthracis,* STI-1, which lacks the pX02 plasmid is licensed for use in Russia and has been used for more than 50 years to vaccinate cattle, as well as humans.(166) Although effective, adverse reactions to live vaccines, which increase in frequency with the number of boosters administered, preclude their use in Western countries. In animal experiments, however, live vaccines confer superior protection to subunit vaccines in terms of time to onset of immunity, number of doses required to achieve effective immunity, the duration of protection, and the breadth of coverage.(187) However, the efficacy of cell-free culture filtrate vaccines (Biothrax®) in animal models other than rabbits and macaques (ie., mice and guinea pigs) has been variable,(57,172) although this variability may be related to differences in vaccine formulation such as certain adjuvants being used.(189) Taken together, these data suggest that research on development of alternate vaccines is warranted. In addition, new vaccines will be required to combat the potential threat of recombinant strains of *B. anthracis.*

Although AVA has proven to be a relatively safe and immunogenic vaccine,(190) adverse reactions to the vaccine have been reported.(191,192) The vaccine is derived from a sterile filtrate of *B. anthracis* adsorbed to aluminum hydroxide adjuvant and contains small quantities of undefined bacterial products.(193) The strain from which the vaccine is made, V770-NP1-R, is fully toxigenic ($pXO1^+$, $pXO2^-$) and thus requires measures to be taken for inactivation of small amounts of lethal and edema factor in the final vaccine product. A next-generation PA-based vaccine has recently been developed that is generated from a recombinant asporogenic, nontoxigenic, and nonencapsulated strain of *B. anthracis*.[194] This vaccine, which contains only purified PA combined with aluminum hydroxide adjuvant, has been shown to be protective in a nonhuman primate model of anthrax infection[193] and elicits toxin-neutralizing antibodies that correlate with protection in a rabbit model of infection.[195] Currently, the recombinant PA vaccine is in the initial stages of human clinical testing.

The major mechanism of protection conferred by vaccination with Biothrax® is to elicit a strong antibody response against PA, the major component of the vaccine. Anti-PA antibodies neutralize the effect of the *B. anthracis* toxins. There are several lines of evidence to date that suggest immune responses to components other than PA may contribute to protective efficacy of a potential anthrax vaccine. For example, in animal studies, live-spore vaccines are more protective than PA-based vaccines.[169,196] Additionally, anti-spore antibodies elicited by vaccination with PA in combination with formaldehyde-inactivated spores provided greater protection against spore challenge in both guinea pigs and mice compared with vaccination with PA alone.[197] Vaccination with spores of an attenuated, recombinant strain expressing PA conferred greater protection than vaccination with vegetative cells of the same strain.[198] As anti-PA antibodies primarily protect against toxemia by neutralizing the effects of the toxin, these data suggest that a vaccine that induces synergistic protective immunity against both toxemia and infection may provide superior overall protection. The demonstration of spore-induced protective immunity to *B. anthracis* has led to renewed interest in identifying spore immunogens that may potentially serve as subunit vaccines. Surface-exposed spore antigens provide the most attractive vaccine targets for this purpose.

Recently, a component of the spore exosporium, BclA was described[53] and experiments on its immunogenicity were conducted.[55] Immunoblotting with a panel of spore-specific monoclonal antibodies showed BclA, and specifically, the protein component of the glycoprotein, to be the immunodominant antigen on the spore surface. In these studies, 60% of the monoclonal antibodies blotted against spore extracts reacted with the BclA protein, suggesting it is the primary immunogenic component. Further development of BclA as a potential contribution to a subunit vaccine is ongoing.

Although antibodies to BclA have not yet been shown to protect animal models, the immune response to spores and specific proteins associated with spores warrants further investigation. It has been shown that PA may be associated with the spore surface.[199,183] In addition, anti-PA antiserum enhances the disposal of spores by promoting their phagocytosis and killing by murine intraperitoneal macrophages.[199,183] It also was shown that anti-PA serum moderately inhibited spore germination *in vitro*. Based on these results, it is reasonable to assume that eliciting high anti-spore-specific antibodies may prove beneficial in inhibiting the initial stages of *B. anthracis* infection. These antibodies bind to ungerminated spores and facilitate their disposal possibly by inhibiting extracellular germination and outgrowth and/or promoting their phagocytosis and killing by macrophages.[200] Thus, a vaccine that elicits both anti-PA and anti-spore immune responses may prove beneficial in combating anthrax infection.

A target for vaccine development that has largely remained uninvestigated because of its low immunogenicity is the poly-D-glutamic capsule that covers the surface of *B. anthracis* vegetative cells. The capsule of *B. anthracis* consists of very high-molecular weight polymerized D-glutamic acid that is synthesized by gene products located on the pX02 plasmid. Its production is regulated by genes located on both pX01 and pX02 and by temperature, bicarbonate, and the presence of CO_2.[121] The capsule is anti-phagocytic *in vitro*[122] and may provide barrier functions against bactericidal serum proteins, including complement. Although antibodies to capsule can be detected after systemic infections,[201] the capsule is poorly immunogenic.[110,201] Recent work showed that opsonizing IgG antibodies can be elicited in mice by conjugating capsule fragments to a protein carrier.[143] In these studies, various length peptides of γDPGA were conjugated at the C- or N- terminus to PA, bovine serum albumin or recombinant *Pseudomonas aeruginosa* exotoxin A and the fusion proteins administered to mice subcutaneously three times at 2-week intervals. High IgG titers could be elicited by this protocol with the highest titers from vaccination with decamers of γDPGA conjugated at the C-terminus to recombinant PA. Schneerson also showed that the IgG antibodies were opsonophagocytic in an *in vitro* assay with neutrophils and that the level of opsonization directly correlated with the IgG titer. Although the protective efficacy of the capsule conjugates was not determined, the results demonstrated the feasibility of eliciting a strong immune response to the *B. anthracis* capsule.[143] Adding capsule antigen to the current anthrax vaccine could potentially enhance the protection afforded by the vaccine and/or reduce the number of doses required by the licensed vaccine, currently an initial series of six boosters over 18 months followed by annual boosters. If the opsonizing activity of the anti-capsule antibodies is shown to be protective in animals, a capsule-containing vaccine may prove to be a valuable tool in combating anthrax.

Live avirulent or attenuated bacterial carriers have also been examined as potential delivery vehicles for anthrax antigens. A plasmid vector, pUB110, carrying the gene encoding PA (*pagA*) was used to transform an asporogenic *B. subtilis* strain. This recombinant strain secreting PA was used to inoculate guinea pigs and mice and elicited high serum anti-PA antibody titers and strong protection against lethal anthrax challenge.[144,163] Attenuated strains of *Salmonella typhimurium* have also been tested as carriers of PA in an oral delivery and challenge model of anthrax. In these studies, an *aroA*-mutant of *S. typhimurium* expressing PA conferred partial protection against challenge with wild type *B. anthracis*.[203] In addition, recombinant strains of *B. anthracis* were examined as potential vaccine candidates. Nonencapsulated and nontoxinogenic strains were transformed with recombinant plasmids expressing PA at various levels.[204] The immunity offered by the live recombinant anthrax vaccine correlated with the amount of PA expressed.[204] Cohen *et al.* used spores from nonencapsulated and nontoxinogenic strains of *B. anthracis* that were engineered to express recombinant PA to vaccinate guinea pigs. This study also suggested that immunity correlated well with the amount of PA expressed by the recombinant strains; however, it was suggested that spore-associated antigens may also be a valuable component of an effective anthrax vaccine.[198] Ivins *et al.* inoculated guinea pigs and mice with recombinant *B. anthracis* strains that were deficient in the synthesis of aromatic compounds. Significant protection was achieved by the live Aro-strains of *B. anthracis*, even when the animals were challenged with the fully virulent *B. anthracis* Ames strain.[163] Although not as common, viral vectors carrying the PA gene have also been tested in animal models of anthrax. Recombinant viral vectors (vaccinia vector and Venezuelan equine encephalitis vector) carrying *pagA* effectively elicited antibodies against PA and afforded protection from *B. anthracis* challenge in mice and guinea pigs.[205,206]

Intramuscular vaccination with DNA encoding PA is another potential strategy to protect against anthrax infection. Mice vaccinated with eukaryotic expression plasmids carrying the genes encoding PA and/or a truncated form of LF were protected against challenge with lethal doses of anthrax toxin.[207,209] These studies found that a combination of PA and LF expressing plasmids elicited far higher titers than either plasmid alone, suggesting that a combination subunit vaccine may offer more protection than one in which PA is the only component. Many strategies have been employed in an attempt to efficiently induce immunity against *B. anthracis*. Although progress has been made, there remains a need for new anthrax vaccine approaches.

Although effective prophylactic vaccines are vital to preventing disease in exposed human populations and in controlling infection with *B. anthracis* in domestic animals, the nonspecific early symptoms, rapid course, and high mortality associated with inhalation anthrax mandate the availability of effective therapeutics. In addition to several antibiotics used to treat anthrax, other strategies to neutralize the bacilli *in vivo* are also being explored. One such approach to combating anthrax infections that has recently been explored is by passive treatment with antibodies specific to *B. anthracis* antigens. The approach is attractive as it is well established that protection against anthrax is provided by a strong humoral immune response, in particular against PA. The protective effects of polyclonal antisera specific to PA, Anthrax Vaccine Absorbed (Biothrax®), or the Sterne spore vaccine have been examined.[208] Only PA-specific antiserum was significantly protective, resulting in a 67% survival rate. In a separate study, the protective effects and toxin neutralizing activity of polyclonal antisera directed against LT components (PA or LF) or against Sterne were evaluated. It demonstrated that while antibodies to LF strongly neutralized the effects of toxin *in vitro*, polyclonal anti-PA-specific serum was the most effective in protecting against a lethal spore challenge. Note that a combination dose of both anti-PA and anti-LF serum provided significantly better protection than anti-PA alone.[210] Thus, an effective passive exposure regimen should include anti-PA or a combination of anti-PA with other *B. anthracis* or toxin specific antisera. A combination of antibiotic therapy combined with active postexposure vaccination may provide the highest probability of successfully treating anthrax bacteremia.[211] Administering polyclonal anti-sera may also be an effective prophylactic treatment in cases where the prolonged vaccine regimen is prohibitive or when exposure to high levels of *B. anthracis* spores or toxins is a threat, as during a biological terrorist attack.

Another strategy to block the effects of anthrax toxin is the development of dominant negative mutants of PA. By constructing mutations in domain II of PA, the domain that is critical for membrane insertion and thus translocation of EF and LF into the target cell cytosol, the authors showed they could inhibit toxin activity *in vitro* and *in vivo.*[212] The mutant form of PA still bound to the cell receptor could thus be used to competitively inhibit wild-type PA in a dominant negative fashion. *In vivo* studies showed that whereas wild-type PA plus LF killed rats within 60 minutes, an equimolar mix of the dominant negative mutant with-wild-type PA and LF protected the animals, even after 48 hours. These data raise the possibility of using mutant PA as a therapy for anthrax infection.

Recent work has focused on the use of bacteriophages as a means of treatment and detection of *B. anthracis.*[213] Bacteriophages produce lytic enzymes necessary to hydrolyze the bacterial cell wall to release progeny phage. PlyG lysin from the γ phage could lyse *B. anthracis* and *B. cereus* strains belonging to the *B. anthracis* cluster. PlyG was lytic to both germinating spores and vegetative bacilli of *B. anthracis.* When mice were infected with a *B. cereus* strain that is closely related to *B. anthracis,* 80% of the mice were protected when supplied with the PlyG lysin 15 minutes after infection.

The potential of using *B. anthracis* as an agent of bioterror has been studied for decades. Although the potential was realized many years ago, it was brought to the forefront most dramatically in the United States in 2001. Although current vaccines and antibiotic therapies have improved estimated survival rates after exposure to *B. anthracis,* new threats may emerge. Antibiotic-resistant strains, super-virulent strains, and engineered strains are likely to be developed. Further studies of the basic biology and pathogenesis of *B. anthracis* leading to new vaccine candidates and novel treatments, both prophylactic and postexposure, are needed to counteract future threats.

REFERENCES

1. Koch, R., 1877, Die aetiologie der Milzbrand-Krankheit, begrundet auf die antwicklungsgeschichte des Bacillus anthracis, *Beritage zur Biologie der Pflanzen.* **2**:277–310.
2. Tigertt, W. D., 1990, Anthrax. William Smith Greenfield, M.D. FCRP. Concerning the priority due to him for the production of the first vaccine against anthrax, *J. Hyg.* **85**:415–420.
3. Pasteur, L., 1881, De l'attenuation des virus et de leur retour a la virulence, *C. R. Acad. Sci.* **92**:429–435.

4. Varkey, P., Poland, G. A., *et al.*, 2002, Confronting bioterrorism: physicians on the front line, *Mayo Clin. Proc.* **77**:661–672.
5. Brook, I., 2002, The prophylaxis and treatment of anthrax, *Int. J. Antimicrob.Agents.* **20**:320–325.
6. Malecki, J., Wiersma, S., *et al.*, 2001, Update: investigation of bioterrorism-related anthrax and interim guidelines for exposure management and antimircobial therapy, *MMWR.* **50**:909–919.
7. Franz, D. R., Jahrling, P. B., *et al.*, 2001, Clinical recognition and management of patients exposed to biological warfare agents, *Clin. Lab. Med.* **21**:435–473.
8. Mock, M., and Fouet, A., 2001, Anthrax, *Annu. Rev. Microbiol.* **55**:647–671.
9. Kassenborg, H., Danila, R., *et al.*, 2000, Human ingestion of *Bacillus anthracis*-contaminated meat–Minnesota, *MMWR.* **49**:813–816.
10. Inglesby, T. V., Henderson, D. A., *et al.*, 1999, Anthrax as a biological weapon: medical and public health management. Working group on civilian biodefense, *JAMA.* **281**:1735–1745.
11. Jernigan, J. A., Stephens, D. S., *et al.*, 2001, Bioterrorism-related inhalational anthrax: the first 10 cases reported in the United States, *Emerg. Infect. Dis.* **7**:933–944.
12. Christopher, G. W., Cieslak, T. J., *et al.*, 1997, Biological warfare. A historical perspective, *JAMA.* **278**:412–417.
13. Meselson, M., Guillemin, J., *et al.*, 1994, The Sverdlovsk anthrax outbreak of 1979, *Science* **266**:1202–1208.
14. WuDunn, S., Miller, J., *et al.*, 1998, How Japan germ terror alerted world, *New York Times* 1–6.
15. Keim, P., Smith, K. L., *et al.*, 2001, Molecular investigation of the Aum Shinrikyo anthrax release in Kameido, Japan, *J. Clin. Micro.* **39**:4566.
16. Helgason, E., Okstad, O. A., *et al.*, 2000, *Bacillus anthracis, Bacillus cereus,* and *Bacillus thuringiensis*—one species on the basis of genetic evidence, *Appl. Environ. Microbiol.* **66**:2627–2630.
17. Read, T. D., Peterson, S. N., *et al.*, 2003, The genome sequence of *Bacillus anthracis* Ames and comparison to closely related bacteria, *Nature* **423**:81–86.
18. Groisman, E. A. and Ochman, H., 1996, Pathogenicity islands: bacterial evolution in quantum leaps, *Cell.* **87**:791–794.
19. Okinaka, R., Cloud, K., *et al.*, 1999a, Sequence, assembly and analysis of pX01 and pX02, *J. Appl. Microbiol.* **87**:261–262.
20. Okinaka, R. T., Cloud, K., *et al.*, 1999b, Sequence and organization of pXO1, the large *Bacillus anthracis* plasmid harboring the anthrax toxin genes, *J. Bacteriol.* **181**:6509–6515.
21. Pannucci, J., Okinaka, R. T., *et al.*, 2002, DNA sequence conservation between the *Bacillus anthracis* pXO2 plasmid and genomic sequence from closely related bacteria, *BMC Genomics.* **3**:34.
22. Dragon, D. C., and Rennie, P. R., 1995, The ecology of anthrax spores: tough but not invincible, *Can. Vet. J.* **36**:295–301.
23. Mignot, T., Mock, M., *et al.*, 2001, The incompatibility between the PlcR- and AtxA-controlled regulons may have selected a nonsense mutation in *Bacillus anthracis, Mol. Microbiol.* **42**:1189–1198.
24. Ireland, J. A., and Hanna, P. C., 2002, Amino acid- and purine ribonucleoside-induced germination of *Bacillus anthracis* DeltaSterne endospores: *gerS* mediates responses to aromatic ring structures, *J. Bacteriol.* **184**:1296–1303.
25. Moir, A., and Smith, D. A., 1990, The genetics of bacterial spore germination, *Annu. Rev. Microbiol.* **44**:531–553.
26. Burbulys, D., Trach,K. A., *et al.*, 1991, Initiation of sporulation in *B. subtilis* is controlled by a multicomponent phosphorelay, *Cell.* **64**:545–552.
27. LeDeaux, J. R., and Grossman, A. D., 1995, Isolation and characterization of *kinC,* a gene that encodes a sensor kinase homologous to the sporulation sensor kinases KinA and KinB in *Bacillus subtilis, J. Bacteriol.* **177**:166–175.

28. LeDeaux, J. R.,Yu, N., *et al.*, 1995, Different roles for KinA, KinB, and KinC in the initiation of sporulation in *Bacillus subtilis*, *J. Bacteriol.* **177**:861–863.
29. Spiegelman, G. B., Bird, T. H., *et al.*, 1995, Transcription regulation by the*Bacillus subtilis* response regulator Spo0A, in *Two-component signal transduction* (J. A. Hoch and T. J. Silhavy, eds.), ASM Press, Washington D.C., pp. 159–179.
30. Stephenson, S. J. and Perego, M., 2002, Interaction surface of the Spo0A response regulator with the Spo0E phosphatase, *Mol. Microbiol.* **44**:1455–1467.
31. Feavers, I. M., Miles, J. S., *et al.*, 1985, The nucleotide sequence of a spore germination gene (*gerA*) of *Bacillus subtilis* 168. *Gene.* **38**:95–102.
32. Irie, R., Okamoto, T., *et al.*, 1986, Characterization and mapping of *Bacillus subtilis gerD* mutants, *J. Gen. Appl. Microbiol.* **32**:303–315.
33. Zuberi, A. R., Moir, A., *et al.*, 1987, The nucleotide sequence and gene organization of the *gerA* spore germination operon of *Bacillus subtilis* 168, *Gene.* **51**:1–11.
34. Clements, M. O., and Moir, A., 1998, Role of the *gerI* operon of *Bacillus cereus* 569 in the response of spores to germinants, *J. Bacteriol.* **180**:6729–6735.
35. Guidi-Rontani, C., Pereira, Y., *et al.*, 1999a, Identification and characterization of a germination operon on the virulence plasmid pXO1 of *Bacillus anthracis, Mol. Microbiol.* **33**: 407–414.
36. Guidi-Rontani, C., Weber-Levy, M., *et al.*, 1999b, Germination of *Bacillus anthracis* spores within alveolar macrophages, *Mol. Microbiol.* **31**:9–17.
37. Weiner, M. A., Read, T. D., *et al.*, 2003, Identification and characterization of the *gerh* operon of *Bacillus anthracis* endospores: A differenetial role for purine nucleosides in germination, *J. Bacteriol.* **185**:1462–1464.
38. Moberly, B. J., Shafa, F., *et al.*, 1966, Structural details of anthrax spores during stages of transformation into vegetative cells, *J. Bacteriol.* **92**:220–228.
39. Welkos, S. L., Cote, C. K., *et al.*, 2004, A microtiter fluorometric assay to detect the germination of *Bacillus anthracis* spores and the germination inhibitory effects of antibodies, *J. Microbiol. Meth.* **56**:253–265.
40. Gerhardt, P., 1967, Cytology of *Bacillus anthracis, Fed. Proc.* **26**:1504–1517.
41. Titball, R. W. and Manchee, R. J., 1987, Factors affecting the germination of spores of *Bacillus anthracis*, *J. Appl. Bacteriol.* **62**:269–273.
42. Hachisuka, Y., 1969, Germination of *B. anthracis* spores in peritoneal cavity of rats and establishment of anthrax, *Jpn. J. Microbiol.* **13**:199–207.
43. Lai, E. M., Phadke, N. D., *et al.*, 2003, Proteomic analysis of the spore coats of *Bacillus subtilis* and *Bacillus anthracis*, *J. Bacteriol.* **185**:1443–1454.
44. Driks, A., 2002, Maximum shields: the assembly and function of the bacterial spore coat, *Trends. Microbiol.* **10**:251–254.
45. Roels, S., Driks, A., *et al.*, 1992, Characterization of *spoIVA*, a sporulation gene involved in coat morphogenesis in *Bacillus subtilis*, *J. Bacteriol.* **174**:575–585.
46. Driks, A., Roels, S., *et al.*, 1994, Subcellular localization of proteins involved in the assembly of the spore coat of *Bacillus subtilis, Genes. Dev.* **8**:234–244.
47. Pogliano, K., Harry, E., *et al.*, 1995, Visualization of the subcellular location of sporulation proteins in *Bacillus subtilis* using immunofluorescence microscopy, *Mol. Microbiol.* **18**: 459–470.
48. Zheng, L. B., Donovan, W. P., *et al.*, 1988, Gene encoding a morphogenic protein required in the assembly of the outer coat of the *Bacillus subtilis* endospore, *Genes. Dev.* **2**:1047–1054.
49. Gerhardt, P., and Ribi, E., 1964, Ultrastructure of the exosporium enveloping spores of *Bacillus cereus*, *J. Bacteriol.* **88**:1774–1789.
50. Matz, L. L., Beaman, T. C., *et al.*, 1970, Chemical composition of exosporium from spores of *Bacillus cereus*, *J. Bacteriol.* **101**:196–201.
51. Garcia-Patrone, M., and Tandecarz, J. S., 1995, A glycoprotein multimer from *Bacillus thuringiensis* sporangia: dissociation into subunits and sugar composition, *Mol. Cell. Biochem.* **145**:29–37.

52. Todd, S. J., Moir, A. J., *et al.*, 2003, Genes of *Bacillus cereus* and *Bacillus anthracis* encoding proteins of the exosporium, *J. Bacteriol.* **185**:3373–3378.
53. Sylvestre, P., Couture-Tosi, E., *et al.*, 2002, A collagen-like surface glycoprotein is a structural component of the *Bacillus anthracis* exosporium, *Mol. Microbiol.* **45**:169–178.
54. Sylvestre, P., Couture-Tosi, E., *et al.*, 2003, Polymorphism in the collagen-like region of the *Bacillus anthracis* BclA protein leads to variation in exosporium filament length, *J. Bacteriol.* **185**:1555–1563.
55. Steichen, C., Chen, P., *et al.*, 2003, Identification of the immunodominant protein and other proteins of the *Bacillus anthracis* exosporium, *J. Bacteriol.* **185**:1903–1910.
56. Little, S. F. and Lowe, J. R., 1991, Location of receptor-binding region of protective antigen from *Bacillus anthracis, Biochem. Biophys. Res. Commun.* **180**:531–537.
57. Welkos, S. L., Lowe, J. R., *et al.*, 1988, Sequence and analysis of the DNA encoding protective antigen of *Bacillus anthracis, Gene.* **69**:287–300.
58. Petosa, C., Collier, R. J., *et al.*, 1997, Crystal structure of the anthrax toxin protective antigen, *Nature.* **385**:833–838.
59. Welkos, S. L., 1991, Plasmid-associated virulence factors of non-toxigenic (pX01-) *Bacillus anthracis, Microb. Pathog.* **10**:183–198.
60. Khanna, H., Chopra, A. P., *et al.*, 2001, Role of residues constituting the 2beta1 strand of domain II in the biological activity of anthrax protective antigen, *FEMS. Microbiol. Lett.* **199**:27–31.
61. Mogridge, J., Mourez, M., *et al.*, 2001, Involvement of domain 3 in oligomerization by the protective antigen moiety of anthrax toxin, *J. Bacteriol.* **183**:2111–2116.
62. Miller, C. J., Elliott, J. L., *et al.*, 1999, Anthrax protective antigen: prepore-to-pore conversion, *Biochem.* **38**(32): 10432–10441.
63. Beauregard, K. E., Collier, J. R., *et al.*, 2000, Proteolytic activation of receptor-bound anthrax protective antigen on macrophages promotes its internalization, *Cell. Microbiol.* **2**:251–258.
64. Mogridge, J., Cunningham, K., *et al.*, 2002a, Stoichiometry of anthrax toxin complexes, *Biochem.* **41**:1079–1082
65. Mogridge, J., Cunningham, K., *et al.*, 2002b, The lethal and edema factors of anthrax toxin bind only to oligomeric forms of the protective antigen, *Proc. Natl. Acad. Sci. U S A* **99**:7045–7048.
66. Chauhan, V., and Bhatnagar, R., 2002, Identification of amino acid residues of anthrax protective antigen involved in binding with lethal factor, *Infect. Immun.* **70**:4477–4484.
67. Cunningham, K., Lacy, D. B., *et al.*, 2002, Mapping the lethal factor and edema factor binding sites on oligomeric anthrax protective antigen, *Proc. Natl. Acad. Sci. U S A.* **99**:7049–7053.
68. Ahuja, N., Kumar, P., *et al.*, 2001, Rapid purification of recombinant anthrax-protective antigen under nondenaturing conditions, *Biochem. Biophys. Res. Commun.* **286**:6–11.
69. Sellman, B. R., Mourez, M., *et al.*, 2001, Dominant-Negative mutants of a toxin subunit: An approach to therapy of anthrax, *Science* **292**:695–697.
70. Ezzell, J. W., Jr., and Abshire, T. G., 1992, Serum protease cleavage of *Bacillus anthracis* protective antigen, *J. Gen. Microbiol.* **138** :543–549.
71. Robertson, D. L., and Leppla, S. H., 1986, Molecular cloning and expression in *Escherichia coli* of the lethal factor gene of *Bacillus anthracis, Gene.* **44**:71–78.
72. Klimpel, K. R., Molloy, S. S., *et al.*, 1992, Anthrax toxin protective antigen is activated by a cell surface protease with the sequence specificity and catalytic properties of furin, *Proc. Natl. Acad. Sci. U S A* **89**:10277–10281.
73. Hammond, S. E. and Hanna, P. C., 1998, Lethal factor active-site mutations affect catalytic activity in vitro, *Infect. Immun.* **66**:2374–2378.
74. Pannifer, A. D., Wong, T. Y., *et al.*, 2001, Crystal structure of the anthrax lethal factor, *Nature* **414**:229–233.
75. Popov, S. G., Villasmil, R., *et al.*, 2002a, Effect of *Bacillus anthracis* lethal toxin on human peripheral blood mononuclear cells, *FEBS. Lett.* **527**:211-5.

76. Popov, S. G., Villasmil, R., *et al.*, 2002b, Lethal toxin of *Bacillus anthracis* causes apoptosis of macrophages, *Biochem. Biophys. Res. Commun.* **293**:349-55.
77. Friedlander, A. M., 1986, Macrophages are sensitive to anthrax lethal toxin through an acid-dependent process, *J. Biol. Chem.* **261**:7123–7126.
78. Singh, Y., Leppla, S. H., *et al.*, 1989, Internalization and processing of *Bacillus anthracis* lethal toxin by toxin-sensitive and -resistant cells, *J. Biol. Chem.* **264**:11099–11102.
79. Hanna, P. C., Acosta, D., *et al.*, 1993, On the role of macrophages in anthrax, *Proc. Natl. Acad. Sci. U S A.* **90**:10198–10201.
80. Vitale, G., Pellizzari, R., *et al.*, 1998, Anthrax lethal factor cleaves the N-terminus of MAP-KKs and induces tyrosine/threonine phosphorylation of MAPKs in cultured macrophages, *Biochem. Biophys. Res. Commun.* **248**:706–711.
81. Duesbery, N. S. and Vande Woude, G. F., 1999, Anthrax lethal factor causes proteolytic inactivation of mitogen-activated protein kinase kinase, *J. Appl. Microbiol.* **87**:289–293.
82. Vitale, G., Bernardi, L., *et al.*, 2000, Susceptibility of mitogen-activated protein kinase kinase family members to proteolysis by anthrax lethal factor, *Biochem. J.* **352**:739–745.
83. Park, J. M., Greten, F. R., *et al.*, 2002, Macrophage apoptosis by anthrax lethal factor through p38 MAP kinase inhibition, *Science* **297**:2048–2051.
84. Friedlander, A. M., Bhatnagar, R., *et al.*, 1993a, Characterization of macrophage sensitivity and resistance to anthrax lethal toxin, *Infect. Immun.* **61**:245–252.
85. Erwin, J. L., DaSilva, L. M., *et al.*, 2001, Macrophage-derived cell lines do not express proinflammatory cytokines after exposure to *Bacillus anthracis* lethal toxin, *Infect. Immun.* **69**:1175–1177.
86. Dang, O., Navarro, L., *et al.*, 2004, Cutting Edge: Anthrax lethal toxin inhibits activation of IFN-Regulatory Factor 3 by lipopolysaccharide, *J. Immunol.* **172**:747–751.
87. Tucker, A. E., Salles, I. I., *et al.*, 2003, Decreased glycogen synthase kinase 3-beta levels and related physiological changes in *Bacillus anthracis* lethal toxin-treated macrophages, *Cell. Microbiol.***5**:523–532.
88. Webster, J. I., Tonelli, L. H., *et al.*, 2003, Anthrax lethal factor represses glucocorticoid and progesterone receptor activity. *Proc. Natl. Acad. Sci. U S A* **100**:5706–5711.
89. Tippetts, M. T., and Robertson, D. L., 1988, Molecular cloning and expression of the *Bacillus anthracis* edema factor toxin gene: a calmodulin-dependent adenylate cyclase, *J. Bacteriol.* **170**:2263–2266.
90. Stanley, J. L., and Smith, H., 1961, Purification of factor I and recognition of a third factor of the anthrax toxin, *J. Gen. Microbiol.* **26**:49–63.
91. Leppla, S. H., 1982, Anthrax toxin edema factor: a bacterial adenylate cyclase that increases cyclic AMP concentrations of eukaryotic cells, *Proc. Natl. Acad. Sci. U S A* **79**:3162–3166.
92. O'Brien, J., Friedlander, A., *et al.*, 1985, Effects of anthrax toxin components on human neutrophils, *Infect. Immun.* **47**:306–310.
93. Kumar, P., Ahuja, N., *et al.*, 2002, Anthrax edema toxin requires influx of calcium for inducing cyclic AMP toxicity in target cells,*Infect. Immun.* **70**:4997–5007.
94. Ulmer, T. S., Soelaiman, S., *et al.*, 2003, Calcium dependence of the interaction between calmodulin and anthrax edema factor, *J. Biol. Chem.* **278**:29261–29266.
95. Kumar, P., Ahuja, N., *et al.*, 2001, Purification of anthrax edema factor from *Escherichia coli* and identification of residues required for binding to anthrax protective antigen, *Infect. Immun.* **69**:6532–6536.
96. Bradley, K. A., Mogridge, J., *et al.*, 2001, Identification of the cellular receptor for anthrax toxin, *Nature.* **414**:225–229.
97. Abrami, L., Liu S., *et al.*, 2003, Anthrax toxin triggers endocytosis of its receptor via a lipid raft-mediated clathrin-dependent process, *J. Cell. Biol.* **160**(3): 321–328.
98. Scobie, H. M., Rainey, G. J., *et al.*, 2003, Human capillary morphogenesis protein 2 functions as an anthrax toxin receptor, *Proc. Natl. Acad. Sci. U S A* **100**:5170–5174.

99. Ristroph, J. D. and Ivins, B. E., 1983, Elaboration of *Bacillus anthracis* antigens in a new, defined culture medium, *Infect. Immun.* **39**:483–486.
100. Bartkus, J. M., and Leppla, S. H., 1989, Transcriptional regulation of the protective antigen gene of *Bacillus anthracis, Infect. Immun.* **57**:2295–2300.
101. Uchida, I., Sekizaki, T., *et al.*, 1985, Association of the encapsulation of *Bacillus anthracis* with a 60 megadalton plasmid, *J. Gen. Microbiol.* **131**:363–367.
102. Dai, Z., Sirard, J. C., *et al.*, 1995, The *atxA* gene product activates transcription of the anthrax toxin genes and is essential for virulence, *Mol. Microbiol.* **16**:1171–1181.
103. Dai, Z. and Koehler, T. M., 1997, Regulation of anthrax toxin activator gene (*atxA*) expression in *Bacillus anthracis*: temperature, not CO_2/bicarbonate, affects AtxA synthesis, *Infect. Immun.* **65**:2576–2582.
104. McIver, K. S., and Myles, R. L., 2002, Two DNA-binding domains of Mga are required for virulence gene activation in the group A streptococcus, *Mol. Microbiol.* **43**:1591–1601.
105. Hoffmaster, A. R., and Koehler, T. M., 1997, The anthrax toxin activator gene *atxA* is associated with CO_2-enhanced non-toxin gene expression in *Bacillus anthracis, Infect. Immun.* **65**:3091–3099.
106. Bourgogne, A., Drysdale, M., *et al.*, 2003, Global effects of virulence gene regulators in a *Bacillus anthracis* strain with both virulence plasmids, *Infect. Immun.* **71**:2736–2743.
107. Hoffmaster, A. R., and Koehler, T. M., 1999a, Autogenous regulation of the *Bacillus anthracis pag* operon, *J. Bacteriol.* **181**:4485–4492.
108. Hoffmaster, A. R., and Koehler, T. M., 1999b, Control of virulence gene expression in *Bacillus anthracis, J. Appl. Microbiol.* **87**:279–281.
109. Mignot, T., Mock, M., *et al.*, 2003, A plasmid-encoded regulator couples the synthesis of toxins and surface structures in *Bacillus anthracis, Mol. Microbiol.* **47**:917–927.
110. Ezzell, J., and Abshire, T., 1999, Encapsulation of *Bacillus anthracis* spores and spore identification, Proceedings of the International Workshop on Anthrax, *Salisbury Medical Bulletin* **87**:42.
111. Saile, E., and Koehler, T. M., 2002, Control of anthrax toxin gene expression by the transition state regulator *abrB, J. Bacteriol.* **184**:370–380.
112. Mesnage, S., Tosi-Couture, E., *et al.*, 1998, The capsule and S-layer: two independent and yet compatible macromolecular structures in *Bacillus anthracis, J. Bacteriol.* **180**:52–58.
113. Makino, S., Watarai, M., *et al.*, 2002, Effect of the lower molecular capsule released from the cell surface of *Bacillus anthracis* on the pathogenesis of anthrax, *J. Infect. Dis.* **186**:227–233.
114. Ivanovics, G., 1937, Unter welchen bedingungen werden bei der nahrbodenzuchtung der milzbandbazillen kapseln gebildet?, *Zentr. Bact. Parasitenk. Orig.* **90**:449–455.
115. Bruckner, V., Kovacs, J., *et al.*, 1953, Structure of poly-D-glutamic acid isolated from capsulated strains of *B. anthracis, Nature.* **172**:508.
116. Ivanovics, G., and Erdos, L., 1937, Ein Beitrag zum Wesen der Kapselsubstnz des Milizbrand-bazillus, *Z. Immunitatsforsch.* **90**:5–19.
117. Zwartouw, H. T., Smith, H., 1956, Polyglutamic acid from *Bacillus anthracis* grown *in vivo*: structure and aggress in activity, *Biochem. J.* **63**:437–454.
118. Thorne, C. B., and Leonard, C. G., 1958, Isolation of D- and L-glutamyl polypeptides from culture filtrates of *Bacillus subtilis, J. Biol. Chem.* **233**:1109–1112.
119. Goodman, J. W., and Nitecki, D. E., 1966, Immunochemical studies on the poly-gamma-D-glutamyl capsule of *Bacillus anthracis*. I. Characterization of the polypeptide and of the specificity of its reaction with rabbit antisera, *Biochemistry* **5**:657–665.
120. Green, B. D., Battisti, L., *et al.*, 1985, Demonstration of a capsule plasmid in *Bacillus anthracis, Infect. Immun.* **49**(2): 291–297.
121. Makino, S., Sasakawa, C., *et al.*, 1988, Cloning and CO2-dependent expression of the genetic region for encapsulation from *Bacillus anthracis, Mol. Microbiol.* **2**:371–376.
122. Makino, S., Uchida, I., *et al.*, 1989, Molecular characterization and protein analysis of the cap region, which is essential for encapsulation in *Bacillus anthracis, J. Bacteriol.* **171**:722–730.

123. Troy, F. A., 1973, Chemistry and biosynthesis of the poly(-D-glutamyl) capsule in *Bacillus licheniformis*. II. Characterization and structural properties of the enzymatically synthesized polymer, *J. Biol. Chem.* **248**:316–324.
124. Troy, F. A., 1985, Capsular poly-gamma-D-glutamate synthesis in *Bacillus licheniformis, Meth. Enzymol.* **113**:146–168.
125. Urushibata, Y., Tokuyama, S., *et al.*, 2002, Characterization of the *Bacillus subtilis ywsC* gene, involved in gamma- polyglutamic acid production, *J. Bacteriol.* **184**:337–343.
126. Uchida, I., Makino, S., *et al.*, 1993a, Identification of a novel gene, *dep*, associated with depolymerization of the capsular polymer in *Bacillus anthracis, Mol. Microbiol.* **9**:487–496.
127. Minami, H., Suzuki, H., *et al.*, 2003, A mutant *B. subtilis* g-glutamyltranspeptidase specialized in hydrolysis activity, *FEMS. Microbiol. Lett.* **224**:169–173.
128. Sterne, M., 1937, Variation in *Bacillus anthracis, Onderstepoort. J. Vet. Sci. Animal. Ind.* **8**: 271–349.
129. Uchida, I., Hornung, J. M., *et al.*, 1993b, Cloning and characterization of a gene whose product is a trans-activator of anthrax toxin synthesis, *J. Bacteriol.* **175**:5329–5338.
130. Thorne, C. B., Gomez, C. G., *et al.*, 1952, Biosynthesis of glutamic acid and glutamyl polypeptide by *Bacillus anthracis*: II. The effect of carbon diolxide on peptide production on solid media, *J. Bacteriol.* **63**:363–368.
131. Meynell, G. G. and Lawn, A. M., 1965, Inheritance of capsule and the manner of cell-wall formation in *Bacillus anthracis, J. Gen. Microbiol.* **39**:423–427.
132. Uchida, I., Makino, S., *et al.*, 1997, Cross-talk to the genes for *Bacillus anthracis* capsule synthesis by *atxA*, the gene encoding the trans-activator of anthrax toxin synthesis, *Mol. Microbiol.* **23**:1229–1240.
133. Fouet, A., and Mock, M., 1996, Differential influence of the two *Bacillus anthracis* plasmids on regulation of virulence gene expression, *Infect. Immun.* **64**:4928–4932.
134. Vietri, N. J., Marrero, R., *et al.*, 1995, Identification and characterization of a trans-activator involved in the regulation of encapsulation by *Bacillus anthracis, Gene.* **152**:1–9.
135. Guignot, J., Mock, M., *et al.*, 1997, AtxA activates the transcription of genes harbored by both *Bacillus anthracis* virulence plasmids, *FEMS Microbiol. Lett.* **147**:203–207.
136. Drysdale, M., Bourgogne, A., *et al.*, 2004, *atxA* control *Bacillus anthracis* capsule synthesis via *acpA* and a newly discovered regulator, *acpB, J. Bacteriol.* **186**:307–315.
137. Leonard, C. G., and Thorne, C. B., 1961, Studies on the nonspecific precipitation of basic serum proteins with g-glutamyl polypeptides, *J. Immunol.* **87**:175–88.
138. Goodman, J. W., and Nitecki, D. E., 1967, Studies on the relation of a prior immune response to immunogenicity, *Immunology* **13**:577–583.
139. Rhie, G., Roehrl, M. H., *et al.*, 2003, A dually active anthrax vaccine that confers protection against both bacilli and toxins, *Proc. Natl. Acad. Sci. U S A* **100**:10925–10930.
140. Maurer, P. H., 1965, Antigenicity of polypeptides (poly alpha amino acids) XIII. Immunological studies with synthetic polymers containing only D- or D- and L-a-amino acids, *J. Exp. Med.* **121**:339–349.
141. Lindberg, A. A., 1999, Glycoprotein conjugate vaccines, *Vaccine.* **17**:S28–S36.
142. Lesinski, G. B., and Westerink, M. A., 2001, Vaccines against polysaccharide antigens, *Curr. Drug. Targets. Infect. Disord.* **1**:325–334.
143. Schneerson, R., Kubler-Kielb, J., *et al.*, 2003, Poly (gamma-D-glutamic acid) protein conjugates induce IgG antibodies in mice to the capsule of *Bacillus anthracis*: a potential addition to the anthrax vaccine, *Proc. Natl. Acad. Sci. U S A* **100**:8945–8950.
144. Ivins, B. E., Ezzell, Jr., J. W., *et al.*, 1986, Immunization studies with attenuated strains of *Bacillus anthracis, Infect. Immun.* **52**:454–458.
145. Sirisanthana, T., Nelson, K. E., *et al.*, 1988, Serological studies of patients with cutaneous and oral-oropharyngeal anthrax from northern Thailand, *Am. J. Trop. Med. Hyg.* **39**:575–581.
146. Gruber, M., and Futaki, K., 1907, Uber die resistenz gegen milzbrand und uber die herkunft der milzbrandfeindlichen stoffe, *Med. Wschr.* **54**:249.

147. Bail, O., 1915, Veranderung der bakterien in tierkorper. Ueber die korrelation zwischen kapselbildung sporenbildung und infektiositat des milzbrnadbazillus, *Zentralbl Bakt Paras Infekt Krank I Orig.* **75**:159–173.
148. Keppie, J., Harris-Smith, P. W., *et al.*, 1963, The chemical basis of the virulence of *Bacillus anthracis.* IX. Its aggressins and their mode of action, *Br. J. Exp. Pathol.* **44**:446–453.
149. Kishore, B. K., Fuming, L., *et al.*, 1996a, Mechanism of the thesaurismosis and altered lysosomal dynamics induced by poly-D-glutamic acid in kidney proximal tubular cells, *Lab. Invest.* **74**:1025–1037.
150. Kishore, B. K., Maldague, P., *et al.*, 1996b, Poly-D-glutamic acid induces an acute lysosomal thesaurismosis of proximal tubules and a marked proliferation of interstitium in rat kidney, *Lab. Invest.* **74**:1013–1023.
151. Read, T. D., Salzberg, S. L., *et al.*, 2002, Comparative genome sequencing for discovery of novel polymorphisms in *Bacillus anthracis, Science* **296**:2028–2033.
152. Kolsto, A. B., Lereclus, D., *et al.*, 2002, Genome structure and evolution of the *Bacillus cereus* group, *Curr. Top. Microbiol. Immunol.* **264**:95–108.
153. Ivanova, N., Sorokin, A., *et al.*, 2003, Genome sequence of *Bacillus cereus* and comparative analysis with *Bacillus anthracis, Nature* **423**:87–91.
154. Margulis, L., Jorgensen, J. Z., *et al.*, 1998, The Arthromitus stage of *Bacillus cereus*: intestinal symbionts of animals, *Proc. Natl. Acad. Sci. U S A* **95**:1236–1241.
155. Jensen, G. B., Hansen, B. M., *et al.*, 2003, The hidden lifestyles of *Bacillus cereus* and relatives, *Environ. Microbiol.* **5**:631–640.
156. Agaisse, H., Gominet, M., *et al.*, 1999, PlcR is a pleiotropic regulator of extracellular virulence factor gene expression in *Bacillus thuringiensis, Mol. Microbiol.* **32**:1043–1053.
157. Klichko, V. I., Miller, J., *et al.*, 2003, Anaerobic induction of *Bacillus anthracis* hemolytic activity, *Biochem. Biophys. Res. Commun.* **303**:855–862.
158. Shannon, J. G., Ross, C. L., *et al.*, 2003, Characterization of anthrolysin O, the *Bacillus anthracis* cholesterol-dependent cytolysin, *Infect. Immun.* **71**:3183–3189.
159. Pomerantsev, A. P., Kalnin, K. V., *et al.*, 2003, Phosphatidylcholine-specific phospholipase C and sphingomyelinase activities in bacteria of the *Bacillus cereus* group, *Infect. Immun.* **71**(11): 6591–606.
160. Fouet, A., Namy, O., *et al.*, 2000, Characterization of the operon encoding the alternative sigma (B) factor from *Bacillus anthracis* and its role in virulence, *J. Bacteriol.* **182**:5036–5045.
161. Pomerantsev, A. P., Staritsin, N. A., *et al.*, 1997, Expression of cereolysine AB genes in *Bacillus anthracis* vaccine strain ensures protection against experimental hemolytic anthrax infection, *Vaccine.* **15**:1846–1850.
162. Papinutto, E., Dundon, W. G., *et al.*, 2002, Structure of two iron-binding proteins from *Bacillus anthracis, J. Biol. Chem.* **277**:15093–15098.
163. Ivins, B. E., Welkos, S. L., *et al.*, 1990. Immunization against anthrax with aromatic compound-dependent (Aro-) mutants of *Bacillus anthracis* and with recombinant strains of *Bacillus subtilis* that produce anthrax protective antigen, *Infect. Immun.* **58**:303–308.
164. Welkos, S. L., Vietri, N. J., *et al.*, 1993, Non-toxigenic derivatives of the Ames strain of *Bacillus anthracis* are fully virulent for mice: role of plasmid pX02 and chromosome in strain-dependent virulence, *Microb. Pathog.* **14**:381–388.
165. Keim, P., Price, L. B., *et al.*, 2000, Multiple-locus variable-number tandem repeat analysis reveals genetic relationships within *Bacillus anthracis, J. Bacteriol.* **182**:2928–2936.
166. Stepanov, A. S., Mikshis, N. I., *et al.*, 1999, Contribution of determinants, located in *Bacillus anthracis* chromosomes, in realizing the pathogenic properties of the pathogen, *Mol. Gen. Mikrobiol. Virusol.* **1**:20–23.
167. Auerbach, S., and Wright, G. G., 1955, Studies on immunity in anthrax. VI. Immunizing activity of protective antigen against various strains of *Bacillus anthracis, J. Immunol.* **75**:129–133.
168. Ward, M. K., McGann, V. G., *et al.*, 1965, Studies on anthrax infections in immunized guinea pigs, *J. Infect. Dis.* **115**:59–67.

169. Little, S. F., and Knudson, G. B., 1986, Comparative efficacy of *Bacillus anthracis* live spore vaccine and protective antigen vaccine against anthrax in the guinea pig, *Infect. Immun.* **52**:509–512.
170. Ivins, B. E., Fellows, P. F., *et al.*, 1994, Efficacy of a standard human anthrax vaccine against *Bacillus anthracis* spore challenge in guinea-pigs, *Vaccine.* **12**:872–874.
171. Coker, P. R., Smith, K. L., *et al.*, 2003, *Bacillus anthracis* virulence in Guinea pigs vaccinated with anthrax vaccine adsorbed is linked to plasmid quantities and clonality, *J. Clin. Microbiol.* **41**:1212–1218.
172. Fellows, P. F., Linscott, M. K., *et al.*, 2001, Efficacy of a human anthrax vaccine in guinea pigs, rabbits, and rhesus macaques against challenge by *Bacillus anthracis* isolates of diverse geographical origin, *Vaccine.* **19**:3241–3247.
173. Cataldi, A., Mock, M., *et al.*, 2000, Characterization of *Bacillus anthracis* strains used for vaccination, *J. Appl. Microbiol.* **88**:648–654.
174. Fasanella, A., Losito, S. *et al.*, 2001, Detection of anthrax vaccine virulence factors by polymerase chain reaction, *Vaccine.* **19**:4214–4218.
175. Adone, R., Pasquali, P., *et al.*, 2002, Sequence analysis of the genes encoding for the major virulence factors of *Bacillus anthracis* vaccine strain 'Carbosap', *J. Appl. Microbiol.* **93**:117–121.
176. Patra, G., Fouet, A., *et al.*, 2002, Variation in rRNA operon number as revealed by ribotyping of *Bacillus anthracis* strains, *Res. Microbiol.* **153**:139–148.
177. Young, G. A., Zelle, M. R., *et al*, 1946, Respiratory pathogenicity of *Bacillus anthracis* spores I. Methods of study and observations on pathogenesis, *J. Infect. Dis.* **79**:233–246.
178. Ross, J. M., 1957, The pathogenesis of anthrax following the administration of spores by the respiratory route, *J. Pathol. Bacteriol.* **73**:485–494.
179. Shafa, F., Moberly, B. J., *et al.*, 1966, Cytological features of anthrax spores phagocytized *in vitro* by rabbit alveolar macrophages, *J. Infect. Dis.* **116**:401–413.
180. Dixon, T. C., Fadl, A. A., *et al.*, 2000, Early *Bacillus anthracis*-macrophage interactions: intracellular survival survival and escape, *Cell. Microbiol.* **2**:453–463.
181. Guidi-Rontani, C., Levy, M., *et al.*, 2001, Fate of germinated *Bacillus anthracis* spores in primary murine macrophages, *Mol. Microbiol.* **42**:931–938.
182. Welkos, S. L., Trotter, R. W., *et al.*, 1989, Resistance to the Sterne strain of *B. anthracis*: phagocytic cell responses of resistant and susceptible mice, *Microb. Pathog.* **7**:15–35.
183. Welkos, S., Friedlander, A., *et al.*, 2002, In-vitro characterization of the phagocytosis and fate of anthrax spores macrophages and the effects of anti-PA antibody, *J. Med. Microbiol.* **51**: 821–831.
184. AVA, 1978, Anthrax vaccine adsorbed (package insert), *Lansing: Michigan Department of Public Health.*
185. Puziss M., Manning, L. C., *et al.*, 1963, Large-scale production of protective antigen from *Bacillus anthracis* anaerobic cultures, *Appl. Microbiol.* **11**:330–334.
186. Puziss, M., and Wright, G. G., 1963, Studies on immunity to anthrax. X. Gel-adsorbed protective antigen for immunization of man, *J. Bacteriol.* **85**:230–236.
187. Friedlander, A. M., Welkos, S. L., *et al.*, 2002, Anthrax vaccines, *Curr. Top. Microbiol. Immunol.* **271**:33–60.
188. Brachman P. S., Plotkin S. A., *et al.*, 1962, Field evaluation of a human anthrax vaccine, *Am. J. Public Health.* **19**:3241–3247.
189. Ivins, B. E., Welkos, S. L., *et al.*, 1992, Immunization against anthrax with *Bacillus anthracis* protective antigen combined with adjuvants, *Infect. Immun.* **60**:662–668.
190. Pittman, P. R., Gibbs, P. H., 2001, Anthrax vaccine: short-term safety experience in humans. *Vaccine.* **20**:972–978.
191. Wasserman, G. M., Grabenstein, J. D., *et al.*, 2003, Analysis of adverse events after anthrax immunization in US Army medical personnel, *J. Occup. Environ. Med.* **45**:222–233.
192. Gaur, R., Gupta, P. K., *et al.*, 2002, Effect of nasal immunization with protective antigen of *Bacillus anthracis* on protective immune response against anthrax toxin, *Vaccine.* **20**:2836–2839.

193. Ivins, B. E., Pitt, M. L., *et al.*, 1998, Comparative efficacy of experimental anthrax vaccine candidates against inhalation anthrax in rhesus macaques, *Vaccine.* **16**:1141–1148.
194. Farchaus, J. W., Ribot, W. J., *et al.*, 1998, Fermentation, purification, and characterization of protective antigen from a recombinant, avirulent strain of *Bacillus anthracis. Appl. Environ. Microbiol.* **64** 982–991.
195. Little, S. F., Ivine, B. E., *et al.*, 2004, Defining a serological correlate of protection in rabbits for a recombinant anthrax vaccine, *Vaccine.* **22**:422–430.
196. Welkos, S. L., and Friedlander, A. M., 1988, Comparative safety and efficacy against *Bacillus anthracis* of protective antigen and live vaccines in mice, *Microb. Pathog.* **5**:127–139.
197. Brossier, F., Levy, M., *et al.*, 2002, Anthrax spores make an essential contribution to vaccine efficacy, *Infect. Immun.* **70**:661–664.
198. Cohen, S., Mendelson, I., *et al.*, 2000, Attenuated nontoxinogenic and nonencapsulated recombinant *Bacillus anthracis* spore vaccines protect against anthrax, *Infect. Immun.* **68**:4549–4558.
199. Welkos, S., Little, S., *et al.*, 2001, The role of antibodies to *Bacillus anthracis* and anthrax toxin components in inhibiting the early stages of infection by anthrax spores, *Microbiology.* **147**:1677–1685.
200. Piris Gimenez, A., Mock, M., *et al.*, 2002, Use of diffusion chamber to explore *Bacillus anthracis* development in the host. *5th International Conference on Anthrax (Abstract)*, Nice, France.
201. Sage H. J., Fasman, G. D., Levine, L., 1964, The serological specificity of the poly-alanine immune system, *Immunochemistry.* **1**:133–134.
202. Ezzell, J. W., Jr., Abshire, T. G., *et al.*, 1990, Identification of *Bacillus anthracis* by using monoclonal antibody to cell wall galactose-N-acetylglucosamine polysaccharide, *J. Clin. Microbiol.* **28**:223–2231.
203. Coulson, N. M., Fulop, M., *et al.*, 1994, Effect of different plasmids on colonization of mouse tissues by the aromatic amino acid dependent *Salmonella typhimurium* SL 3261, *Microb. Pathog.* **16**:305–311.
204. Barnard, J. P., and Friedlander, A. M., 1999, Vaccination against anthrax with attenuated recombinant strains of *Bacillus anthracis* that produce protective antigen, *Infect. Immun.* **67**:562–567.
205. Iacono-Connors, L. C., Welkos, S. L., *et al.*, 1991, Protection against anthrax with recombinant virus-expressed protective antigen in experimental animals, *Infect. Immun.* **59**:1961–1965.
206. Lee, J. H., Hadjipanayis, A. G., and Welkos, S. L., 2003, Venezuealn equine virus-vectored vaccines protect mice against anthrax spore challenge, *Infect. Immun.* **71**:1491–1496.
207. Gu, M. L., Leppla, S. H., *et al.*, 1999, Protection against anthrax toxin by vaccination with a DNA plasmid encoding anthrax protective antigen, *Vaccine.* **17**:340–344.
208. Little, S. F., Ivins, B. E., *et al.*, 1997, Passive protection by polyclonal antibodies against *Bacillus anthracis* infection in guinea pigs, *Infect. Immun.***65**:5171–5175.
209. Price, B. M., Liner, A. L., *et al.*, 2001, Protection against anthrax lethal toxin challenge by genetic immunization with a plasmid encoding the lethal factor protein, *Infect. Immun.* **69**(7): 4509-15.
210. Kobiler, D., Gozes, Y., *et al.*, 2002, Efficiency of protection of guinea pigs against infection with *Bacillus anthracis* spores by passive immunization, *Infect. Immun.* **70**:544–560.
211. Friedlander, A. M., Welkos, S. L., *et al.*, 1993b, Postexposure prophylaxis against experimental inhalation anthrax, *J. Infect. Dis.* **167**:1239–1243.
212. Singh, Y., Khanna, H., *et al.*, 2001, A dominant negative mutant of *Bacillus anthracis* protective antigen inhibits anthrax toxin action *in vivo, J. Biol. Chem.* **276**:22090–22094.
213. Schuch, R., Nelson, D., *et al.*, 2002, A bacteriolytic agent that detects and kills *Bacillus anthracis, Nature.* **418**:884–889.

7

Tularemia Pathogenesis and Immunity

STEPHEN L. MICHELL, KATE F. GRIFFIN, and RICHARD W. TITBALL

1. INTRODUCTION

Francisella tularensis, the etiological agent of tularemia, is one of the most infectious pathogens known. Human cases of the disease occur infrequently in the northern hemisphere, mainly in some parts of Scandanavia and in Russia.(1) It is probably the high infectivity, ease of culture, and low levels of natural immunity to the bacterium that originally attracted interest in *F. tularensis* as a bioweapon.(2) During the 1930s and 1940s the bacterium was evaluated by Japanese germ warfare units. Later, both the former Soviet Union (fSU) and the USA reportedly produced weapons capable of disseminating the bacterium.(2) The programme to develop biological weapons in the USA was abandoned in 1969. In other countries the status of the programme is not clear, and there are some suggestions that strains which are resistant to commonly available antibiotics have been developed for use as bioweapons.(2)

The severity of disease caused by *F. tularensis* is highly dependent on the causative strain and the route of entry of the bacterium into the host. Currently, there are four accepted subspecies (Table I), and *F. tularensis* subsp. *tularensis* is the most virulent in humans. Most naturally acquired cases of disease in humans are the consequence of a bite from an arthropod vector that has previously fed on an infected animal.(1) Ulceroglandular tularemia is the usual form of disease

STEPHEN L. MICHELL, KATE F. GRIFFIN, and RICHARD W. TITBALL • Defence Science and Technology Laboratory, Porton Down, Salisbury, Wiltshire, SP4 0JQ, UK.

TABLE I
Some Properties of the Four Subspecies of *F. tularensis*

		Identification and diagnostic tests		
Subspecies	50% Lethal dose in humans	Citrulline ureidase	Glycerol fermentation	Glucose fermentation
Tularensis	10–50 cfu[3]	+	+	+
Holarctica	<10^3 cfu[3]	−	−	+
Mediaasiatica	NR	+	+	−
Novicida	>10^3 cfu[3]	NR	NR	+

cfu: colony forming units.

that develops, and is severely debilitating, but not often fatal.[1,2,4] Ticks such as *Dermacentor reticulatus* and *Ixodes ricinus* are the most frequent vectors, and mammals such as ground squirrels, rabbits, hares, voles, water rats and other rodents are believed to be the usual reservoirs of infection.

The inhalation of *F. tularensis* can result in the development of pneumonic disease. Naturally occurring cases of primary pneumonic tularemia are infrequent, and are usually a consequence of the inhalation of dusts from hay contaminated from infected rodents.[5–7] Most of the information on pneumonic tularemia comes from the infrequent cases that occur naturally and from trials with human volunteers in the USA during the 1950s.[8,9] Two naturally occurring outbreaks have attracted particular attention. Firstly, a number of cases of pneumonic tularemia were reported in Sweden during 1966–1967.[7] The disease was contracted by those working in farming communities, and the available evidence indicates that the bacteria were inhaled in dusts generated when contaminated hay was moved from storage sites in fields into barns. Secondly, there have been a number of cases of pneumonic tularemia on Martha's Vineyard in the USA.[10] The etiology of these cases is somewhat unusual, being associated with lawn mowing or brush cutting activities that resulted in the generation of airborne bacteria from the remains of rabbits that had died from tularemia.[10]

F. tularensis used as a bioweapon would be expected to be delivered by the aerosol route, and would most likely cause pneumonic tularemia.[2] Previous human volunteer studies in the USA have shown that the infectious dose of *F. tularensis* subsp. *tularensis* by the airborne route is between 10 and 50 cfu.[8] The World Health Organization has used this information to predict the number of casualties following a bioweapon attack with *F. tularensis* (Fig. 1).[11] Providing appropriate medical care for the large number of incapacitated casualties would pose significant logistical problems. Based on these predictions, the centers for disease control and prevention estimated that the cost to society of an airborne exposure to *F. tularensis* would be $5.4 billion for every 100,000 persons exposed.[2]

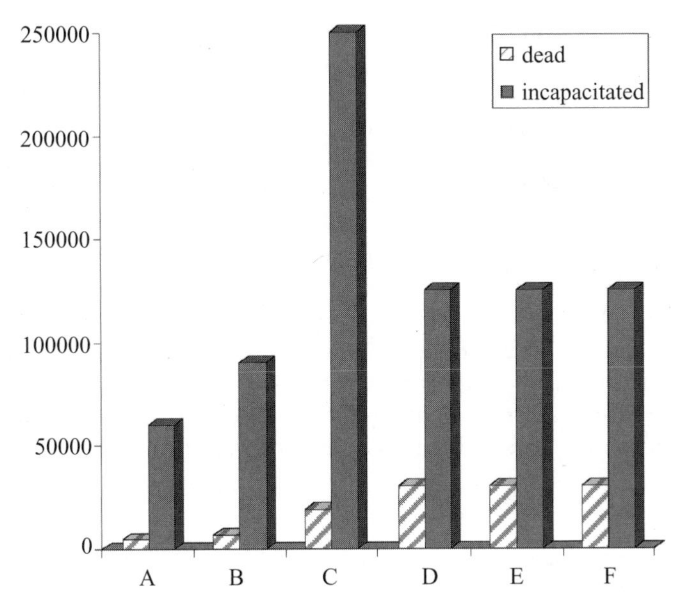

FIGURE 1. Number of deaths and number of cases of incapacitating disease following an airborne attack with 50 kg of dried *F. tularensis*. Exposure of an urban population of 500,000 (A), 1,000,000 (B) or 5,000,000 (C) in an economically developed country. Exposure of an urban population of 500,000 (D), 1,000,000 (E) or 5,000,000 (F) in a developing country. Data taken from *Health Aspects of Chemical and Biological Weapons,* 1970, World Health Organisation, Geneva.(11)

2. PATHOGENESIS

2.1. Human Disease

F. tularensis is able to enter the host after vector-borne delivery (i.e., across the skin) or by crossing a mucosal surface.(1) In the case of vector-borne delivery the most likely outcome is ulceroglandular tularemia. The typical incubation period for ulceroglandular tularemia is 3–6 days, with the subsequent formation of an ulcer at the site of infection (i.e., the vector bite).(1,2,4) The patient experiences sudden onset of flu-like symptoms including fever (38–40°C), headache, chills, and generalized aches.(1,2,4) Often swollen lymph nodes develop which resemble the bubo's associated with bubonic plague. Disease without the development of an ulcer is termed glandular tularemia, while disease without either the development of an ulcer or lymphadenopathy is usually referred to as typhoidal or septicaemic tularemia. Oculoglandular, oropharyngeal, gastrointestinal, or pneumonic tularemia are all rare forms of the disease, which occur as a consequence of entry into the host via the relevant mucosal surface.(1)

Septicemic and pneumonic tularemia, when caused by *F. tularensis* subsp. *tularensis*, represent the most severe forms of disease with a typical mortality rate of 30–60%.[1,2] Both of these forms of disease may develop from ulceroglandular or glandular tularaemia, but primary pneumonic disease is the consequence of the inhalation of bacteria. The clinical features of pneumonic tularemia are quite variable and the disease may present without obvious signs of a pneumonia.[5] Human volunteer studies during the 1950s have provided data on the development of disease in a controlled environment. In these studies the symptoms of pneumonic tularemia developed within 3–5 days of exposure to airborne bacteria and disease was characterized by a fever of up to 40°C. The signs and symptoms of primary pneumonic tularemia include brachycardia, chills, dyspnea, and a nonproductive cough.[5] There may also be headache, sore throat, myalgia, and nausea. There is often hemorrhagic inflammation of the airways that may progress to a bronchopneumonia. Pleuritis is a common feature and enlargement of the hilar lymph nodes is a common radiological feature.[2] Clinical disease may last from a few days to several weeks.[5]

2.2. Animal Models

A number of animal species have been investigated for susceptibility to *F. tularensis* infection. A study published in 1946 reported that *F. tularensis* subsp. *tularensis* (strain SchuS4) was virulent in the mouse, guinea pig, hamster, rabbit, and the cotton rat.[1] The cotton rat was shown to have a high degree of host variation and consequently has received little further attention. A number of subsequent studies have evaluated the infectivity of the four subspecies of *F. tularensis* in mice, guinea pigs, and rabbits. Mice and guinea pigs have been shown to be susceptible to acute disease caused by *F. tularensis* subsp. *tularensis* and subsp. *holarctica.* Rabbits are most susceptible to strains of subspecies *tularensis.*[1] However, the animal species of choice for most studies to date has been inbred mouse strains with BALB/c or C57BL/6 (and genetic mutants of each) mice being most commonly used for pathogenesis and protection studies.

BALB/c and C57BL/6 mice are susceptible to infection with subsp. *tularensis* or subsp. *holarctica.* However, these strains show significant differences in susceptibilities to challenge with the live vaccine strain of *F. tularensis* (LVS).[12] The virulence of *F. tularensis* LVS in mice is also dependent on the route of delivery. This strain is fully virulent when delivered intraperitoneally, but is attenuated when delivered intradermally or subcutaneously.[13] Because *F. tularensis* LVS is virulent in mice when given by some routes, this strain has been used extensively for many studies on the pathogenesis of tularaemia. Conversely, since the intradermal or subcutaneous routes of challenge are analogous to the most frequent route of natural infections, this has led to some debate over the applicability of the model to accurately mimic natural disease caused by fully virulent strains of *F. tularensis.*

The administration of *F. tularensis* LVS into mice by the intradermally or subcutaneous route can result in the induction of protective immunity. However, the nature of this protective response is also dependent on the mouse strain used. Immunized BALB/c mice have been shown to be protected against a subsequent challenge with either *F. tularensis* subsp. *holarctica* or *F. tularensis* subsp. *tularensis* strains, while immunized C57BL/6 mice show little protection against the latter strains.[14,15] The reasons for this have not yet been identified.

2.3. Cellular Pathogenesis

The pathogenesis of tularemia is poorly characterized. However, many insights have come from studies of *F. tularensis* infection of mice, a model generally considered to represent tularemia in humans.[13,16] A major finding was the demonstration that *F. tularensis* is an intracellular pathogen with the ability to replicate within macrophages.[17] In cases of ulceroglandular tularemia it is believed that initial replication of the bacteria occurs locally in the skin within polymorphonuclear leukocytes (neutrophils or PMNs), attracted by chemokines resulting from a pronounced inflammatory response, and resident macrophages. The bacteria are rapidly transported to regional lymph nodes and disseminated by leukocytes[18] by systemic circulation to other organs, especially the spleen, kidney, and liver.[2]

In addition to their ability to parasitize macrophages and neutrophils, it has been demonstrated that *F. tularensis* can also replicate within hepatocytes upon arrival in the liver. The inflammatory response to this foci of infection leads to the recruitment of activated macrophages, NK cells, monocytes, and T cells.[19] These immune cells function to destroy the infected hepatocytes and clear the released bacteria by ingestion of activated macrophages, forming granulomas in the process. The pathogenesis of pneumonic tularaemia is less well understood. However, studies in non-human primates have provided some insight into the likely pathogenesis of the disease. The bacteria are initially confined to the bronchial lymph nodes, and replication appears to occur at this site.[20] Within a few days bacteria are disseminated to the spleen and liver where pyogranulomatous lesions are observed,[21] an outcome also seen following intradermal infection. Similarly, autopsies in fatal cases of tularemia revealed the presence of necrotic granulomas in several tissues including the spleen and lymph nodes.[22] Necrosis of the lung and spleen is also seen in mice following aerosol infection with virulent *F. tularensis,* thus supporting the mouse as a model of tularemia in man.[19]

The mechanism(s) by which *F. tularensis* causes death of some infected individuals is unknown. However, death is often a consequence of organ failure, sepsis, with the subsequent development of systemic inflammatory response syndrome, disseminated intravascular coagulation, and acute and respiratory distress syndrome.[2,5]

2.4. Molecular Pathogenesis

The intracellular niche adopted by *Francisella*, while affording protection from serum immune responses, results in the pathogen's exposure to the potent antimicrobial activity of immune effector cells. However, before entry into phagocytic cells the bacteria must first evade killing by innate serum components such as complement. It has been shown that a capsule deficient mutant of *F. tularensis* LVS is susceptible to the bactericidal effect of nonimmume human sera, although uptake by PMNs of complement opsonized *Francisella* prevents induction of the respiratory burst and killing of the bacteria. In contrast, uptake of *F. tularensis* strain LVS by PMNs leads to killing of greater than 75% of the phagocytosed bacteria.[23] If and when virulent strains of *F. tularensis* express this capsule during infection, and subsequently the mechanism of opsonization and uptake by PMNs and macrophages, remains to be determined.

Bacterial lipopolysaccharide (LPS) is a potent mediator of the proinflammatory response causing infected cells to release cytokines and chemokines such as TNFα, IFNγ, IL-12, CXCL8 and CCL2, and the activation of other innate immune cells. LPS from *F. tularensis* strain LVS is intriguing in its inability to elicit these classic proinflammatory responses. This may be due to an empirical lack of immunostimulatory properties or as a result of binding to a host cell receptor that fails to initiate the production of an inflammatory response. In support of the latter, Telepnev *et al.*[24] have shown that *F. tularensis* LPS does not act as a competitive inhibitor of *E. coli* LPS for the Toll-like receptor (TLR4). Notwithstanding the latter observations, reflection to the gross pathology of *F. tularensis* infection reveals a rapid pronounced inflammatory response, suggesting the existence of an alternative T-cell-independent activation mechanism of macrophages.

The determination of the role of LPS in disease may be further complicated by the observation that some strains of the bacteria can display two types of LPS. The predominant type is a nonstimulatory chemotype (FT LPS) and a second chemotype resembles the LPS produced by *Francisella tularensis* subsp. *novicida* (FN LPS).[25] Analysis of this latter chemotype has revealed that its immunobiological activities are similar to classic immunocompetent bacterial LPS molecules, inducing robust amounts of IL12 and TNFα from mouse macrophages.[26] This raises the possibility that expression of FN LPS during infection may contribute to the pathogenesis of tularemia. However, it has yet to be demonstrated that FN LPS has a similar immunostimulatory effect on human macrophages. Also, concurrent with the requirements to determine the role of capsule in pathogenesis, there is a need to determine when, during infection, the various chemotypes of LPS are expressed.

F. tularensis has been shown to enter macrophages via a cytochalasin B-insensitive pathway with the result that the respiratory burst is not activated.[27] Opsonized bacteria taken up by neutrophils do activate the respiratory burst and while *F. tularensis* LVS succumbs to this bactericidal action, fully virulent strains of *F. tularensis* subsp. *holarctica* are able to survive. This difference is probably due to the resistance of fully virulent strains to hypochlorous acid, a potent product of the hydrogen peroxide-myeloperoxidase-chloride system.[28] Another antibacterial mechanism of phagocytic cells is the production of nitric oxide (NO). The different chemotypes of LPS can reportedly affect NO production in peritoneal macrophages,[25] suggesting that phase variation of LPS by *Francisella* may modulate this innate immune response. The importance of *Francisella* LPS as a modulator of the immune response and potential virulence factor is supported by studies suggesting that the *Francisella* ABC transporter, *valA,* is important in the assembly of LPS.[29] Furthermore, mutation of this locus renders bacteria susceptible to killing by serum and restricted for growth within macrophages.[30]

Another significant antimicrobial mechanism of phagocytes is the fusion of the lysosome, an organelle containing numerous enzymes capable of degrading a range of macromolecules, with the bacteria-laden phagosome. Other pathogens either prevent phagolysosme fusion, such as *Mycobacteria,* or as in the case of *Listeria monocytogenes,* escape from the phagosome before fusion. *Mycobacteria* prevent fusion of the lysosome by excluding a host vesicular proton ATPase from the phagosome resulting in lack of acidification of the phagosome.[31] In contrast, *Francisella* requires acidification of the phagosome for the sequestration of iron.[32] It has been proposed that this acidification of the phagosome may be a prerequisite for the induction of *Francisella* virulence factors that lead to escape from the phagosome to the cytosol.[33] The mechanism by which Francisella escapes the phagosome is as yet undetermined, but is thought to be distinct from the mechanisms employed by several other intracellular pathogens. *L. monocytogenes* escapes the phagosome by producing a pore-forming listeriolysin; however, no such homologue of this virulence factor has been identified in the genome sequence of *F. tularensis* strain SchuS4 (subspecies *tularensis*).

Analysis of the protein profile of *Francisella* expressed within macrophages identifies very few proteins that are upregulated, suggesting that this pathogen has evolved to be tolerant of the hostile intracellular environment of host phagocytic cells.[34] Of the four proteins that do show upregulation during growth in macrophages, a 23-kDa protein was also shown to be upregulated in response to exposure to oxidative stress, suggesting that this protein's function is related to the adaptation to an intracellular environment. It is interesting to note that there are two copies of the gene encoding this protein, *iglC,* in both *F. tularensis* subspecies *tularensis* and *F. tularensis* LVS, reinforcing the hypothesis that this gene is essential for the intracellular growth of *Francisella.* This hypothesis has been further substantiated by the finding that a derivative of *F. tularensis* LVS, containing mutations in the genes encoding this protein, shows impaired multiplication in a macrophage cell line. This mutant is also attenuated in a mouse model of infection.[35] Another genetic loci that has been implicated as necessary for intracellular growth of *Francisella* is the *mglAB* locus.[36] It has been proposed that MglAB may be a transcriptional regulator given its high similarity to the *E. coli* regulator SspAB. This idea is corroborated by the observation that a strain of *F. tularensis* subspecies *novicida* harboring a mutation in *mglAB* results in a change in expression of several proteins and precludes intracellular growth.

As with some of the mechanisms of cellular pathogenicity discussed above, *F. tularenis* appears to have a distinct method of ultimately killing its host. It has been shown that the bacteria must multiply intracellularly to induce cytopathogenicity and host cell apoptosis.[37] The mechanism by which apoptosis is effected by *Francisella* is similar to that of the intrinsic apoptotic pathway involving the release of cytochrome C from the mitochondria.[38] However, the upstream mechanisms leading to this programmed cell death remain to be determined.

3. IMMUNITY

3.1. Natural Infection and Immunity

It is generally accepted that recovery from tularemia results in long-lived immunity, with re-infection reported very infrequently.[39] Agglutinating antibodies in serum appear during the 2nd or 3rd week of disease,[5,39] reaching a maximum several weeks later, and remaining detectable in some individuals for at least 10 years after infection.[39] The antibody response that develops after infection is primarily directed toward lipopolysaccharide.[39–41] It is clear that antigens other than LPS are recognized during infection, but some studies have shown that surface proteins may be partially masked by surface polysaccharide.[39] Over the past 15 years a number of protein antigens that are recognized by convalescent sera have been identified,[42,43] work that has been supported recently by the development of proteomic approaches to the identification of immunoreactive antigens (Table II). Antibody to the heat shock protein components Hsp 60 and Hsp10 reportedly predominate,[44] and surprisingly many of the other immunoreactive proteins would also be considered to be cytoplasmically located. The antibody responses that develop to these proteins might be used as the basis of future diagnostic tests for tularemia. However, it is not clear at this stage whether these immunoreactive proteins might be exploited as components of a subunit vaccine.

In parallel, there has been some work to identify antigens able to activate T-cells. At least four outer membrane proteins are able to stimulate proliferation of αβT-cells taken from individuals who had previously been vaccinated with the live vaccine strain (LVS) of *F. tularensis* or who had previously contracted tularemia.[45,46] These proteins generally appear to stimulate the proliferation of $CD4^{+}$ T-cells rather than $CD8^{+}$ T-cells.[47] One of the membrane

TABLE II
Proteins Antigens Reported to be Recognized by Antisera from Individuals Previously Infected with *F. tularensis*

Protein
43 kDa outer membrane protein[42]
Chaperone DnaK[43]
Hsp60[43]
Hsp10[43]
17 kDa lipoprotein (Tul4)[43]
Elongation factor TU[43]
Glycine cleavage system T1 protein[43]
Hypothetical protein[43]
Oxidoreductase[43]
Biotin carrier protein[43]
50S ribosomal protein[43]
Probable bacterioferritin[43]
3-dehydroquinase[43]
Histone-like protein[43]

proteins that is capable of inducing proliferation has been identified as the 17 kDa TUL4 lipoprotein.[48] Hsp10, Hsp60, and DnaK are also capable of causing proliferation of $\alpha\beta$T-cells from individuals who have recovered from tularemia.[47] These proteins are associated with the general stress responses of bacteria and are normally considered to be cytoplasmically located. All of these proteins are also recognized by convalescent sera.[43]

There is also sufficient evidence that $\gamma\delta$ T-cells are activated in individuals suffering from tularemia.[49] The identity of the antigen(s) which stimulates $\gamma\delta$ T-cell activation is not known. In other intracellular pathogens such as *Mycobacterium tuberculosis* nonpeptidic phosphoesters are implicated in this response.[50] Experimental evidence indicates that phosphoantigens also play a role in the activation of $\gamma\delta$ T-cells in tularemia patients.[49] Only limited activation of $\gamma\delta$ T-cells was seen in individuals who had been immunized with *F. tularensis* LVS, leading to the suggestion that the activation of $\gamma\delta$ T-cells may be linked to the virulence of the infecting strain. The significance of the activation of $\gamma\delta$ T-cells is not clear. However, the long-lasting recall responses of $\gamma\delta$ T-cells appears to be minimal, suggesting that these cells may not contribute to long-term protection against re-infection with *F. tularensis.*[47]

3.2. Live Vaccines

Following Pasteur's demonstration that attenuated viruses could be used as effective vaccines, numerous researchers employed this strategy to the development of vaccines for other pathogens. Perhaps the most notable success of this approach is the vaccine against tuberculosis, bacille Calmette-Guérin (BCG). Before World War II, similar approaches were undertaken in the former Soviet Union for the development of a vaccine against tularemia. In the 1930s El'bert *et al.* demonstrated protection in a small animal model against a virulent culture of *F. tularensis* following immunisation with an attenuated strain.(51) In 1942, an attenuated strain of *F. tularensis* strain Moscow, was administered to humans with effective protection demonstrated. Development of live vaccines continued in the former Soviet Union, several of which were received by the United States in the 1950s. From one of these strains, a subspecies *holarctica* (Type B), the live vaccine strain (LVS) was developed.(52)

Initial studies with LVS demonstrated that this vaccine was more efficacious in a small animal model when administered as a viable culture,(52) as is the case for BCG. The reason for the increased protection observed with live attenuated intracellular pathogens is not fully understood, although it has been proposed for BCG that active secretion of proteins is required for protection.(53) Although the infectious dose of *F. tularensis* strain SchuS4 is reported to be between 10 and 50 cfu,(8) volunteers immunized with *F. tularensis* LVS were protected against an aerosol challenge with 200 cfu of strain SchuS4.(54) Other studies have shown that *F. tularensis* LVS administered by the respiratory route affords better protection against an aerogenic challenge than intradermal immunisation,(55) a finding also observed with the BCG vaccine.(56) However, *F. tularensis* LVS is usually administered by scarification.

At present LVS, although an effective vaccine against tularemia, is not currently licensed for use. Reasons for this may include mixed colonial morphology and variable immunogenicity, and not least a lack of understanding of the mechanisms of attenuation and protection. However, the finding that an attenuated strain of *F. tularensis* can provide protective immunity suggests that genetically defined and rationally attenuated mutants are a feasible prospect. Such a mutant should be avirulent, be able to replicate *in vivo*, but have a limited ability to survive, ensuring that a protective immune response develops without causing disease. Indeed, for other pathogens, strains containing mutations in genes of essential biosynthetic pathways are already being considered as potential vaccines.(57) The generation of rationally attenuated auxotrophic mutants is favorable, as it has been proposed that their limited replication would allow their administration to immunocompromized hosts without the threat of disease.(58) Many investigators have targeted genes involved in the purine biosynthesis pathways for the construction of rationally attenuated mutants. Analysis of the *F. tularensis* SchuS4 genome sequence indicates that genes encoding all of the enzymes in this pathway are present, but the functionality of this pathway has not been confirmed experimentally. Other genes, which play a role in the growth of *F. tularensis* in macrophages, might also be targeted for the construction of rationally attenuated mutants and a more detailed analysis of the genome sequence may reveal other gene targets for inactivation.

3.3. Subunit Vaccines

At around the same time as the Soviet Union was developing live attenuated vaccines, researchers in the United States suggested that immune serum could be used as a prophylactic treatment of Tularemia in humans. Subsequently, Lee Foshay investigated the possibility of using killed *F. tularensis* cells as a vaccine by virtue of its ability to induce a humoral immune response. Studies in mice, nonhuman primates, and also in humans did demonstrate low level protection against disease,[59] although the reactogenicity of killed whole cell vaccines and the more favorable protection studies with live attenuated strains stemmed further research into the development of killed whole cell vaccines. Nevertheless, the identification of the components of *F. tularensis* responsible for the induction of a protective response, either after immunisation with LVS or natural infection, has been the focus of several studies over the past 50 years.

To date the only protective antigen of *F. tularensis* identified is LPS. Immunization with LPS provides protection against low virulence strains of *F. tularensis*, but is less effective against *F. tularensis* subsp. *tularensis.*[60,61] The lack of protection against high virulence strains following immunization with LPS is thought to result from the requirement of T-cell-mediated immunity for protection.[39] Thus, the development of a subunit vaccine against tularemia can be envisaged as containing LPS coupled with antigen that is capable of eliciting cellular immunity. The number of *Francisella* antigens reported as being able to induce a cellular response is limited. T-cells taken from humans that have been immunized with the LVS vaccine showed proliferation to polypeptides of *Francisella*, having relative molecular weights of 61, 40, 37, 32, 17, and 17.5 kDa.[45] Only the 17-kDa protein and FopA, a 43-kDa protein recognized by convalescent sera, have been evaluated as protective subunits in the murine model of disease. Although both are immunogenic, this response did not provide protection against disease.[62,63]

A method currently employed for the identification of protective subunits uses a novel *in silico* approach for the identification of putative vaccine antigens from genome sequence data.[64] Similar approaches have successfully been used for the identification of potential subunit vaccines for extracellular pathogens. These approaches coupled with the recent completion of the *F. tularensis* strain SchuS4 genome sequence raises the possibility of the identification of new proteins that could be included in a subunit vaccine. In addition, for many pathogens there is a dichotomy that virulence factors are also protective antigens. Identification of virulence determinants of *Francisella* may also add to the arsenal of potential subunit vaccine candidates. How these antigens should be delivered is a major factor contributing to the development of subunit vaccines. Classically, the protein antigen is purified from host bacteria expressing the protein from a plasmid containing its corresponding gene, the purified protein is then administered with a suitable adjuvant. Developments, to enhance immunogenicity, include administering the gene of the antigen on a plasmid that is recognized by the vaccinee, with subsequent expression of the antigen *in vivo.*(65) Similarly, the gene subunit antigen may be administered in the context of a live attenuated vaccine that invokes an immune response similar to that required for protection against *Francisella.*(66)

3.4. Mechanisms of Protection in Adaptive Immunity

The design of effective new tularemia vaccines requires an understanding of the mechanisms of adaptive immunity that contribute to protection. In humans immunization with LVS leads to protection against virulent tularemia infection and, although the immune responses stimulated have been studied,(67,68) those mechanisms essential for protection are unknown. A murine model of immunization provides a convenient experimental system that can be manipulated to identify these essential protective components in a mammalian system.

The role that antibody plays in protection against disease remains controversial. The adoptive transfer of antibodies has been shown to protect mice against attenuated strains of either *F. tularensis* subsp. *tularensis*(69) or the attenuated *F. tularensis* subsp. *holarctica* strain LVS.(16,70,71) In contrast, no protection has been seen against fully virulent strains of *F. tularensis* subsp. *tularensis.*(39,69) However, in experiments using a low dose challenge of a virulent isolate of *F. tularensis* subsp. *holarctica,* a reduced bacterial burden in the liver and spleens of B-cell-deficient mice following administration of LVS-specific antibody has been demonstrated.(71) Antibodies to LPS have conferred passive protection in mice against challenge with *F. tularensis* LVS, but not against the *F. tularensis* subsp. *tularensis* strain SchuS4.(60) The utility of anti-LPS antibodies was also seen in immunization trials with the O-antigen of *F. tularensis* LVS, which successfully protected against challenge with a fully virulent strain of subsp. *holarctica,* but gave no protection against challenge with a subsp. *tularensis* strain.(61) The role of specific antibody in protection against intracellular pathogens has traditionally been regarded as limited due to the protection from effector mechanisms afforded by the intracellular niche of the pathogen. However, more recent papers review several mechanisms by which antibody may act on intracellular pathogens.(72,73) Various mechanisms by which antibody may protect against tularemia infection have been suggested. LVS and a virulent strain of *F. tularensis* subsp. *holarctica* have both been shown to be susceptible to opsonin-dependent intracellular killing by human polymorphonuclear leukocytes in an *in vitro* assay.(74) In the LVS challenge model, efficacy of passive antibody protection has been shown to be dependent on cellular immunity since no protection was observed in mice deficient in interferon gamma, CD4+ or CD8+ T cells.(75) Thus, the evidence so far suggests that at least in mice antibody is a mechanism of protection against attenuated strains and virulent strains of subspecies *holarctica,* but not against strains of subspecies *tularensis.* The role of antibody in protection of humans should not be discounted. Nonetheless, as for other intracellular pathogens, T-cell effector functions are likely to be the major component of resistance to infection.

The role of T cells in protection against tularemia is dependent on the animal model used. Several studies have demonstrated that mice immunized with either LPS or *F. tularensis* LVS can survive a subsequent challenge with LVS after depletion of CD4+ and/or CD8+ T cells, although T cells are required for clearance of the challenge.[60,76,77] The role of T cells, including the Thy1.2+CD4-CD8- population, in this model has been reviewed extensively by Elkins *et al.*[78] However, mice immunized with LPS followed by an LVS boost and challenged with the fully virulent *F. tularensis* subsp. *tularensis* strain SchuS4 did not survive when depleted of CD4+ and/or CD8+ T cells.[60] The absolute requirement for T cells in this latter experiment illustrates the difficulty of assessing the importance of a mechanism of protection when using attenuated strains.

CONCLUSIONS

Although *F. tularensis* is one of the most infectious pathogens known, very little is known about the pathogenesis of disease or virulence mechanisms. The origins of this pathogen are not clear—there are apparently no close relatives and there may therefore be few parallels with other pathogens which can be drawn on to inform future work. Notwithstanding this, it is likely that significant progress will be made in understanding the biology of this organism in the near future. The determination of the genome sequence of this bacterium, coupled with the development of methods for the construction of defined allelic replacement mutants will support this work. Several important questions need to be addressed. Do similar mechanisms of virulence operate in disease caused by inhalation and vector-borne delivery of the bacteria into the host? What are the mechanisms that allow the bacteria to grow within host cells and to spread from cell to cell, and what is the molecular basis of the clear differences in virulence of the four subspecies of the bacterium? Two approaches to the development of a vaccine seem feasible. Firstly, it may be possible to construct rationally attenuated mutants. The feasibility of this approach is supported by previous clinical experiences with the LVS strain in humans. However, a longer-term goal may be to devise a subunit vaccine.

REFERENCES

1. Ellis, J., Oyston, P. C. F., Green M., *et al.*, 2002, Tularemia, *Clin. Microbiol. Rev.* **15**:631–646.
2. Dennis, D. T., Inglesby, T. V., Henderson, D. A., *et al.*, 2001, Tularemia as a biological weapon–Medical and public health management, *JAMA*. **285**:2763–2773.

3. Eigelsbach, H. T., and McGann, V. G., 1984, in: *Bergey's Manual of Systematic Bacteriology*, Vol. 1 (N. R. Kreig and J. G. Holt, eds.), Williams and Wilkins, Baltimore, pp. 394–399.
4. Evans, M. E., Gregory, D. W., Schaffner W. *et al.*, 1985, 1997, Tularemia: a 30-year-experience with 88 cases, *Medicine.* **64**:251–69.
5. Gill V., and Cunha, B. A., 1997, Tularemia pneumonia, *Sem. Resp. Infect.* **12**:61–67.
6. Stewart, S. J., 1996,Tularemia: association with hunting and farming, *FEMS Immunol. Med. Microbiol.* **13**:197–199.
7. Tarnvik, A., Sandstrom, G., and Sjostedt, A., 1996, Epidemiological analysis of tularemia in Sweden 1931–1993, *FEMS Immunol. Med. Microbiol.* **13**:201–204.
8. McCrumb, F. R., 1961, Aerosol infection of man with *Pasteurella tularensis, Bacteriol. Revs.***25**:262–267.
9. Saslaw, S., Eigelsbach, H. T., Prior, J. A. *et al.*, 1961, Tularemia vaccine study. II. Respiratory challenge, *Arch. Intern. Med.* **107**:702–714.
10. Feldman, K. A., Enscore, R.E. Lathrop, S. L., *et al.*, 2001, An outbreak of primary pneumonic tularemia on Martha's vineyard, *N. Engl. J. Med.* **345**:1601–1637.
11. Anon, 1970, *Health Aspects of Chemical and Biological Weapons,* World Health Organisation, Geneva.
12. Anthony L. S. D., and Kongshavn, P. A. L., 1988, H-2-restriction in acquired cell mediated immunity to infection with *Francisella tularensis* LVS, *Infect. Immun.* **56**:452–456.
13. Elkins, K. L., Rhinehart-Jones, T. R., Culkin, S. J., *et al.*, 1996, Minimal requirements for murine resistance to infection with *Francisella tularensis* LVS, *Infect. Immun.* **64**:3288–3293.
14. Chen, W. X., Shen, H. Webb, A. *et al.*, 2003, Tularemia in BALB/c and C57BL/6 mice vaccinated with *Francisella tularensis* LVS and challenged intradermally, or by aerosol with virulent isolates of the pathogen: protection varies depending on pathogen virulence, route of exposure, and host genetic background, *Vaccine.* **21**:3690–3700.
15. Green, M., Choules, G., Rogers D., *et al.*, 2004, Efficacy of the live attenuated *Francisella tularensis* vaccine (LVS) in a murine model of disease, *Vaccine.* **23**:2680–2686.
16. Anthony, L. S., and Kongshavn, P. A., 1987, Experimental murine tularemia caused by *Francisella tularensis,* live vaccine strain: a model of acquired cellular resistance, *Microb. Pathog.* **2**:3–14.
17. Anthony, L. S. D., Burke R. D., and Nano, F. E., 1991, Growth of *Francisella* spp. in rodent macrophages, *Infect. Immun.* **59**:3291–3296.
18. Long, G. W., Oprandy, J. J., Narayanan, R. B., *et al.*, 1993, Detection of *Francisella tularensis* in blood by polymerase chain reaction, *J. Clin. Microbiol.* **31**:152–154.
19. J. W. Conlan, W. Chen, H. Sheh, A. Webb and R. KuoLee, 2003, Experimental tularemia in mice challenged by aerosol or intradermally with virulent strains of *Francisella tularensis*: bacteriologic and histopathologic studies, *Microb. Pathog.* **34**:239–248.
20. R. L. Schricker, H. T. Eigelsbach, J. Q. Mitten and W. C. Hall, 1972, Pathogenesis of tularemia in monkeys aerogenically exposed to *Francisella tularensis* 425, *Infect. Immun.* **5**:734–744.
21. A. Baskerville and P. Hambleton, 1976, Pathogenesis and pathology of respiratory tularemia in the rabbit, *Br. J. Exp. Pathol.* **57**:339–347.
22. E. W. Goodpasture and S. J. House, 1928, The pathologic anatomy of tularemia in man, *Am. J. Pathol.* **4**:213–226.
23. G. Sandstrom, S. Lofgren and A. Tarnvik, 1988, A capsule-deficient mutant of *Francisella tularensis* LVS exhibits enhanced sensitivity to killing by serum but diminished sensitivity to killing by polymorphonuclear leukocytes, *Infect. Immun.* **56**:1194–1202.
24. M. Telepnev, I. Golovliov, T. Grundstrom, A. Tarnvik and A. Sjostedt, 2003, *Francisella tularensis* inhibits Toll-like receptor-mediated activation of intracellular signalling and secretion of TNF-alpha and IL-1 from murine macrophages., *Cell. Microbiol.* **5**:41–51.
25. S. C. Cowley, S. V. Myltseva and F. E. Nano, 1996, Phase variation in *Francisella tularensis* affecting intracellular growth, lipopolysaccharide antigenicity and nitric oxide production, *Mol. Microbiol.* **20**:867–874.

26. T. L. Kieffer, S. Cowley, F. E. Nano and K. L. Elkins, 2003, *Francisella novicida* LPS has greater immunobiological activity in mice than *F. tularensis* LPS, and contributes to *F. novicida* murine pathogenesis, *Microbes. Infect.* **5**:397–403.
27. A. H. Fortier, S. J. Green, T. Polsinelli, T. R. Jones, R. M. Crawford, D. A. Leiby, K. L. Elkins, M. S. Meltzer and C. A. Nacy, 1994, Life and death of an intracellular pathogen: *Francisella tularensis* and the macrophage, *Immunol. Ser.*, **60**:349–361.
28. S. Lofgren, A. Tarnvik, M. Thore and J. Carlsson, 1984, A wild and an attenuated strain of *Francisella tularensis* differ in susceptibility to hypochlorous acid—a possible explanation of their different handling by polymorphonuclear leukocytes, *Infect. Immun.* **43**:730–734.
29. M. K. McDonald, S. C. Cowley and F. E. Nano, 1997, Temperature-sensitive lesions in the *Francisella novicida valA* gene cloned into an *Escherichia coli msbA lpxK* mutant affecting deoxycholate resistance and lipopolysaccharide assembly at the restrictive temperature. *J. Bacteriol.* **179**:7638-7643.
30. K. E. Mdluli, L. S. Anthony, G. S. Baron, M. K. McDonald, S. V. Myltseva and F. E. Nano, 1994, Serum-sensitive mutation of *Francisella novicida*: association with an ABC transporter gene, *Microbiology* **140**:3309–3318.
31. S. Sturgill-Koszycki, P. H. Schlesinger, P. Chakraborty, P. L. Haddix, H. L. Collins, A. K. Fok, R. D. Allen, S. L. Gluck, J. Heuser and D. G. Russell, 1994, Lack of acidification in *Mycobacterium tuberculosis* phagosomes produced by exclusion of the vesicular proton-ATPase, *Science* **263**:678–681.
32. A. H. Fortier, D. A. Leiby, R. B. Narayanan, E. Asafoadjei, R. M. Crawford, C. A. Nacy and M. S. Meltzer, 1995,Growth of *Francisella tularensis* LVS in macrophages: the acidic intracellular compartment provides essential iron required for growth, *Infect. Immun.* **63**:1478–1483.
33. I. Golovliov, V. Baranov, Z. Krocova, H. Kovarova and A. Sjostedt, 2003, An attenuated strain of the facultative intracellular bacterium *Francisella tularensis* can escape the phagosome of monocytic cells, *Infect. Immun.* **71**:5940–5950.
34. I. Golovliov, M. Ericsson, G. Sandstrom, A. Tarnvik and A. Sjostedt, 1997, Identification of proteins of *Francisella tularensis* induced during growth in macrophages and cloning of the gene encoding a prominently induced 23-kilodalton protein, *Infect. Immun.* **65**: 2183–2189.
35. I. Golovliov, A. Sjostedt, A. N. Mokrievich and V. M. Pavlov, 2003, A method for allelic replacement in *Francisella tularensis*, *FEMS Microbiol. Lett.* **222**:273–280.
36. G. S. Baron and F. E. Nano, 1998, MglA and MglB are required for the intramacrophage growth of *Francisella novicida*, *Mol. Microbiol.* **29**:247–259.
37. X. H. Lai, I. Golovliov and A. Sjostedt, 2001, *Francisella tularensis* induces cytopathogenicity and apoptosis in murine macrophages via a mechanism that requires intracellular bacterial multiplication, *Infect. Immun.* **69**:4691–4694.
38. X.-H. Lai and A. Sjostedt, 2003, Delineation of the molecular mechanisms of *Francisella tularensis*-induced apoptosis in murine macrophages, *Infect. Immun.* **71**:4642–4646.
39. A. Tarnvik, 1989, Nature of protective immunity to *Francisella tularensis*, *Revs. Infect. Dis.* **11**:440–451.
40. H. E. Carlsson, A. A. Lindberg, G. Lindberg, B. Hederstedt, K. A. Karlsson and B. O. Agell, 1979, Enzyme-linked immunosorbent assay for immunological diagnosis of human tularemia, *J. Clin. Microbiol.* **10**:615–621.
41. G. Sandstrom, 1994, The tularemia vaccine, *J. Chem. Tech. Biotechnol.* **59**:315–320.
42. L. Bevanger, J. A. Maeland and A. I. Naess, 1988, Agglutinins and antibodies to *Francisella tularensis* outer membrane antigens in the early diagnosis of disease during an outbreak of Tularemia, *J. Clin. Microbiol.* **26**:433–437.
43. J. Havlasova, L. Hernychova, P. Halada, V. Pellantova, J. Krejsek, J. Stullk, A. Macela, P. R. Jungblut, P. Larsson and M. Forsman, 2002, Mapping of immunoreactive antigens of *Francisella tularensis* live vaccine strain, *Proteomics* **2**:857–867.

44. L. Hernychova, J. Stulik, P. Halada, A. Macela, M. Kroca, T. Johansson and M. Malina, 2001, Construction of a *Francisella tularensis* two-dimensional electrophoresis protein database, *Proteomics* **1**:508–515.
45. G. Sandstrom, A. Tarnvik and H. Wolf-watz, 1987, Immunospecific T-lymphocyte stimulation by membrane proteins from *Francisella tularensis, J. Clin. Microbiol.* **25**:641–644.
46. A. Sjostedt, G. Sandstrom and A. Tarnvik, 1990, Several membrane polypeptides of the live vaccine strain *Francisella tularensis* LVS stimulate T cells from naturally infected individuals, *J. Clin. Microbiol.* **28**:43–48.
47. M. Ericsson, M. Kroca, T. Johansson, A. Sjostedt and A. Tarnvik, 2001, Long-lasting recall response of CD4(+) and CD8(+) alpha beta T cells, but not gamma delta T cells, to heat shock proteins of *Francisella tularensis, Scand. J. Infect. Dis.* **33**:145–152.
48. A. Sjostedt, G. Sandstrom, A. Tarnvik and B. Jaurin, 1989, Molecular cloning and expression of a T-cell stimulating memebrane protein of *Francisella tularensis, Microb. Pathog.* **6**: 403–414.
49. Y. Poquet, M. Kroca, F. Halary, S. Stenmark, M. A. Peyrat, M. Bonneville, J. J. Fournie and A. Sjostedt, 1998, Expansion of V gamma 9V delta 2 T cells is triggered by *Francisella tularensis*-derived phosphoantigens in tularemia but not after tularemia vaccination, *Infect. Immun.* **66**:2107–2114.
50. P. Constant, F. Davodeau, M. A. Peyrat, Y. Poquet, G. Puzo, M. Bonneville and J. J. Fournie, 1994, Stimulation of human gamma-delta T-cells by nonpeptidic Mycobacterial ligands, *Science* **264**:267–270.
51. W. D. Tigertt, 1962, Soviet viable *Pasteurella tularensis* vaccines: a review of selected articles, *Bacteriol. Rev.* **26**:354–373.
52. H. T. Eigelsbach and C. M. Downs, 1961, Prophylactic effectiveness of live and killed tularemia vaccines. I, *J. Immunol.* **87**:415–425.
53. P. Andersen, 1994, Effective vaccination of mice against *Mycobacterium tuberculosis* infection with a soluble mixture of secreted mycobacterial proteins, *Infect. Immun.* **62**:2536–2544.
54. H. T. Eigelsbach, R. B. Hornick and J. J. Tulis, 1967, Recent studies on live tularemia vaccine, *Med. Ann. Dist. Columbia.* **36**:282–286.
55. R. B. Hornick, A. T. Dawkins, H. T. Eigelsbach and J. J. Tulis, 1966, Oral tularemia vaccine in man, *Antimicrob. Agents Chemother.* **6**:11–14.
56. M. L. Cohn, C. L. Davis and G. Middelbrook, 1958, Airborne immunization against tuberculosis, *Science* **128**:1282–1283.
57. P. V. Scott, J. F. Markham and K. G. Whithear, 1999, Safety and efficacy of two live pasteurella multocida aro-A mutant vaccines in chickens, *Avian Dis.* **43**:83–88.
58. I. Guleria, R. Teitelbaum, R. A. McAdam, G. Kalpana, W. R. J. Jacobs and B. R. Bloom, 1996, Auxotrophic vaccines for tuberculosis, *Nat. Med.* **2**:334–337.
59. L. Foshay, Tularemia, 1950, *Ann. Rev. Microbiol.* **4**:313–330.
60. M. Fulop, P. Mastroeni, M. Green and R. W. Titball, 2001, Role of antibody to lipopolysaccharide in protection against low- and high-virulence strains of *Francisella tularensis, Vaccine* **19**:4465–4472.
61. J. W. Conlan, H. Shen, A. Webb and M. B. Perry, 2002, Mice vaccinated with the O-antigen of *Francisella tularensis* LVS lipopolysaccharide conjugated to bovine serum albumin develop varying degrees of protective immunity against systemic or aerosol challenge with virulent type A and type B strains of the pathogen, *Vaccine* **20**:3465–3471.
62. I. Golovliov, M. Ericsson, L. Akerblom, G. Sandstrom, A. Tarnvik and A. Sjostedt, 1995, Adjuvanticity of iscoms incorporating a T-cell-reactive lipoprotein of the facultative intracellular pathogen *Francisella tularensis, Vaccine* **13**:261–267.
63. M. Fulop, R. Manchee and R. Titball, 1995, Role of lipopolysaccharide and a major outer-membrane protein from *Francisella tularensis* in the induction of immunity against tularemia, *Vaccine* **13**:1220–1225.

64. C. Mayers, M. Duffield, S. Rowe, J. Miller, B. Lingard, S. Hayward and R. W. Titball, 2003, Analysis of known bacterial protein vaccine antigens reveals biased physical properties and amino acid composition, *J. Comp. Funct. Genomics* **4**:468–478.
65. H. S. Garmory, K. A. Brown and R. W. Titball, 2003, DNA vaccines: improving expression of antigens., *Gen. Vac. Ther.* **1**:2.
66. F. Bowe, D. J. Pickard, R. J. Anderson, P. Londono-Arcila and G. Dougan, 2003, Development of attenuated *Salmonella* strains that express heterologous antigens, *Methods. Mol. Med.* **87**:83–100.
67. D. W. Waag, K. T. McKee, G. Sandstrom, L. L. K. Pratt, C. R. Bolt, M. J. England, G. O. Nelson and J. C. Williams, 1995, Cell-mediated and humoral immune responses after vaccination of human volunteers with the live vaccine strain of *Francisella tularensis*, *Clin .Diagn. Lab. Immunol.* **2**:143–148.
68. D. M. Waag, A. Galloway, G. Sandstrom, C. R. Bolt, M. J. England, G. O. Nelson and J. C. Williams, 1992, Cell mediated and humoral immune responses induced by scarification vaccination of human volunteers with a new lot of the live vaccine strain of *Francisella tularensis*, *J. Clin. Microbiol.* **30**:2256–2264.
69. W. Allen, 1962, Immunity against tularemia: passive protection of mice by transfer of immune tissues, *J. Exp. Med.* **115**:411–420.
70. J. J. Drabick, R. B. Narayanan, J. C. Williams, J. W. Leduc and C. A. Nacy, 1994, Passive protection of mice against lethal *Francisella tularensis* (Live tularemia vaccine strain) infection by the sera of human recipients of the live tularemia vaccine, *Am. J. Med. Sci.* **308**:83–87.
71. S. Stenmark, H. Lindgren, A. Tarnvik and A. Sjostedt, 2003, Specific antibodies contribute to the host protection against strains of *Francisella tularensis* subspecies holarctica, *Microb. Pathog.* **35**:73–80.
72. A. Casadevall, 1998, Antibody-mediated protection against intracellular pathogens, *Trends Microbiol.* **6**:102–107.
73. A. Casadevall and L. A. Pirofski, 2003, Antibody-mediated regulation of cellular immunity and the inflammatory response, *Trends. Immunol.* **24**:474–478.
74. S. Lofgren, A. Tarnvik, G. D. Bloom and W. Sjoberg, 1983, Phagocytosis and killing of *Francisella tularensis* by human polymorphonuclear leukocytes, *Infect. Immun.* **39**:715–720.
75. T. R. Rhinehart-Jones, A. H. Fortier and K. L. Elkins, 1994, Transfer of immunity against lethal murine *Francisella* infection by specific antibody depends on host gamma interferon and T cells, *Infect. Immun.* **62**:3129–3137.
76. D. Yee, T. R. Rhinehart-Jones and K. L. Elkins, 1996, Loss of either CD4+ or CD8+ T cells does not affect the magnitude of protective immunity to an intracellular pathogen, *Francisella tularensis* strain LVS, *J. Immunol.* **157**:5052–5048.
77. J. W. Conlan, A. Sjostedt and R. J. North, 1994, CD4+ and CD8+ T-cell-dependent and -independent host defence mechanisms can operate to control and resolve primary and secondary *Francisella tularensis* LVS infection in mice, *Infect. Immun.* **62**:5603–5607.
78. K. L. Elkins, S. C. Cowley and C. M. Bosio, 2003, Innate and adaptive immune responses to an intracellular bacterium, *Francisella tularensis* live vaccine strain, *Microbe. Infect.* **5**:135–142.

8

Brucella and Bioterrorism

MICHELLE WRIGHT VALDERAS and R. MARTIN ROOP II

1. INTRODUCTION

The *Brucella* spp. are included on the class B list of select agents as defined by the Centers for Disease Control and Prevention.[1] Human brucellosis is rarely fatal and the disease is not transmitted from person to person.[2] Nevertheless, the brucellae are considered to be incapacitating agents capable of rendering an opposing military force severely debilitated when used as a bioweapon,[3] or overwhelming hospitals and other medical care facilities if used as an agent of bioterrorism in an urban setting.[4] The fact that successful treatment of human brucellosis with antibiotics can be problematic and no safe and effective vaccine exists for use in humans[2,5] enhances the risk associated with the potential use of the *Brucella* spp. as agents of bioterrorism.

2. BRUCELLOSIS: A ZOONOTIC DISEASE

The *Brucella* spp. are Gram-negative bacteria that, like other notable agents of biological warfare such as *Bacillus anthracis*, *Yersinia pestis*, and *Francisella tularensis*, naturally cause zoonotic disease.[6] Six species of *Brucella*—*B. melitenis*, *B. abortus*, *B. suis*, *B. canis*, *B. ovis*, and *B. neotomae* are formally recognized,[7] and two provisional species designations, *B. cetaceae* and *B. pinnipediae*, have been given to the brucellae isolated from marine mammals.[8] For practical purposes, the brucellae are classified into these nomenspecies based on their phenotypic and metabolic characteristics and their host specificity,[9,7,10] but at the genetic level all of these strains represent a single genospecies.[11] The brucellae predominately cause abortion and infertility in their natural hosts, which include domesticated mammals such as goats, cattle, sheep, pigs,

MICHELLE WRIGHT VALDERAS and R. MARTIN ROOP II • Department of Microbiology and Immunology, East Carolina University School of Medicine, Greenville, NC 27858-4354, USA.

TABLE I
Host Specificity of the Currently Recognized *Brucella* sp. and Their Infectivity for Humans

Strain	Natural hosts	Human infectivity
B. abortus	Cattle, bison, and elk	Yes
B. canis	Dogs	Yes
B. cetaceae	Dolphins and porpoises	Unknown
B. melitensis	Sheep and goats	Yes
B. neotomae	Desert wood rat	Unknown
B. ovis	Sheep	No
B. pinnipediae	Seals and otters	Yes
B. suis	Pigs, reindeer, and caribou	Yes

and dogs, and wild mammals such as bison, elk, and the marine mammals (Table I).[6,12,13] The economic impact of brucellosis in food animals can be significant to both the owners of the livestock and to the countries where they are located due to the imposition of trade barriers.

Humans contract brucellosis through direct contact with infected animals or their products, through exposure to the brucellae in a laboratory setting, or through accidental inoculation with the live attenuated vaccine strains used in food animals.[12] Numerous countries including the United States, Canada, and parts of Europe have been successful in controlling the incidence of brucellosis in food animals through concerted vaccination, surveillance, quarantine, and slaughter programs, and in these areas human infection is uncommon and has become predominately an occupational hazard for animal handlers, slaughterhouse workers, veterinarians, and laboratory personnel. The practice of pasteurizing milk and other dairy products has also played a significant role in reducing the incidence of human brucellosis. In areas of the world where brucellosis in food animals is still endemic, however, human brucellosis remains a serious public health concern. The risk of human infection in these regions is intensified if the consumption of unpasteurized dairy products is a common practice. Cases of human brucellosis almost exclusively result from infection by *B. melitensis, B. abortus,* or *B. suis.*[2] *B. canis* and the brucellae from marine mammals have also been implicated in zoonotic infections in humans, but these cases appear to be rare.[14–16]

3. CLINICAL SYMPTOMS OF HUMAN BRUCELLOSIS

Human brucellosis was originally described in British soldiers serving in the Mediterranean during the Crimean war, who fell ill with a chronic relapsing febrile illness that was eventually known as Mediterranean Fever or Malta

Fever. Sir David Bruce discovered the causative agent of this disease in 1886, and Themistocles Zammit subsequently established the epidemiological link between the consumption of unpasteurized goats milk and Malta Fever.[17,18] In humans, brucellosis can manifest itself in a variety of ways ranging from asymptomatic subclinical infection to lethal endocarditis.[2] Although the symptoms of human brucellosis are similar in nature regardless of the bacterial strain involved, the severity of these symptoms can vary greatly, with *B. melitensis* strains producing the most severe symptoms followed in severity by *B. suis*, *B. abortus*, and *B. canis* strains in descending order.[2] Generally, the disease presents as a protracted debilitating illness characterized by intermittent fever, chills, myalgia, and malaise[2] (Table II). Unresolved infections can also lead to focal abscesses in organs of the reticuloendothelial system, and joint infection is common. Rarely endocarditis, encephalitis, and meningitis occur, and in the small percentage (2% or below) of deaths that occur, mortality is usually attributed to endocarditis.

TABLE II
Symptoms and Clinical Findings in 480 Patients Diagnosed with Brucellosis in a Turkish Clinic Between 1989 and 1998[a]

	No. of patients	(%) Percentage
Symptoms		
Malaise	432	90
Sweating	405	84.4
Arthralgia	393	81.9
Fever	383	79.8
Back pain	281	58.5
Myalgia	236	49.2
Weight loss	213	44.4
Anorexia	198	41.3
Nausea	155	32.3
Vomiting	104	21.7
Abdominal pain	101	21
Headache	91	19
Findings		
Fever	187	39
Hepatolomegaly	102	21.3
Osteoarticluar involvement	91	19
Splenomegaly	68	14.2
Neurological involvement	31	6.5
Genitourinary involvement	5	1
Endocarditis	2	0.4
Peritonitis	2	0.4
Cutaneous involvement	2	0.4
Pneumonia	1	0.2

[a]Modified from Doganay and Aygen (70).

4. LIFE WITHIN THE MACROPHAGE AND SUBVERSION OF HOST IMMUNE RESPONSES

The predominant host cell for the brucellae in humans is the macrophage.[19,20] Indeed, it is the remarkable capacity of these bacteria to survive and replicate in these host phagocytes that is responsible for the chronicity of brucellosis. Following ingestion by host macrophages, the brucellae initially withstand acidification of the phagosome and exposure to the oxidative burst of these host phagocytes. Shortly after entry into the acidified phagosome, the brucellae employ their Type IV secretion machinery[21] to elaborate as yet unidentified effector molecules into the host macrophage that divert the *Brucella*-containing vacuole from the endolysosomal pathway into an alternative maturation pathway. This pathway ultimately leads to the brucellae residing in an intracellular compartment bounded by a membrane that is formed by continual interactions with the endoplasmic reticulum of the host cell.[22] Although this compartment, known as the replicative phagosome, represents a favorable environment for intracellular survival and replication of the brucellae, experimental evidence indicates that these bacteria must make significant physiological adaptations to withstand the environmental stresses encountered during long-term residence in this intracellular niche.[19,20]

It is well established that cell-mediated immunity is required for the resolution of *Brucella* infections[23,24] and this is consistent with the intracellular niche occupied by these bacteria. Specifically, it appears that Th1-type cellular immune responses and the resulting IFN-γ mediated activation of macrophages plays a crucial role in protective immunity. Experimental evidence also suggests that the induction of IgG-type antibodies play a role in protective immunity and it is likely that the opsonic activity of these antibodies enhances the brucellacidal activity of activated macrophages.

Not surprisingly, the brucellae appear to be able to evade specific components of the host immune response and prolong the life span of their host cell. Undoubtedly, these properties contribute significantly to their ability to maintain long-term residence in host macrophages and correspondingly produce chronic infection. The O-polysaccharide of the smooth lipopolysaccharide (LPS) of the *Brucella* spp., for instance, binds to lipid rafts on the surface of host macrophages, and this interaction appears to play an important role in the initial stages of development of the replicative phagosome prior to the participation of the Type IV secretion machinery.[19] The smooth LPS also interferes with the capacity of infected macrophages to present *Brucella* antigens via their MHC class II molecules, which leads to a diminished capacity of these phagocytes to activate *Brucella* antigen-specific CD4^{+} T cells.[25] It has long been known that the lipid A component of the *Brucella* LPS displays a much lower level of biological activity than lipid A molecules from other gram-negative bacteria and it has been proposed that this reduced biological activity allows the brucellae to avoid inducing a full-blown inflammatory response in the infected host.[26] Experimental evidence also suggests that the brucellae are able to inhibit TNF-α production by the human monocytic cell line THP-1 and inhibit apoptosis of these cells in culture.[27] Peripheral blood monocytes from human brucellosis patients are also more resistant to spontaneous or ligand-induced apoptosis than monocytes from uninfected patients, and this resistance disappears shortly after successful treatment with antibiotics.[28]

5. HISTORY OF THE USE OF *BRUCELLA* AS AN AGENT OF BIOLOGICAL WARFARE

There are several biological and pathogenic properties of the brucellae that make them useful as agents of biological warfare.[3,4,5,29] First, they are highly infectious via the aerosol route, with an infectious dose estimated at approximately 10–100 organisms.[5] It has been estimated that the release of 50 kg of *B. suis* from a plane along a 2 km line 10 km upwind of a population center of 500,000 people would result in 500 deaths and 125,000 people being incapacitated.[3] Second, human brucellosis is a notoriously debilitating disease,[2] and it is considered an incapacitating agent rather than a lethal agent.[3] Consequently, the logistics of caring for a large number of patients with undulant fever can soon become overwhelming. Third, successful treatment of human brucellosis requires prolonged antibiotic treatment (i.e., doxycycline and rifampin for 6 weeks)[2] and relapse rates of 5–10% after apparently successful treatment are not uncommon.[30,31] Finally, there is no safe, effective vaccine available for use in humans.

During World War II it became evident through intelligence received by the United States and other allied countries that Japan and Germany were actively researching the use of bioweaponry.[3,32] Consequently, President Roosevelt ordered the creation of the War Research Service in 1942 to explore the potential use of biological weapons and to develop strategies for defending against the use of these agents by an opposing military force. This Executive Action led to the construction and operation of a number of facilities, including Camp Detrick (which later became known as Fort Detrick), and Dugway Proving Ground. Following the Korean War, the United States government expanded its offensive biowarfare capabilities and in 1954, *B. suis* was weaponized at the Pine Bluff Arsenal in Arkansas. During this period of active research, *B. suis* was considered to be one of the major debilitating agents available for use in the U.S. bioweapons arsenal.

In 1969, new political attitudes about the morality of use of biological agents for warfare prevailed and the United States destroyed its stockpile of biological weapons and ceased their production altogether. In 1972 it was one of 140 nations that signed a treaty at the Biological Weapons Convention agreeing "not to develop, produce, stockpile, or otherwise acquire or retain" biological agents for military use. In 1992, however, U.S. intelligence officials received word of the existence of genetically altered, multiply drug resistant *Brucella* strains that had been developed by Biopreparat, the agency responsible for the development of bioweapons in the Former Soviet Union.[33] During this same time period, military intelligence obtained during and after Operation Desert Storm indicated that bioweapons (including possibly *Brucella* spp.) had been stockpiled in Iraq.[32,34] These incidents resurrected the specter of bioweapons as a potential threat to U.S. military forces and served as a reminder of the ill-prepared nature of these forces to protect themselves against deliberate exposure to the *Brucella* spp. in a battlefield setting. Consequently, scientists at Walter Reed Army Institute of Research were given the task of developing a vaccine that could be safely used to protect troops against aerosolized *Brucella.*[35] The resulting studies have greatly enhanced our understanding of the protective immunity against brucellosis and several attractive vaccine candidates identified during the course of these studies have either been evaluated in nonhuman primates,[36] or are presently under evaluation in this model.

6. IMPACT OF AN ATTACK USING BRUCELLA AS AN AGENT OF BIOTERRORISM

The same properties that make the brucellae useful as an agent of biological warfare make them a potential terrorist threat. As opposed to the use of agents such as *Bacillus anthracis* and *Yersinia pestis* for such purposes, in which case many deaths would be expected,[4] the use of *Brucella* spp. for bioterrorist purposes would produce a more insidious and long-term effect. It might take weeks before the exposed individuals develop full-blown undulant fever,[5] at which time effective resolution of the infection would require lengthy antibiotic treatment and associated medical care.[2,4] In many cases repeat treatment would be necessary due to the high percentage of relapses. Indeed, the insidious nature of the infection would likely complicate the process of identifying and appropriately treating exposed individuals, especially if the exact time and place of the exposure were unknown. Due to the risk of these bacteria being used as agents of bioterrorism, *B. melitensis, B. suis,* and *B. abortus* are included on the class B list of select agents as defined by the Center for Disease Control,[1] and accordingly these strains are included in the National Institute of Allergy and Infectious Disease's Strategic Plan for Biodefense Research.[37]

Over the last 50 years, the United States and many other industrialized nations have spent billions of dollars on vaccination, quarantine, and slaughter programs aimed at controlling brucellosis in food animals.[38] Consequently, another important consideration with regard to the use of the *Brucella* spp. as agents of bioterrorism is the potential impact on the agricultural community resulting from the re-introduction of these agents into the food animals populations of countries where brucellosis has been considered to be "eradicated".

7. DIAGNOSIS AND TREATMENT OF BRUCELLOSIS

Although the symptoms of human brucellosis are relatively nonspecific,[2] physicians may have a high index of suspicion for this disease in geographic areas where *Brucella* infections are endemic in food animals. In regions where the disease is rare in food animals and the use of pasteurized dairy products is widespread, however, human brucellosis is uncommon and can easily be mistaken for other febrile illnesses in the absence of a patient history that includes exposure to infected animals or in the laboratory. This low index of suspicion in areas where brucellosis in food animals is uncommon could potentially delay the recognition of an outbreak of this disease in humans associated with a bioterrorist event. It is notable in this regard that all of the bacterial agents that are considered to be potential bioweapons are natural pathogens for animals that produce zoonotic diseases in humans. Thus, one of the potential "warning signs" of the use of these agents for bioterrorism is the occurrence of clusters of ordinarily zoonotic infections such as brucellosis, tularemia, anthrax, or plague affecting patients with no obvious history of exposure to infected animals or occupational exposure to these agents.

Culture of the brucellae from blood or other tissue is the most definitive means of diagnosing human brucellosis,[2,39] but isolation of these bacteria from patients can be problematic. The biphasic culture method described by Castañeda[40] is considered to be the best traditional approach for isolating the *Brucella* spp., but regardless of the culture medium employed, it is important that cultures be maintained for at least 28 days before they are considered negative. Automated blood culture systems (e.g., BACTEC) also appear to be efficient for isolation of the *Brucella* spp. and in some cases may be faster than traditional culture methods. Polymerase chain reaction (PCR)-based assays have been designed for rapid detection of the *Brucella* in patient tissue samples, and in some cases primers specific for these bacteria have been included in multiplex PCR-based assays designed to screen for the presence of multiple biowarfare agents.[41] Antigen detection-based biosensors have also been developed to detect the *Brucella* spp. in the environment.[42]

Although culture of the brucellae is confirmatory, serology is the most widely used method for diagnosing human brucellosis.[2,39] The presence of IgG-type antibodies specific for the O-chain of the LPS of *B. abortus*, *B. melitensis*, and *B. suis* in a patient's serum is considered to be diagnostic for an active infection. The serum agglutination test (SAT) employing killed *Brucella* cells is the most widely used serological assay for diagnosing human infection, and IgM and IgA can be removed from patient's sera prior to evaluation in the SAT by treatment with 2-mercaptoethanol or Rivanol to provide a more accurate assessment of IgG levels. The Rose Bengal card test can be used for screening patients' sera, but positive reactions in this test must be confirmed by the SAT. Serologic cross reactivity between the *Brucella* spp. and

Francisella tularensis, another potential bioterrorism agent, has been reported in the SAT, but the titers of brucellosis patients are higher against *Brucella* antigens than they are against the *Francisella tularensis* antigens in this assay and the reverse is true for patients with tularemia. Enzyme-linked immunosorbent assays (ELISA) are also available for use in the serologic diagnosis of human brucellosis, but they are not used as routinely as the SAT. Rapid dipstick serologic tests have recently been developed for field diagnosis of human brucellosis[43] and these assays may be particularly useful for screening individuals for exposure to the *Brucella* spp. during a suspected incident of bioterrorism.

Proper treatment of brucellosis is imperative as too short a course of antibiotics or use of a single antibiotic has been associated with increased risk of relapse.[2] The combination of doxycycline for 6 weeks and streptomycin for 14–21 days or gentamicin for 7–10 days has recently become the primary treatment regimen recommended by the World Health Organization[44] for human brucellosis. Streptomycin and gentamicin are administered parenterally (i.e., intramuscularly or intravenously), however, and consequently, doxycyline in combination with rifampin (both of which can be taken orally) is considered to be the principal alternative therapy for treatment of human brucellosis. Fluoroquinolones or trimethoprim/sulfamethoxazole in combination with doxycyline or rifampin are recommended as secondary alternative therapies. The duration of treatment with these antibiotic combinations may have to be extended in cases where complications such as meningitis or endocarditis occur, and in the latter case surgical intervention (e.g., valve replacement) may also be necessary. Alternative drugs may also have to be added to the regimen in meningitis cases because the tetracyclines and aminoglycosides do not penetrate the blood/brain barrier well. Because the tetracyclines cannot be used in pregnant women and children, trimethoprim/sulfamethoxazole in combination with gentamicin or rifampin have been recommended for use in pregnant women and trimethoprim/sulfamethoxazole in combination with streptomycin or gentamicin for use in children less than 8 years of age.

Even in cases where recommended treatment regimens are completed and patients appear to have resolved their symptoms, relapse rates of 5–10% are not uncommon.[30,31] This potential difficulty associated with the effective treatment of human brucellosis is one of the major reasons that this agent is considered an attractive bioweapon. Accordingly, because the mortality rate of human brucellosis is very low, the development of chemotherapeutic strategies that would reliably shorten the duration of the overt illness with a shorter course of treatment and less potential for relapse would certainly lessen the potential impact on both civilian populations and military forces resulting from the use of the *Brucella* spp. as agents of bioterrorism or biowarfare.

8. VACCINE DEVELOPMENT: HISTORICAL PERSPECTIVES AND CONSIDERATIONS FOR THE FUTURE

No safe, effective human brucellosis vaccine presently exists, but live, attenuated strains of *B. abortus*, *B. melitensis*, and *B. suis* have been used effectively as vaccines in food animals.[45,46] Although these strains either retain their virulence in humans or display other phenotypic properties that preclude them from consideration as candidates for a human vaccine, use of these strains as effective vaccines in their natural hosts has provided us with valuable information concerning protective immunity that can potentially be applied toward the development of a safe and effective human vaccine.

B. abortus Strain 19[47] has been widely used in cattle as a live, attenuated vaccine, and the efficacy of this strain is best exemplified by the success of eradication programs in which it has been used to control the incidence of bovine brucellosis. Most notably, S19 was the centerpiece of the Brucellosis Eradication Program administered by USDA/APHIS[38] that resulted in the incidence of bovine brucellosis in the United States being reduced from approximately 124,000 infected herds in 1957 to 2 infected herds in 2003. Despite its utility as a bovine vaccine, however, *B. abortus* S19 remains fully virulent in humans[47] and thus presents risk for occupational exposure to veterinarians. This strain can also occasionally induce abortion if administered to pregnant cattle. From a practical standpoint, however, probably the most significant problem associated with the use of S19 in cattle was the fact that the administration of this vaccine to adult cattle produced antibody responses that were indistinguishable from those present in naturally infected cattle in the standard serologic tests. In 1996, RB51, an attenuated strain of *B. abortus* that lacks the O-polysaccharide component of its LPS,[48] was licensed in the United States as an alternative live vaccine for use in cattle. This strain has subsequently replaced S19 as the official bovine vaccine strain in the United States and its efficacy is presently being examined in other geographic areas.[46] Since RB51 lacks the major antigenic component that elicits the antibodies detected in standard serologic assays, RB51-vaccinated and naturally infected cattle can be easily distinguished by routine serology. *B. melitensis* Rev1[49] is an attenuated strain that has been effectively used as a live, attenuated vaccine in programs aimed at controlling the incidence of brucellosis in sheep and goats in areas of the world where these animals represent major food animals.[50] *B. suis* strain 2 has also been used as a live attenuated vaccine in cattle, goats, sheep, and pigs in China,[51] but the use of this strain has not been as widespread as the use of strains S19, Rev1, or more recently RB51. Experimental studies with these vaccine strains have clearly shown that cellular immunity is essential for protection against *Brucella* infection. Moreover, they have established that the use of live, attenuated *Brucella* strains is the most reliable way to induce protective immunity in the natural host.

Both live attenuated strains and cellular subunit preparations have been tested as experimental vaccines against human brucellosis. A derivative of *B. abortus* Strain 19, designated S19-BA, was used extensively to immunize humans in the Soviet Union during a period beginning in the 1940s and extending through the 1960s,[52] and an attenuated *B. abortus* strain known as 104M has also been used in China to immunize farmers and agricultural workers during brucellosis outbreaks in rural areas.[52] Although both *B. abortus* S19-BA and 104M provided protective immunity when administered as live vaccines in humans, the level of virulence retained by these strains or the extent of the side-effects associated with their use in humans is difficult to assess from the literature. A phenol-insoluble subcellular fraction prepared from *B. melitensis* M15[54] has also been evaluated as a vaccine candidate in humans, but this preparation appeared to be ineffective at eliciting protective immunity.[55]

Current efforts aimed at developing a safe and effective brucellosis vaccine for use in humans are primarily taking two approaches.[35] First, *Brucella* strains with defined mutations that display the desired level of attenuation and immunogenicity in the mouse model are being evaluated for efficacy in providing protective immunity in nonhuman primates. The most promising of these strains examined to date is a *purE* mutant derived from virulent *B. melitensis* 16M, which is unable to synthesize purines *de novo* during residence in host macrophages.[56] Other *Brucella* genes that may offer attractive targets for the construction of live vaccine candidates based on the attenuation of the corresponding mutants in the murine model are listed in Table III. The second approach that is being used to formulate human vaccine candidates is the identification of specific *Brucella* antigens that elicit protective cellular immune responses and the design of antigen delivery systems (including DNA vaccine and recombinant DNA-based strategies) that can be used to elicit *Brucella* antigen-specific immune responses in the host.[46,57] *Brucella* antigens that have been identified as being capable of eliciting protective cellular immune responses in the mouse model are shown in Table IV.

TABLE III
Selected *Brucella* Genes that May Serve as Viable Targets for the Construction of Novel Live Vaccine Candidates

Gene(s)	References
aroC	Foulongne *et al.*,[58]
bvrRS	Sola-Landa *et al.*,[59]
cydB	Endley *et al.*,[60]
hfq	Robertson and Roop,[61]
virB1-12	O'Callaghan *et al.*[62]

TABLE IV
Selected *Brucella* Cell Components that have the Capacity to Induce Protective Immunity in the Mouse Model when Administered as Subunit Preparations or Produced from Recombinant Genes *In Situ* in the Host

Cu/Zn superoxide dismutase	Onate *et al.*,[71,72]
Lumazine synthase	Velikovsky *et al.*,[65,66]
Ribosomal protein L7/L12	Oliveira and Splitter,[67]
Smooth LPS O-polysaccharide	Winter *et al.*,[68]
P39	Al-Mariri *et al.*,[63]

9. SUMMARY

Brucellosis has long been a major concern of the agricultural community worldwide, and an important public health issue in geographic areas where this disease is endemic in food animals. The biological properties of the brucellae and the nature of the disease they cause in humans also make them a threat for use as an agent of bioterrorism. Our understanding of the basic biology of these bacteria and their interactions with their mammalian hosts has improved greatly over the last few years and very powerful genetic, molecular, biochemical, and microscopic approaches can now be used to gain even better insight into the pathobiology of *Brucella* infections. Information derived from such studies will hopefully provide the basis for the development of safe and effective human brucellosis vaccines and improved chemotherapeutic strategies for treating this disease. Because human brucellosis is generally nonlethal and naturally occurring human disease is only common in geographic regions where brucellosis is endemic in food animals, a human brucellosis vaccine will likely have its greatest impact on military personnel and individuals with occupational or geographically associated risk of exposure. If chemotherapeutic agents are developed or existing agents identified that reliably lead to the rapid resolution of human infection, however, these agents will be broadly beneficial for the prophylaxis and treatment of human brucellosis regardless of whether or not the infection is acquired naturally or as the result of intentional exposure through an act, or war, or bioterrorism.

ACKNOWLEDGMENTS

The authors greatly acknowledge Ottorino Cosivi of the World Health Organization for allowing them access to the most recent WHO guidelines on brucellosis while this document was still in the draft stage; thanks also to Col. David Hoover of the Walter Reed Army Institute of Research for providing us with unpublished and unclassified information regarding the development of a human vaccine candidate. Work in the laboratory of R.M.R. II was funded by grants from the National Institute of Allergy and Infectious Disease (AI 48499) and the United States Department of Agriculture's National Research Initiative's Competitive Grants Program (02-02215).

REFERENCES

1. United States Department of Health and Human Services, 2002a, 42 CFR Part 1003 Possession, use and transfer of select agents and toxins, *Fed. Reg.* **240**:76886–76905.
2. Young, E. J., 2000, *Brucella* species, in: *Principles and Practice of Infectious Disease*, 5th ed. (Mandel, G. L., Bennett, J. E., and Dolin, R., eds.), Churchill-Livingstone, Philadelphia, PA, pp. 2386–2393.
3. Christopher, G. W., Cieslak, T. J., Pavlin J. A., *et al.*, 1997, Biological warfare. A historical perspective, *JAMA*. **279**:412–417.
4. Kaufmann, A. F., Meltzer, M. I., and Schmid G. P., 1997, The economic impact of a bioterrorist attack: are prevention and post-attack intervention programs justifiable? *Emerg. Infect. Dis.* **3**:83–94.
5. Franz, D. R., Jahrling, P. B., McClain, D. J., et al., 2001, Clinical recognition and management of patients exposed to biological warfare agents, *Clin. Lab. Med.* **21**:435–473.
6. Nicoletti, P., 1989, Relationship between animals and human disease, in: *Brucellosis: Clinical and Laboratory Aspects* (Young, E. J. and M. J. Corbel, eds.), CRC Press, Boca Raton, FL, pp. 41–51.
7. Corbel, M. J., and Brinley-Morgan, W. J., 1984, Genus *Brucella*, in: *Bergey's Manual of Systematic Bacteriology*, vol. 1 (N. R. Krieg and J. G. Holt, eds.), Williams & Wilkins, Baltimore, pp. 377–388.
8. Cloeckart, A., Verger, J. M., Grayon, M., *et al.*, 2001, Classification of *Brucella* spp. isolated from marine mammals by DNA polymorphism at the *omp2* locus, *Microbes Infect.* **3**:729–738.
9. Alton, G. G., Jones, L. M., Angus R. D., *et al.*, 1988, Techniques for the brucellosis laboratory, Institute National de la Recherche Agronomique, Paris.
10. Jahans, K.L., Foster, G., and Broughton E.S., 1997, The characterization of *Brucella* strains isolated from marine mammals, *Vet. Microbiol.* **57**:373–382.
11. Verger, J. M., F. Grimont, P. A. D. Grimont, and M. Grayon. 1985. *Brucella*: a monospecific genus as shown by deoxyribonucleic acid hybridization, *Int. J. Syst. Bacteriol.* **35**:292–295.
12. Acha, P. N., and Szyfres, B., 1980, *Zoonoses and Communicable Diseases Common to Man and Animals*, Pan American Health Organization, Washington, DC., pp. 28–45.
13. Foster, G., MacMillan, A. P., Godfroid, J., *et al.*, 2002, A review of *Brucella* sp. infection of sea mammals with particular emphasis on isolates from Scotland, *Vet. Microbiol.* **90**:563–580.
14. Polt, S. S., W. E. Dismukes, A. Flint, and J. Schaefer. 1982. Human brucellosis caused by *Brucella canis*: clinical features and immune response, *Ann. Intern. Med.* **97**:717-719.
15. Brew, S. D., Perrett, L. L., Stack, J. A., *et al.*, 1999, Human exposure to *Brucella* recovered from a sea mammal, *Vet. Rec.* **144**:483.
16. Sohn, A. H., W. S. Probert, C. A. Glaser, N. Gupta, A. W. Bollen, J. D. Wong, E. M. Grace, and W. C. McDonald. 2003, Human neurobrucellosis with intracerebral granuloma caused by a marine mammal *Brucella* spp. *Emerg. Infect. Dis.* **9**:485-488.
17. Hall, W. H., 1989, History of *Brucella* as a human pathogen, in: *Brucellosis: Clinical and Laboratory Aspects* (E. J. Young., and M. J. Corbel., eds.), CRC Press, Boca Raton, FL, pp. 1–9.
18. Nicoletti, P., 2000, A short history of brucellosis, *Vet. Microbiol.* **90**:5–9.
19. Köhler, S., Michaux-Charachon, S. Porte, F., *et al.*, 2003, What is the nature of the replicative niche of a stealthy bug named *Brucella*? *Trends Microbiol.* **11**:215–219.

20. Roop, R. M. II, B. H. Bellaire, M. W. Valderas, and J. A. Cardelli, 2004, Adaptation of the brucellae to their intracellular niche. *Mol. Microbiol.* **52**:621–630.
21. Boschiroli, M. L., Ouahrani-Bettache, S., Foulongne, V., *et al.*, 2002, Type IV secretion and *Brucella* virulence, *Vet. Microbiol.* **90**:341–348.
22. Celli, J., de Chastellier, C., Franchini, D. M., *et al.*, 2003, *Brucella* evades macrophage killing via VirB-dependent sustained interactions with the endoplasmic reticulum, *J. Exp. Med.* **198**:545–556.
23. Baldwin, C. L., and Parent, M., 2002, Fundamentals of host immune response against *Brucella abortus*: what the mouse model has revealed about control of infection, *Vet. Microbiol.* **90**:367–382.
24. Doganay, M., and Aygen, B., 2003, Human brucellosis: an overview, *Int. J. Infect. Dis.* **7**:173–182.
25. Forestier, C., Deleuil, F., Lapaque, N., *et al.*, 2000, *Brucella abortus* lipopolysaccharide in murine peritoneal macrophages acts as a down-regulator of T cell activation, *J. Immunol.* **165**:5202–5210.
26. Rasool, O., E. Freer, E. Moreno, and C. Jarstrand, 1992, Effect of *Brucella abortus* lipopolysaccharide on oxidative metabolism and lysozyme release by human neutrophils, *Infect. Immun.* **60**:1699-1702.
27. Dornand, J., Gross, A., Lafont, V., *et al.*, 2002, The innate immune response against *Brucella* in humans, *Vet. Microbiol.* **90**:383–394.
28. Tolomeo, M., P. Di Carlo, V. Abbadessa, L. Titone, S. Miceli, E. Barbusca, G. Cannizo, S. Mancuso, S. Arista, and F. Scarlata, 2003, Monocyte and lymphocyte resistance in acute and chronic brucellosis and its possible implications in clinical management, *Clin. Infect. Dis.* **36**:1533–1538.
29. Huxsoll, D. L., Patrick III, W. C., and Parrott, C. D., 1987, Veterinary services in biological disasters, *J. Am. Vet. Med. Assoc.* **190**:714–722.
30. Hall, W. H., 1990, Modern chemotherapy for brucellosis in humans, *Rev. Infect. Dis.* **12**:1060–1099.
31. Memish, Z., Mah, M. W., Al Mahmoud, S., *et al.*, 2000, *Brucella* bacteraemia: clinical and laboratory observations in 160 patients, *J. Infect.* **40**:59–63.
32. Miller, J., Engelberg, S., and Broad W., 2002, *Germs–Biological Weapons and America's Secret War*, Simon & Schuster, New York.
33. Alibek, K., and Handelman, S., 1999, *Biohazard: The Chilling True Story of the Largest Covert Biological Weapons Program in the World—Told From the Inside by the Man Who Ran It*, Random House, New York.
34. Zilinskas, R. A. 1997. Iraq's biological weapons – the past as future? *J. Amer. Med. Assoc.* **278**:418–424.
35. Hoover, D. L., and Borschel R. H., 2004, Medical protection against brucellosis, in: *Biological Weapons Defense: Infectious Diseases and Counterbioterrorism* (L. E. Lindler, F. J. Lebeda, and G. Korch, *eds.*). Humana Press, Totowa, NJ.
36. Hoover, D., Izadjoo, M., Borschel, R. *et al.*, 2003, Oral immunization with WR210, a live, attenuated purine auxotrophic strain of *B. melitensis* protects mice and nonhuman primates against respiratory challenge with *B. melitensis* 16M, in: *Abstract, Proc. Brucellosis 2003 International Research Conference*, vol. **7**, p. 79.
37. United States Department of Health and Human Services, 2002b, NIAID strategic plan for biodefense research. NIH Publication No. 03-5306
38. Ragan, V. E., 2002, The Animal and Plant Health Inspection Service (APHIS) brucellosis eradication program in the United States, *Vet. Microbiol.* **90**:11–18.
39. Chu, M. C., and Weyant, R. S., 2003, *Francisella* and *Brucella*, in: *Manual of Clinical Microbiology*, 8th Edition (P. R. Murray, E. J. Baron, J. H. Jorgensen, *et al.*, eds.), American Society for Microbiology, Washington, DC., pp. 789–808.
40. Castañeda, M. R., 1947, A practical method for routine blood cultures in brucellosis, *Proc. Soc. Exp. Biol. Med.* **64**:114–115.

41. McDonald, R., Cao, T., and Borschel R., 2001, Multiplexing for the detection of multiple biowarfare agents shows promise in the field, *Mil. Med.* **166**:237–239.
42. Lee, W. E., Thompson, H. G., Hall, J. G., *et al.*, 2000, Rapid detection and identification of biological and chemical agents by immunoassay, gene probe assay and enzyme inhibition using a silicon-based biosensor, *Biosens. Bioelectron.* **14**:795–804.
43. Altuglu, I., Zeytinoglu, A., Bilgic, A., *et al.*, 2002, Evaluation of *Brucella* dipstick assay for diagnosis of acute brucellosis, *Diagn. Microbiol. Infect. Dis.* **44**:241.
44. World Health Organization, 2004, Brucellosis in humans and animals: WHO guidance. Department of Communicable Disease Surveillance and Response, World Health Organization, Geneva (draft document; Available at: http://www.brucella.org/cost/dynamicfiles/memberpages/148_WHO%20guid%20Web-2.pdf)
45. Nicoletti, P., 1990, Vaccination, in: *Animal Brucellosis* (K.Nielsen and J. R.Duncan, eds.), CRC Press, Boca Raton, FL, pp. 283–299.
46. Schurig, G. G, N. Sriranganathan, and M. J. Corbel. 2002. Brucellosis vaccines: past, present and future, *Vet. Microbiol.* **90**:479–496.
47. Buck, J. M., 1930, Studies of vaccination during calfhood to prevent bovine infectious abortion, *J. Agric. Res.* **41**:667–689.
48. Spink, W. W., J. W. Hall III, J. Finstad, and E. Mallet, 1962, Immunization with viable *Brucella* organisms. Results of a safety test in humans. *Bull. World Health Organ.* **26**:409-419.
49. Schurig, G. G., R. M. Roop II, T. Bagchi, S. Boyle, D. Buhrman, and N. Sriranganathan. 1991. Biological properties of RB51: a stable rough strain of *Brucella abortus. Vet. Microbiol.* **28**:171–188.
50. Elberg, S. S., and Faunce, K., 1957, Immunization against *Brucella* infection. VI. Immunity conferred on goats by a non-dependent mutant from a streptomycin-dependent mutant strain of *Brucella melitensis, J. Bacteriol.* **73**:211
51. Alton, G. G., 1990, *Brucella melitensis*, in: *Animal Brucellosis* (K.Nielsen, and J. R. Duncan, eds.), CRC Press, Boca Raton, FL, pp. 383–409.
52. Xin, X., 1986, Orally administrable brucellosis vaccine: *Brucella suis* strain 2 vaccine, *Vaccine* **4**:212–216.
53. Kolar, J., 1989, *Brucellosis* in Eastern European countries, in: *Brucellosis: Clinical and Laboratory Aspects* (E. J.Young, and M. J. Corbel, eds.), CRC Press, Boca Raton, FL. pp. 163–172.
54. Lu, S. L. and Zhang, J. L., 1989, *Brucellosis* in China, in: *Brucellosis: Clinical and Laboratory Aspects* (E. J.Young, and M. J. Corbel, eds.), CRC Press, Boca Raton, FL, pp. 173–180.
55. Roux, J., J. Asselineau, A. Serre, and C. Lacave. 1967. Immunological properties of a phenol-insoluble extract from *Brucella melitensis* (P.I. fraction), *Ann. Inst. Pasteur (Paris)* **113**:411–423.
56. Hadjichristodoulou, C., Voulgaris, P. Toulieres, L., *et al.*, 1994, Tolerance of the human brucellosis vaccine and the intradermal reaction test for brucellosis, *Eur. J. Clin. Microbiol. Infect. Dis.* **13**:129–134.
57. Hoover, D. L., Crawford, R. M., Van der Verg, L. L., *et al.*, 1999, Protection of mice against brucellosis by vaccination with *Brucella melitensis* WR201 (16MΔ*purE*), *Infect. Immun.* **67**:5877–5884.
58. Ko, J., and Splitter G. A., 2003, Molecular host-pathogen interaction in brucellosis: current understanding and future approaches to vaccine development for mice and humans, *Clin. Microbiol. Rev.* **16**:65–78.
59. Foulongne, V., Walravens, K., Bourg, G., *et al.*, 2001, Aromatic compound-dependent *Brucella suis* is attenuated in both cultured cells and mouse models, *Infect. Immun.* **69**:547–550.
60. Sola-Landa, A., J. Pizarro-Cerdá, M. J. Grilló, E. Moreno, I. Moriyón, J. M. Blasco, J. P. Gorvel, and I. López-Goñi. 1998. A two-component regulatory system playing a critical role in plant pathogens and endosymbionts is present in *Brucella abortus* and controls cell invasion and virulence, *Mol. Micro.* **29**:125–138.
61. Endley, S., McMurray, D., and Ficht T. A., 2001, Interruption of the *cydB* locus in *Brucella abortus* attenuates intracellular survival and virulence in the mouse model of infection, *J. Bacteriol.* **183**:2454–2462.

62. Robertson, G. T. and R. M. Roop II. 1999, The *Brucella abortus* host factor I (HF-I) protein contributes to stress resistance during stationary phase and is a major determinant of virulence in mice, *Mol. Micro.* **34**:690–700.
63. O'Callaghan, D., C. Cazevieille, A. Allardet-Servent, M. L. Boschiroli, G. Bourg, V. Foulongne, P. Frutos, Y. Kulakov, and M. Ramuz, 1999, A homologue of the *Agrobacterium tumefaciens* VirB and *Bordetella pertussis* Ptl type IV secretion systems is essential for intracellular survival of *Brucella suis*, *Mol. Microbiol.* **33**:1210–1220.
64. Al-Mariri, A., Tibor, A., Mertens, P., *et al.*, 2001, Protection of BALB/c mice against *Brucella abortus* 544 challenge by vaccination with bacterioferritin or P39 recombinant proteins with CpG oligodeoxynucleotides as adjuvant, *Infect. Immun.* **69**:4816–4822.
65. Al-Mariri, A., Tibor, A., Mertens, P., *et al.*, 2001, Induction of immune response in BALB/c mice with a DNA vaccine encoding bacterioferritin or P39 of *Brucella* spp., *Infect. Immun.* **69**:6264–6270.
66. Velikovsky, C. A., J. Cassataro, G. H. Giambartolomei, F. A. Goldbaum, S. Estein, R. A. Bowden, L. Bruno, C. A. Fossati, and M. Spitz, 2002, A DNA vaccine encoding lumazine synthase from *Brucella abortus* induces protective immunity in BALB/c mice, *Infect. Immun.* **70**:2507–2511.
67. Velikovsky, C. A., F. A. Goldbaum, J. Cassataro, S. Estein, R. A. Bowden, L. Bruno, C. A. Fossati, and G. H. Giambartolomei, 2003, *Brucella* lumazine synthase elicits a mixed Th1-Th2 immune response and reduces infection in mice challenged with *Brucella abortus* 544 independently of the adjuvant formulation used, *Infect. Immun.* **71**:5750–5755.
68. Oliveira, S. C., and G. A. Splitter. 1996. Immunization of mice with recombinant L7/L12 ribosomal protein confers protection against *Brucella abortus* infection, *Vaccine* **14**:959–962.
69. Winter, A. J., G. E. Rowe, J. R. Duncan, M. J. Eis, J. Widom, B. Ganem, and B. Morein. 1988. Effectiveness of natural and synthetic complexes of porin and O polysaccharide as vaccines against *Brucella abortus* in mice, *Infect. Immun.* **56**:2808–2817.
70. Doganay, M., and B. Aygen. 2003. Human brucellosis: an overview. *Int. J. Infect. Dis.* **7**:173–182.
71. Onate, A. A., R. Vemulapalli, E. Andrews, G. G. Schurig, S. Boyle, and H. Folch. 1999. Vaccination with live *Escherichia coli* expressing *Brucella abortus* Cu/Zn superoxide dismutase protects mice against virulent *B. abortus*. *Infect. Immun.* **67**:986–988.
72. Onate, A. A., S. Cespedes, A. Cabrera, R. Rivers, A. Gonzalez, C. Munoz, H. Folch, and E. Andrews. 2003. A DNA vaccine encoding Cu, Zn superoxide dismutase of *Brucella abortus* induces protective immunity in BALB/c mice. *Infect. Immun.* **71**:4857–4861.

9

Pneumonic Plague

DAVID L. ERICKSON and B. JOSEPH HINNEBUSCH

1. INTRODUCTION

Yersinia pestis, the agent of plague, is one of the most feared pathogens in human history. By some estimates, plague has killed up to 200 million people during three major pandemics.[1] The first recorded pandemic was the Justinian plague, which originated in Africa and spread around the Mediterranean during the sixth century. The second occurred in Europe during the fourteenth and fifteenth centuries and is sometimes referred to as the Black Death. The third pandemic began in China in the nineteenth century, spread throughout India and Asia, and eventually reached the Americas and other continents. Between these pandemics, the disease has manifested itself in smaller periodic outbreaks. During the last 50 years, the worldwide annual number of human plague cases documented by the World Health Organization has ranged from about 200 to 6000.[2] Plague remains a significant public health threat today, due to natural outbreaks of the disease, the emergence of antibiotic resistant strains, and its possible use as a biological weapon.[3,4]

Y. pestis is a member of the Enterobacteriaceae, the family of Gram-negative rod-shaped bacteria that includes *Escherichia coli*, *Salmonella*, and *Shigella*. Although humans are highly susceptible to infection, plague is a zoonosis that primarily affects rodents. *Y. pestis* circulates within wild rodent populations in permanent natural foci of disease scattered throughout every continent except Australia (Figure 1), and is transmitted primarily by fleas.[1,5] Over 200 rodent species have been implicated as being competent hosts, although their individual importance as reservoirs is mostly unknown.[2,5]

Although outbreaks of human disease are geographically tied to rodent plague foci, humans are of little or no significance in the ecology of plague,

DAVID L. ERICKSON and B. JOSEPH HINNEBUSCH • Laboratory of Zoonotic Pathogens, Rocky Mountain Laboratories, National Institute of Allergy and Infectious Diseases, National Institutes of Health, 903 S. 4th St., Hamilton, MT, USA 59840.

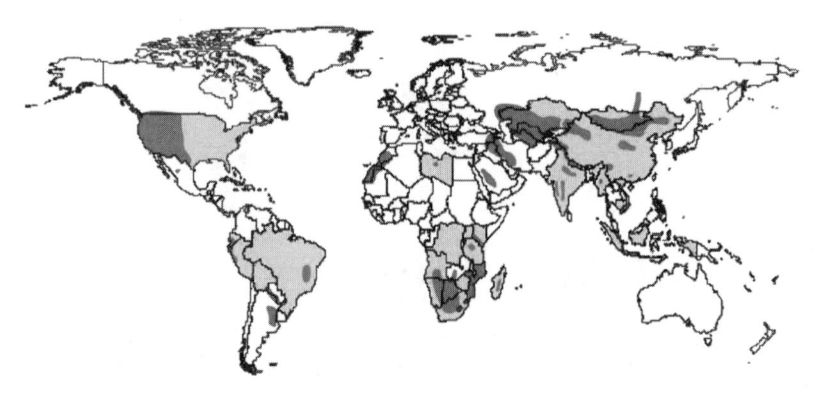

FIGURE 1. World distribution of plague, 1970–2003. Countries reporting human cases of plague are shaded light gray; regions with enzootic plague in wild rodent populations are shaded dark gray. Map courtesy of the Centers for Disease Control and Prevention.

and usually become infected when they come into contact with infected domestic or wild rodents or their fleas. Infected rodents, even those with lung involvement, do not generate aerosolized respiratory droplets, precluding direct transmission to other rodents or to humans.[6] Thus, among the many mammals that are susceptible to plague, direct transmission by the aerosol route appears to be an important means of transmission only in humans.

Despite being the least common form of plague in humans, pneumonic plague rightly commands our attention. Few infectious diseases are uniformly fatal when untreated, but pneumonic plague is one of them. If not diagnosed and treated rapidly, death can occur as early as 24 hours after the onset of the disease. A single case of human pneumonic plague also represents a public health emergency, because in that form the disease is directly transmissible. Furthermore, if plague were used as a bioweapon, it would likely be in the form of aerosolized *Y. pestis*, with the intent of causing a large outbreak of primary pneumonic plague.[3,7] This chapter will discuss the microbiology, pathogenesis, diagnosis, treatment, and epidemiology of pneumonic plague, drawing upon reports of human outbreaks as well as studies incorporating animal models of the disease.

2. PLAGUE IN HUMANS

2.1. Flea-Borne Plague

The three major clinical forms of plague are bubonic, septicemic, and pneumonic plague.[8] Other forms do occur, but they are rare. The different forms are not always distinct, but often overlap, with one form developing into the next during the course of the disease. Bubonic plague, the most common type of plague in humans, occurs when bacteria are transmitted by fleabite. From the fleabite site, the bacteria enter the lymphatics and migrate to the regional lymph nodes where they multiply, causing severe lymphadenopathy. This results in edema and swelling of the infected lymph nodes (the "buboes" which characterize bubonic plague). Transient bacteremia can develop when the bacteria escape the lymph nodes and migrate to other organs, such as the liver and the spleen. If properly diagnosed, bubonic plague can be successfully treated with appropriate antibiotics at this stage of the disease. Frequently, however, dissemination of the bacteria leads to high-density bacteremia (septicemic plague), producing endotoxic shock and disseminated intravascular coagulation, often resulting in death in spite of antibiotic treatment.

Sometimes, infection with *Y. pestis* results in high bacteremia without evidence of a palpable lymphadenopathy. This is referred to as primary septicemic plague. Clinically, this often resembles other Gram-negative septicemias. Patients may suffer fever, chills, headache, vomiting, and/or diarrhea. Primary septicemic plague may be more dangerous because a patient with undifferentiated sepsis might be treated with antibiotics that are not effective against *Y. pestis.*

In some bubonic and septicemic plague cases, the bacteria are able to invade the pulmonary tissue from the blood, inducing a neutrophilic inflammatory response.[8–10] As they multiply, the bacteria sometimes enter the alveolar spaces, resulting in secondary pneumonic plague. For secondary pneumonic plague to develop, the patient is usually severely ill for several days, and often the patient dies before invasion of the alveolar spaces can occur. Patients who do survive long enough for a well-developed pneumonia are often very weak, and therefore their cough reflex is not vigorous enough to produce finely aerosolized droplets. Sputum production also tends to be scant and viscous,[10] which also reduces the frequency of subsequent transmission. The pathology of secondary pneumonic plague shares many features with primary pneumonic plague (discussed below), but the secondary form can sometimes be distinguished based on: (i) more widely scattered, diffuse foci of infection in the lung tissue that show greater inflammation and necrosis; (ii) less coughing, and production of less copious, more viscous sputum that contains fewer bacteria; (iii) less mucosal and submucosal hemorrhage in the airways and more bronchial ulceration; and (iv) more evidence of disease in other organs, such as the presence of buboes, tonsillar infection, or pharyngitis, which are usually seeded from cervical buboes.[10] Secondary pneumonic plague patients are able to spread the infection via respiratory droplets to their close contacts, who may develop primary pneumonic plague. Essentially, all naturally occurring pneumonic plague outbreaks arise from cases of bubonic plague with secondary lung involvement.

2.2. Aerosol-Transmitted Plague

2.2.1. Clinical Features and Pathology

In contrast to secondary pneumonic plague, primary pneumonic plague results from inhalation of infected respiratory droplets directly into the lower respiratory tract, and the lung is the initial site of infection. In humans, the incubation period for primary pneumonic plague is usually between two and four days, but can be as short as 24 hours.[11] The first clinical signs are not easily detected and are nonspecific, and may include fever, chills, headache, dizziness, body pains, and chest discomfort. Gastrointestinal symptoms are also common. This initial stage of the disease, characterized only by general signs, typically lasts 24–36 hours. The second stage of the disease begins with a persistent dry cough, and sputum production is limited initially.[9] In these early stages, the bacteria are primarily in the bronchioles, but also in the peribronchial lymph spaces, alveoli, the interlobar septa, and beneath the pleura.[11] They trigger congestion in the nearby blood vessels and soon the epithelium of the alveoli becomes swollen and an exudate containing erythrocytes, leukocytes, and bacteria appears in the lumen. As shown in Figure 2A, polymorphonuclear leukocytes (PMNs) are also recruited to the area and in some parts of the lung become the predominant cell type. PMN recruitment is sometimes accompanied by a reduction in the number of bacteria at these sites. In contrast to secondary pneumonic plague, desquamated alveolar epithelial cells are rarely observed.[9]

The disease progresses quickly, and by the second day there is an increase in chest pain, accompanied by cough, tachypnea, and dyspnea. The bacteria rapidly fill the alveolar spaces and can move from one alveolus to the next through the intraalveolar pores (Figure 2B). Abundant sputum is produced that is thin and syrupy, sometimes frothy, bloody, or tinged with blood.[9] Bloody sputum contains large numbers of *Y. pestis*, nearly in pure culture, and it is during this stage that the patient is most infectious. At this stage of primary pneumonic plague, the sputum is rarely mucopurulent, in contrast to the early stages or in secondary pneumonic plague.

As the infection grows, there is often segmental consolidation followed by bronchopneumonic spread to other parts of the same or opposite lung.[12] The damage can be so severe as to cause liquefactive necrosis and/or cavitation, which can leave severe scarring should the patient survive. Lung sounds typical of pneumonia are apparent on auscultation, but are variable and may not be pronounced despite the severity of infection. As with other forms of plague, the bacteria are adept at entering the bloodstream and spreading to other organs.[10] The case fatality rate of primary pneumonic plague is greater than 95% without antibiotic therapy, and the mean time to death is only three to four days after infection (range two to nine days).[11] At the terminal stage of disease, patients become cyanotic and assume a dazed or anxious appearance.

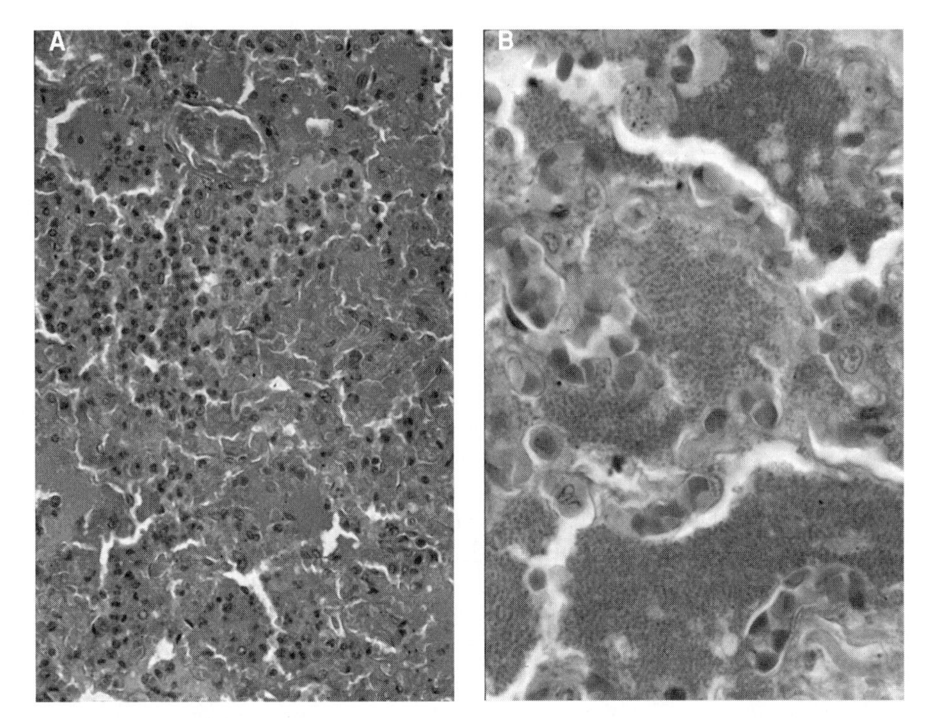

FIGURE 2. Human lung tissue showing typical histopathology of primary pneumonic plague. (A) Dense proteinaceous exudate with infiltration of inflammatory cells, primarily neutrophils (PMNs) with occasional macrophages and lymphocytes, (hemotoxylin and eosin stain); (B) Higher magnification image showing dense fields of *Y. pestis* (stained dark brown). The bacteria fill the alveolar spaces and spill into adjacent alveoli through the intraalveolar pores, (Brown–Hopps Gram stain). Photographs courtesy of Dr. Douglas Wear, Armed Forces Institute of Pathology.

Delirium or coma sometimes occurs, but patients usually remain conscious until death is near.[11] In most instances death results from overwhelming sepsis and cardiac failure, often sudden and brought about by slight exertion.

A less common form of primary pneumonic plague, termed pulmonary plague, can result when the onset of septicemia is particularly rapid.[9,11] In these cases, even though infection is by the respiratory route and the primary site of infection is the lungs, death from septicemia occurs before the alveolar congestion and edema typical of pneumonia can fully develop. Therefore, pulmonary plague is rapidly fatal but presents little risk of transmission, because no sputum or cough is induced.

In most fatal cases of pneumonic plague in humans, necrotic pneumonia, hemorrhage, pleural effusion, fluid in the airways, suppuration, pleuritis, and numerous bacteria in most of the alveoli is evident at autopsy.[8–10] Fibrin deposits are often located in the areas of pneumonia. Changes in the upper respiratory tract, however, are usually less dramatic. In most instances, there is congestion of the tracheal and bronchial mucosa and sometimes a bloody and frothy serous fluid covers the bronchial walls.

Lesions in other organs that follow seeding of the blood stream are similar to those that occur in bubonic plague. In some cases, there may be fibrin thrombi in the renal glomerular capillaries,[13] likely as a result of disseminated intravascular coagulation.[14] In the lymph nodes that drain the infected lobes, there is often liquefactive necrosis, hemorrhage, and large numbers of bacilli. Bacteria may also be found in the spleen, along with cytolysis in the red pulp, hemorrhage, necrosis, and fibrin deposits.

2.2.2. *Transmission*

The aerobiology underlying person-to-person transmission of pneumonic plague is not particularly well characterized. The size range of the particles expelled by a coughing patient is an important determinant of the infectivity of pneumonic plague. Respiratory droplets less than 5 μm in size are most likely to be deposited and retained in the lower respiratory tract and alveoli, large particles greater than 10 μm are mostly trapped in the nasal turbinates, and intermediate-sized particles in the trachea and bronchioles.[15] Strong and Teague[16] and Strong, Cromwell, and Teague,[13] who were the first to demonstrate airborne spread of the disease, estimated that coughing patients might propel *Y. pestis*-laden droplets up to several yards at most, suggesting that primarily coarse particles were generated. They also observed that dispersal of the bacteria occurred more readily during coughing than during talking or breathing. However, the distribution of particle sizes in the aerosol cloud generated by a human with pneumonic plague has never been carefully analyzed, and may range from fine to coarse particles.

Those exposed may develop a local infection, depending on whether the bacteria are deposited on the eyes, throat, skin, or upper respiratory tract. In cases of peripheral infection, lymphadenopathy, and subsequent bacteremia may develop, similar to flea-borne transmission of plague. Coarse droplets are likely to lodge on the upper respiratory tract, and may result in tonsillar or septicemic plague, whereas finer droplets inhaled into the lower respiratory tract cause primary bronchopneumonia.

2.2.3. *Clinical Laboratory Diagnosis*

Since the initial symptoms of primary pneumonic plague are similar to many other viral and bacterial respiratory tract infections, it is often difficult to diagnose. Direct smears of blood or sputum often show large numbers of Gram-negative bacteria and PMNs, but this is not specific for plague. Wright's, Giemsa's, or Wayson's stains of sputum or blood smears sometimes show the typical bipolar or safety-pin staining pattern that is typical of *Y. pestis.*[2] A presumptive diagnosis of pneumonic plague can be made by detection of

Y. pestis-specific antigen or nucleic acid in clinical specimens. For example, the F1 capsular antigen that surrounds *Y. pestis* can be specifically visualized in sputum and blood smears by direct fluourescent antibody assay using fluorescein-labeled anti-F1 antibody.[10] A more rapid method, suitable for use in facilities that lack fluorescence microscopy capability, uses an anti-F1 impregnated dipstick and a colorimetric readout to detect *Y. pestis* in clinical samples.[17] Although these tests specifically detect *Y. pestis*, they are limited in that the F1 capsule is produced primarily at 37°C; therefore, refrigeration of samples for more than 30 hours may result in false-negative results.[1] Additionally, F1-negative *Y. pestis* strains have been reported, which, although rare, are capable of causing pneumonic infection.[18] Polymerase chain reaction (PCR) methods have also been developed to rapidly amplify and detect *Y. pestis*-specific DNA sequences in sputum samples.[19]

A definitive diagnosis of plague can be obtained by culture and identification of *Y. pestis* from sputum or blood, and by serology. *Y. pestis* grows well in most standard biological culture media and in automated blood culture systems. In addition to sputum culture, three blood cultures should be drawn over a 45-minute period. *Y. pestis* colonies, which usually appear within 48 hours on solid media, are opaque and smooth, with irregular edges and a raised center.[1] In liquid, growth is nonturbid, and usually forms a floccular deposit.[5] Biochemical tests[20] and PCR methods[21–23] can be used to identify isolated bacteria. Recently, additional rapid tests to identify *Y. pestis* cultures have been developed, such as a fiber optic biosensor assay to detect the F1 antigen,[24] and fluorescent monoclonal antibodies raised against the Pla protease.[25] Lysis by specific bacteriophages can also be used to definitively identify *Y. pestis* cultures.[1]

Although not rapid, documentation of a significant (fourfold) rise in antibody titer against *Y. pestis* antigens by serological testing is used to confirm a plague diagnosis. Seroconversion usually occurs at the end of the first week of the infection. Several approaches can be used to detect and quantify antibodies raised against the F1 antigen in patient sera, such as the passive hemagglutination test and the more sensitive and specific enzyme-linked immunosorbent assay (ELISA).[1,2]

2.2.4. *Treatment*

The World Health Organization[2] recommends that when plague is suspected based on clinical and/or epidemiological observations, the patient should be started on antimicrobial therapy immediately, without waiting for laboratory confirmation (Table I). There is a lack of published trials evaluating different antibiotics in treating human plague. The antibiotics that are most commonly recommended include the aminoglycosides streptomycin and gentamicin. Streptomycin has historically been the preferred choice for treating plague, and is effective when administered early. However, it is in short supply in

TABLE I
Pneumonic Plague Treatment Guidelines[a]

Drug	Dosage	Interval (hours)	Route[b]
Streptomycin			
Adults	2 g/day	12	IM
Children	30 mg/kg/day	12	IM
Pregnant women	Not recommended		
Gentamicin			
Adults	5 mg/kg/day	8	IM or IV
Children	6.0–7.5 mg/kg/day	8	IM or IV
Pregnant women	5 mg/kg/day	8	IM or IV
Doxycycline			
Adults	200 mg/day	12 or 24	PO or IV
Children	4.4 mg/kg/day	12 or 24	PO or IV
Pregnant women	200 mg/day	12 or 24	PO or IV
Ciprofloxacin			
Adults	1000 mg/day	12	PO or IV
Children	40 mg/kg/day	12	PO or IV
Pregnant women	1000 mg/day	12	PO or IV
Chloramphenicol			
Adults	50 mg/kg/day	6	PO or IV
Children (>2 yrs)	50 mg/kg/day[c]	6	PO or IV
Pregnant women	50 mg/kg/day[c]	6	PO or IV

[a] Adapted from Dennis *et al.*[2] and Inglesby *et al.*[3]
[b] IM = Intramuscular; IV = Intravascular; PO = Orally.
[c] Concentration should be maintained between 5 and 20 μg/ml to prevent bone marrow suppression.

the United States whereas gentamicin is generally available and is likely at least as effective. These drugs may not be ideal in a widespread epidemic because they cannot be given orally. Tetracycline, doxycycline, and chloramphenicol have been proposed for treatment and prophylaxis in such instances.[3] Fluoroquinolones such as ciprofloxacin have been shown to be effective in treating mice with pneumonic plague,[26] but have not been tested in human cases. These drugs can also be orally administered.

3. BACTERIAL PATHOGENESIS

3.1. Experimental Models of Pneumonic Plague

A variety of mammals have been used as animal models to study the pathogenesis of pneumonic plague, including guinea pigs,[27,28] mice,[26,29–31] rats,[32] small and large marmots,[33,34] rabbits,[35] cats,[36] camels,[11] and several nonhuman primates.[18,37–42] The mouse, guinea pig, and nonhuman primate models are the best characterized and closely resemble human pneumonic plague in general pathology. Various infectious challenge methods have also been developed, including nose or head-only aerosol exposure, whole-body aerosol exposure, and delivery of bacterial suspensions either through intranasal or intratracheal instillation.

In mice, infection with small, aerosolized particles consistently results in the development of primary pneumonic plague.(31) Infection with larger or variably sized particles can lead to other additional forms of plague that involve the cervical lymph nodes and bacteremia, but does not always result in pneumonia.(6) Mice infected by intranasal instillation also develop pneumonic plague.(29) The disease progression in mice and humans is similar. Within 12–24 hours there is focal cellular infiltration of the peribronchial lymphatics. Between 24 and 36 hours lobular bronchial pneumonia develops, and an exudate may be present in the alveoli, which contains numerous bacteria. At approximately 36 hours, septicemic spread of the organism leads to colonization of other organs; at this time bacteria can be readily cultured from the spleen and bone marrow.(29)

Guinea pigs were used in early studies of pneumonic plague.(16,43) Suspensions of bacteria were sprayed into the air, which resulted in infections resembling those seen in mice exposed to large particles; i.e., cervical lymph node and laryngeal involvement with secondary pneumonic infection. Druett *et al.*(27) later systematically evaluated the effect of aerosol particle size on pneumonic infection in guinea pigs. Only infectious particles smaller than 1 μm caused bronchopneumonia; particles greater than 4 μm did not reach the lung. Larger particles were deposited in the upper airways and caused infection of the cervical nodes and larynx. In later studies, intratracheal or intranasal instillation of culture suspensions also successfully produced pneumonic plague in guinea pigs. It was observed that some of the guinea pigs infected by these methods transmitted the disease to their cage mates, which has not been observed with mice.(29) Other differences with the guinea pig model were that the infection remained confined to the respiratory tract, and that the animals did not become visibly ill until "sudden death", typically 72–96 hours after infection.(29)

The nonhuman primate species that have been most frequently used in pneumonic plague studies are the African green monkey (*Chlorocebus (Cercopithecus) aethiops*) and various species of macaques (*Macaca fascicularis, M. mulatta,* and *M. rhesus*). Progression of the disease in monkeys is also very similar to human pneumonic plague. Clinical signs after infection include fever and tachypnea, anorexia, dyspnea, and rales.(37,38,44) Pathological changes also resemble those in human disease, and may include pulmonary alveolar flooding, hemorrhage, fibrinous pleuritis, and fibrin deposits in the blood vessels. Bacteremic spread is typical and bacteria may be found in the spleen. As with the small-animal models of pneumonic plague, infection of monkeys via aerosol or intranasal instillation can result in pneumonia, but nonpneumonic forms may also develop if the bacteria invade via the upper respiratory tract.[6,16] Alternatively, bacteria can be introduced directly by laryngoscopy using a catheter, to ensure that the bacteria reach the lung.[38,44] As with guinea pigs, cross-infection of contacts of monkeys with pneumonic plague does occur, even when the animals are individually caged so that no physical contact but free passage of exhaled air is allowed.[6]

Cats are unique among other predators in their susceptibility to plague. They also have a propensity to develop lung infection secondary to lymphadenopathy in submandibular or cervical nodes, or tonsils infected as a result of feeding on infected rodents.[36,45] Cats infected by needle inoculation develop a diffuse interstitial pneumonia, with large areas of necrosis and infiltrates similar to humans with secondary pneumonic plague.[36] Although cats with secondary pneumonic plague can readily transmit the infection to humans, it is not known whether cats themselves are susceptible to primary pneumonic infection.

3.2. *Y. pestis* Virulence Factors

Plague is an acute, fulminant, and often rapidly fatal disease, and *Y. pestis* is justifiably one of the most feared of all bacterial pathogens. The extreme virulence of *Y. pestis* depends on bacterial factors that effectively eliminate the host's ability to localize the infection and to initiate an appropriate immune response, thus allowing unchecked bacterial replication and systemic invasion. The bacterial factors important in *Y. pestis* disease are listed in Table II. Chief among them is a sophisticated, multicomponent Type 3 secretion system that functions to inject cytotoxic proteins directly into host phagocytic and immune cells. All three pathogenic *Yersinia* species have this system, which is encoded on a highly conserved ~70-kb virulence plasmid (pYV). Virulence factors encoded by pYV prevent the host from responding appropriately to infection. This is accomplished partly by preventing phagocytosis directly, and partly by inhibiting the inflammatory response that leads to activation of other phagocytes and immune cells. The cytotoxic effector proteins, termed Yops (Yersinia outer proteins) are injected directly into the host cell cytosol following contact with the eukaryotic cell. Once inside, they block or redirect eukaryotic cell signaling cascade pathways to disrupt the normal response to infection (reviewed in Cornelis, 2002).[46] YopE, YopT, YpkA, and YopH all target the actin cytoskeleton. For example, YopE disrupts actin microfilaments and thus causes rounding of target cells by inhibiting Rho GTPases. YopT disrupts the host cell cytoskeleton by cleaving RhoA, Rac-1, and Cdc42 from the cell membrane. YpkA also targets RhoA and Rac1, through its serine/threonine kinase activity. YopH prevents assembly of macrophage focal adhesions and blocks calcium signaling in neutrophils through its protein-tyrosine phosphatase activity. Other Yops do not target the actin cytoskeleton per se, but inhibit normal

TABLE II
***Y. pestis* Virulence Factors and Their Role in Pneumonic Plague**

Virulence Factor	Mechansim of action	Function	Virulence of mutant in pneumonic infection
Type III secretion system	Injection of effector Yop proteins	Interference with innate immunity	Not tested
YopE	Inhibits Rho GTPases	Disrupts actin filaments	Not tested
YopT	Cleaves RhoA, Rac-1, Cdc42	Disrupts actin filaments	Not tested
YpkA	Inhibits RhoA, Rac-1	Disrupts actin filaments	Not tested
YopH	Protein tyrosine phosphatase, blocks calcium signaling	Disrupts focal adhesion complexes and oxidative burst of phagocytes	Not tested
YopJ	Binds MAPK kinases, IKKβ	Inhibits TNF-α, triggers apoptosis	Not tested
YopM	Unknown	Reduces NK cells	Not tested
LcrV	Regulates Yop expression, triggers IL-10	Yop delivery, immunosuppression	Not tested
Ybt locus	Synthesis and uptake of yersiniabactin siderophore	Iron acquisition	Slightly attenuated (*pgm-*)
Yfe	ABC transporter	Iron acquisition	Not tested
Psa	Binds to glycolipids	Fimbrial-like adhesin	Not tested
Fraction 1 (capsule)		Inhibits phagocytosis; adhesin?	Fully virulent
Plasminogen activator	Cell-surface protease that cleaves fibrin and complement	Dissemination from peripheral sites, adherence to basement membranes	Slightly attenuated (*pgm-pla* double mutant is avirulent)

cytokine signaling patterns. YopJ binds to mitogen-activated protein kinases, downregulating TNF-α production. It also prevents NFκB activation, thereby interfering with antiapoptotic pathways in macrophages.[47,48] YopM may reduce the number of natural killer cells by interfering with IL-15 signaling.[49] Thus, the Yop effector proteins act at multiple levels, to kill phagocytes, inhibit their phagocytic activity, and prevent recruitment of additional immune cells.

The multifunctional LcrV protein is also encoded by pYV. LcrV acts as a regulator of Yop expression and secretion,[50] and also as a virulence factor.[51] LcrV, together with the YopB and YopD proteins, facilitates translocation of the effector Yops, by forming a channel through which the effector proteins pass.[52] Extracellular LcrV released by the bacteria into infected tissues is also predicted to have an antiinflammatory effect because it induces the production of cytokine IL-10, which inhibits expression of the proinflammatory cytokines

TNF-α and IFN-γ.[51] The ensuing down-regulation of the inflammatory response would reduce the recruitment of professional phagocytes and other immune cells to the site of infection.

In addition to the pYV virulence plasmid, *Y. pestis* contains two additional plasmids that are not found in other *Yersinia* species.[1] The ~100-kb pFra contains the *ymt* phosopholipase D gene required for survival in the flea as well as the *caf* locus, which encodes the genes required for synthesis of an extracellular protein capsule.[53] The capsule is a gel-like aggregate of multimeric fraction 1 capsule protein. The role of capsule in pathogenesis of plague is not known. It may prevent phagocytosis,[54] but could also prevent secretion of Yop proteins by the bacteria. Presence of the capsule affects the time of onset of disease, but is not essential for virulence in mice, guinea pigs, or monkeys,[55,56] and strains lacking capsule have been isolated from natural infections.[57] No other genes on pFra have been shown to affect the virulence of *Y. pestis* in mammals or its transmission by fleas.

The other *Y. pestis*-unique plasmid, pPla, contains the *pla* gene, encoding the outer membrane plasminogen activator protein.[1] Pla has protease activity against a number of different substrates, including fibrin and complement.[58] It also may play a role in adherence to host basement membranes.[59] In the mammalian host, Pla is necessary for efficient dissemination from peripheral injection sites, but not when injected directly into the bloodstream.[58]

Several chromosome-encoded functions also affect the virulence of *Y. pestis.* These include mechanisms to scavenge iron within the mammalian host and the pH 6 antigen. The yersiniabactin siderophore-dependent system (Ybt) encoded on the high-pathogenicity island[60] and the siderophore-independent ABC transporter system (Yfe)[61] have been shown to be required for full virulence in mice. A third iron-transport-system (Yfu) does not affect virulence in mice by needle inoculation.[62] The pH 6 antigen (Psa) is a fimbria-like protein that is expressed at a pH below 6.7 and at temperatures above 35°C.[1] Inactivation of Psa leads to reduced virulence.[63] Psa has been shown to bind to β1-linked galactosyl residues in glycosphingolipids[64] but its role in plague pathogenesis has not been determined.

3.3. Role of *Y. pestis* Virulence Factors in Pneumonic Plague

Despite the fact that many of the *Y. pestis* virulence factors described above have been known for many years, with a few exceptions their function has not been explicitly tested in pneumonic plague models (Table II). In fact, very little is known about the molecular mechanisms of pathogenesis of pneumonic plague in any host. For example, the chromosomal *pgm* locus contains the high-pathogenicity island that includes the Ybt iron acquisition and transport proteins. Strains lacking this ~102-kb region have been tested in mice and in African green monkeys.[42] Although Pgm-negative strains are considered avirulent when injected subcutaneously, allowing them to be used as live vaccines, they appear to be less attenuated by aerosol infection. Half of monkeys challenged with Pgm-negative *Y. pestis* aerosols died, although the pathology of the infection differed slightly from infection with wild-type bacteria.[42] This observation implies that the Ybt iron acquisition system contributes to but is not absolutely required for pneumonic infection, but this has not been confirmed directly.

The role of the plasminogen activator protein (Pla) has also been investigated. Infection of mice with aerosols of wild-type *Y. pestis* CO92 or derivatives cured of the plasmid pPst containing the *pla* gene showed that the mutant strain was only slightly attenuated, as the LD_{50} of the pPst- strain was just four times higher.[42] Conversely, a Pgm- version of *Y. pestis* CO92 with a deletion/frameshift in the *pla* gene (Pgm-, Pla-) was unable to kill any African green monkeys at any dosage tested.[42] Significantly, some natural Pestoides isolates of *Yersinia pestis* (strains believed to be more ancient than the other *Y. pestis* biotypes) that lack *pla* also appear to be fully virulent in mice infected by aerosol or subcutaneous injection.[65] Pestoides strains may have another gene that is functionally equivalent to *pla*.

The contribution of the capsule (F1 antigen) to pneumonic infection has also been addressed. Infection of African green monkeys with either *Y. pestis* CO92, its derivative containing nonsense mutation in the *caf* locus, or a naturally occurring F1-negative strain resulted in similar mortality rates, time to death, and pathologic effects.[18] Hence, capsule is not required for pneumonic plague, at least in African green monkeys.

It is almost certain that LcrV and at least some of the Yop effector proteins are necessary for pneumonic plague infection, but their role in lung infection has not been directly tested. However, mice infected by *Y. pestis* CO92 aerosols and then treated with antibiotics produce antibodies against LcrV, YopH, and YopM, and to a lesser extent YopB, YopD, and YopK, but not to YpkA, YopN, or YopE.[66] This suggests that LcrV and some Yops are expressed in pneumonic infections. Likewise, it is also probable that one or more of the iron acquisition systems is necessary for pneumonic infection, but this also has not been tested.

Other possible virulence factors for pneumonic plague include one of the three quorum-sensing systems (2 LuxI/R systems plus a LuxS homologue). Quorum-sensing systems contribute to the virulence of many lung pathogens, including *Pseudomonas aeruginosa*,[67] *Burkholderia cenocepacia*[68] *B. mallei*[69] and *B. pseudomallei*.[70] The *Y. pestis* quorum-sensing systems do not affect virulence in mice infected by needle injection, or transmission by fleas,[71] but may be involved in lung infection. For instance, the effect of *B. mallei* quorum sensing systems on virulence appeared to be greater in mice challenged by aerosols than in hamsters challenged by intraperitoneal injection,[69] which could be due to host-specific factors, but could also be related to the route of infection. In addition to quorum sensing, leukocyte adhesion systems might be important in pneumonic plague. For instance, the *Y. pestis* genome contains paralogues of the filamentous hemagglutinin (FHA) adhesin of *Bordetella pertussis.*[72] Tight adhesion to host cells is required for Yop translocation and secretion, and the bacterial structures involved have not been identified, particularly those for pneumonic infections.

4. BACTERIAL GENETICS

The genus *Yersinia* is composed of 10 species of Gram-negative coccobacilli. Only three of them, *Y. enterocolitica, Y. pseudotuberculosis,* and *Y. pestis,* are pathogenic to humans. Both *Y. enterocolitica* and *Y. pseudotuberculosis* cause self-limiting food- and water-borne enteric diseases, whereas *Y. pestis* causes highly lethal invasive disease and is primarily arthropod-borne. Despite their vastly different pathogenic and epidemiological features, *Y. pestis* and *Y. pseudotuberculosis* are nearly identical genetically.[73] It has been proposed that *Y. pestis* evolved from its *Y. pseudotuberculosis* ancestor only within the last 20,000 years.[74] The diseases caused by these two organisms do share some similarity in that they both have a propensity to cause lymphadenitis, but *Y. pseudotuberculosis* is much less invasive and rarely causes bacteremia. *Y. pseudotuberculosis* can cause abscesses in the lungs and other internal organs as a complication of bacteremia,[75] but no instances of pneumonic transmission have been reported.

Traditionally, *Y. pestis* has been divided into three biovars based on the ability to ferment glycerol and arabinose, and to reduce nitrate.[76] These metabolic properties do not correlate with virulence. It was suggested by Devignat that the three biovars were responsible for the different epidemics; that is, biovar antiqua is descended from bacteria that caused the Justinian plague, biovar mediaevalis from strains that caused the Black Death, and biovar orientalis is associated with modern plague, but there is no direct evidence for this.[76] Analyses of genetic diversity within the species suggest that biovar orientalis is uniform, most mediaevalis strains are similar, while there is greater diversity within antiqua strains.[77] Recently, a fourth biovar (microtus) has been proposed for a group of strains isolated in China which are not pathogenic to humans.[77] Considerable genetic diversity has also been observed in strains circulating in Asia.[78]

Comparisons of the genome sequences of a *Y. pestis* biovar orientalis strain,[53] a biovar mediaevalis strain[79] and a *Y. pseudotuberculosis* serotype I strain[73] reveal that the major genetic difference among them is their plasmid content. There are relatively few *Y. pestis*-specific genes, although compared to *Y. pseudotuberculosis, Y. pestis* has a higher number of pseudogenes, insertion sequences, and numerous chromosomal rearrangements that could alter gene function.[73] Comparative genomics analyses of *Y. pestis* antiqua, mediavalis, and orientalis isolates have revealed strain-related differences in genetic content and arrangement.[80] To date, none of these genetic differences have been shown to be responsible for differences in host range or pathogenicity, but this will likely become a subject of active research.

Because of the overall genetic homogeneity among *Y. pestis* populations worldwide, strain discrimination requires high-resolution molecular techniques. Molecular strain-typing methods include ribotyping,[81] insertion sequence (IS) mapping,[82] and profiling hypervariable regions on the *Y. pestis* chromosome known as variable-number tandem repeats.[83] These methods would be useful in epidemiological studies to provide a molecular fingerprint of the *Y. pestis* strain, or strains circulating during a natural or bioterrorist-generated plague outbreak.

5. EPIDEMIOLOGY OF PNEUMONIC PLAGUE

Outbreaks of primary pneumonic plague in human communities typically arise from a prior outbreak of flea-borne bubonic plague that generates index cases of secondary pneumonia. Seeding of the lungs by hematogenous spread is fairly common in many animals with bubonic and septicemic plague, but the extent of lung infection is usually limited and does not regularly induce the production of infectious aerosols.[5,9] In humans, transmissible secondary plague pneumonia is a relatively rare complication that develops in less than 5% of bubonic and septicemic plague patients.[8] Unlike humans, most rodents with lung infection do not transmit pneumonic plague, because their respiratory tract anatomy is not conducive to the exhalation of infectious particles.[6] Cats, however, develop transmissible secondary pneumonic plague, and several cases of primary pneumonic plague in humans have been traced to veterinary examination of infected cats.[45]

Historical records suggest that each of the three great plague pandemics consisted of a mixture of both flea-borne bubonic plague and human-to-human transmitted pneumonic plague, as do plague epidemics in modern times.[11] Thus, naturally occurring outbreaks begin with bubonic plague, which sets up the potential for a subsequent, self-sustaining pneumonic plague outbreak that is independent of rodents and fleas. However, the proportion of primary pneumonic to bubonic plague cases varies widely from outbreak to outbreak, and the incidence of pneumonic plague can also vary greatly during the course of a single epidemic. During the last pandemic, for example, pneumonic plague accounted for only a small percentage of the millions of cases in southern India, southern China, and Indonesia, but large epidemics consisting entirely of pneumonic plague occurred in Manchuria. In addition, pneumonic plague is thought to have been prevalent only during certain periods of the medieval European pandemic.[11] Of the 390 cases of plague reported in the United States from 1947 to 1996, only 2% were pneumonic plague.[3]

5.1. Biological and Epidemiological Determinants of Pneumonic Plague Epidemics

Several factors have been proposed to account for the disparity in pneumonic plague incidence rates that is observed among outbreaks in different plague foci. One early theory was that different *Y. pestis* strains have intrinsic differences in pathogenicity or tropism for lung tissue. However, *Y. pestis* isolated from pneumonic plague patients do not show enhanced virulence or pneumotropism in animal models compared to isolates from bubonic cases.(6,11) Although no evidence for intrinsic strain differences relevant to pneumonic plague transmission or pathogenesis has yet been reported, the theory is still raised, particularly by Russian investigators who detect considerable strain heterogeneity among *Y. pestis* strains circulating in central Asian plague foci.(78) For example, the claim has been made that a *Y. pestis* strain or subspecies associated with marmots in the former Soviet Union is more apt to cause pneumonic plague in humans than are other subspecies.(84) Regardless of whether or not there are genetic differences among strains that relate to pneumotropism, it is likely that *Y. pestis* growing in the lungs express a phenotype that optimizes survival in aerobic conditions and enhances subsequent aerosol transmission and primary infectivity in the lung environment.

There is likewise no evidence for human genetic, age, or sex differences in susceptibility to pneumonic plague.(11,5) Thus, whether or not pneumonic plague spreads or not depends primarily on socioeconomic, climactic, and other extrinsic factors, and not on factors intrinsic to the bacteria or human host. Pneumonic plague has a reputation for being highly contagious, and in the right conditions it is so. However, the estimated transmission rate of primary pneumonic plague is rather low compared to other infectious diseases. The average number of new cases generated per primary case (R_0) in eight pneumonic plague outbreaks from 1907 to 1997 was 1.3.(85) Correspondingly, the number of people who develop primary plague pneumonia after exposure to an infectious patient is usually low. For example, only about 8% of all contacts and 22% of persons with prolonged contact with sick patients developed the disease during the 1920–1921 Manchurian pneumonic plague epidemic.(11) The short infectious period of pneumonic plague, the virtual absence of asymptomatic carriers, and the necessity for close person-to-person contact all contribute to the modest transmission rate by the airborne route. Transmission from an index case of secondary pneumonic plague is probably even less efficient than subsequent transmission of primary pneumonic plague, because patients with secondary pneumonic plague exhibit more variation in extent of lung involvement, produce a more viscous sputum, and cough less than patients with primary pneumonic plague.(10,11) However, plague transmission dynamics would be completely different in the case of illegitimate release of an aerosolized *Y. pestis* bioweapon. It has been estimated that airborne release of 50 kg of *Y. pestis* over a large city could infect 150,000 people.(3)

Pneumonic plague transmission depends on inhalation of airborne particles containing *Y. pestis* into the lower respiratory tract. The cough of a pneumonic plague patient constitutes the major infectious threat, because large numbers of plague bacilli are disseminated during coughing, even when no visible sputum is expelled.(11)

Determining the number of *Y. pestis* that must be inhaled to cause pneumonia in 50% of susceptible animals (the ID_{50}) is difficult, but estimates vary from 10^2 to 10^4.(7,39,42) *In vitro*-cultured bacteria are typically used for these challenges, however, and it is possible that the number may be lower if *Y. pestis* exhaled from an infected lung are pre-adapted for growth in the respiratory tract. Another factor that tends to limit airborne transmission is that aerosolized *Y. pestis* in aqueous diluents quickly lose viability through dessication, especially at low relative humidity.(86,87) *Y. pestis* does remain viable in drying sputum for 2 to 3 days if not exposed to sunlight, but contact with sputum-soiled clothes or bedding does not appear to be a major mode of infection.(11,88)

The most important risk factors for transmission, which have been well-documented during twentieth century pneumonic plague outbreaks, are overcrowding and close proximity to an infected person in an enclosed space with poor ventilation. Thus, pneumonic plague transmission occurs primarily indoors, and much epidemiological evidence indicates that close contact is critical. The aerosol cloud containing *Y. pestis*-laden droplets that is expelled by a coughing patient appears to extend a few yards at most, and rapid diffusion in air currents and the fragile nature of *Y. pestis* in small particle aerosols tends to limit transmission.(11,16,88,89) The observation has been made that simply turning the patient to face the wall is an effective way to reduce infections among health care workers.(9,11)

Pneumonic plague is often considered to be a disease of cold climates, and to be rare in warm climates. Likely reasons for this correlation include the fact that the saturation deficit of air (the difference between the saturation water vapor pressure and the actual pressure) decreases with air temperature for any given relative humidity. Dry air at 10°C actually has less drying power than air at 37°C and 80% relative humidity.(89) Thus, rates of dessication and loss of viability of *Y. pestis* are relatively greater at higher temperatures. Nevertheless, pneumonic plague outbreaks have occurred regularly in regions with warm climates, such as Upper Egypt and West Africa.(11) The major epidemiological factor related to cold weather may be that it drives people indoors, often in poorly ventilated, overcrowded rooms, designed to conserve heat conditions, which are conducive to aerosol transmission.

5.2. Management and Control of Pneumonic Plague Outbreaks

Much of our knowledge of pneumonic plague epidemiology and control comes from experience gained during two major epidemics of pneumonic plague in Manchuria, which claimed over 40,000 lives in the fall and winter of 1910–1911 and over 9000 in 1920–1921. Careful studies of the epidemiology, microbiology, and public health control measures were published in a series of reports by the North Manchurian Plague Prevention Service, an international team headed by the Cambridge-educated physician Wu Lien-teh. His detailed summary treatise on pneumonic plague (1926) should be required reading for public health officials concerned with responses to pneumonic plague. Experience gained in Manchuria established a model for management of a pneumonic plague epidemic. One of the many important practical measures pioneered during these epidemics was the design and use of an effective face mask used by medical and public health personnel for personal prophylaxis. The major elements of a public health program to control an outbreak of pneumonic plague include:

- Detection and diagnosis of cases
- Treatment and hospitalization of cases under respiratory isolation
- Quarantine and prophylactic treatment of contacts
- Disinfection of dwellings and personal effects of patients
- Public education campaigns
- Controls on public transportation to prevent spread of the epidemic

The extent to which individual control measures are taken depends on the severity of the outbreak. For example, in the city of Harbin, the central focus of the large Manchurian epidemics, entire facilities were designated as plague hospitals and isolation hospitals, a military cordon was drawn around the city to control traffic, and railroad passengers were carefully screened and quarantined if deemed necessary.[90,91] Reviews containing updated recommendations for medical and public health management of pneumonic plague outbreaks have appeared recently, indicative of the growing concern regarding the potential use of *Y. pestis* as an agent of bioterrorism.[3,7] Plague, along with yellow fever and cholera, is one of three diseases subject to mandatory quarantine control measures under international law.

5.3. Lessons from Modern Pneumonic Plague Outbreaks

> "There is probably no infectious disease which, theoretically, is so easy to suppress as lung plague."
>
> Wu Lien-teh[9]

This conclusion, made in the preantibiotic era by one of the great plague fighters of the twentieth century, is logical given the brief period during which the disease is infectious, the proven effectiveness of respiratory isolation of patients, segregation of contacts, prophylactic antibiotic treatment in breaking the transmission chain, and the absence of asymptomatic chronic carriers. Most pneumonic plague outbreaks are in fact quite circumscribed and self-limiting, even with minimal intervention, but in certain conditions explosive epidemics break out, as in Manchuria and fourteenth century Europe.[11,89] Two examples serve to illustrate some of the practical problems that can complicate control of an outbreak.

Manchuria, 1920. As was the case for the great 1910–1911 epidemic, the 1920 pneumonic plague epidemic began in late summer with a small cluster of bubonic plague cases in northwestern Manchuria. These cases probably originated from the neighboring Transbaikal region of Siberia, where a permanent focus of flea-borne bubonic plague exists in wild marmots. When pneumonic plague cases subsequently appeared in the fall, public health authorities acted. Using the control measures listed above and the experience gained during the first epidemic a decade earlier, the initial outbreak was contained and limited to 52 cases. However, nine contacts under quarantine observation forcibly escaped and fled. Two returned to their jobs at a coal mine 100 miles away, where they lived in semiunderground barracks with no sunlight or ventilation. With outside temperatures at −20°C, 60 to 80 men were crowded together in each 20 × 60 foot barracks. These conditions proved ideal for transmission, and over 1000 of the 4000 miners died of pneumonic plague. This reignited the epidemic, which in spite of rigid control measures spread to other towns along railway lines, eventually claiming 9300 lives before subsiding in the spring of 1921.[11,92]

India, 1994. In August 1994, an outbreak of human bubonic plague occurred in Maharashtra State in western India following a plague epizootic in the local rat population. In September, an increasing number of patients with fatal pneumonia were seen in Surat, the capital of the neighboring state of Gujarat. A presumptive diagnosis of pneumonic plague was made for some of these cases, control measures were implemented, and both outbreaks quickly abated.[93,94] Except for the fact that it occurred in a densely populated city and involved more people, the epidemiology of this outbreak was similar to several other outbreaks that have been documented around the world, most recently in South America and Africa.[11,95,96] What made the 1994 Indian outbreak unusual was its psychosocial impact. The lack of specific and rapid diagnostic capabilities led physicians to label many patients with respiratory infection and fever as suspected plague cases. The resulting overreporting and uncertainty over the extent of the outbreak led to anxiety and flight, always a problem in plague control.[97,98] An estimated 300,000 to 500,000 people fled the area, despite the widespread availability and use of effective antibiotic prophylaxis, underscoring the fact that the psychological and social effects of plague can be greater than the medical ones.[93,97] Fortunately, this mass exodus did not result in further spread of the outbreak, even from incubating cases that fled to other large cities. The international response included termination of air transport to and from India by some carriers and increased surveillance of passengers and goods by others.[93,99]

6. CURRENT AND FUTURE NEEDS

Renewed interest in the clinical and public health management of pneumonic plague has come from the recognized potential of aerosolized *Y. pestis* as a bioterrorism weapon.(100) Although much is known about the microbiology, pathology, treatment, and epidemiology of plague, improved medical and public health countermeasures are needed to meet the challenge of pneumonic plague outbreaks, whether naturally occurring or instigated by bioterrorists. All levels of the public health infrastructure are being evaluated and new recommendations have been formulated to optimally respond to acts of biological terrorism.(101) Because it is a zoonosis, routine surveillance programs designed to monitor the incidence and distribution of plague in peridomestic rodent populations can provide a first alert to an increased risk of human plague in an area.(2) The most critical medical needs for better pneumonic plague control include rapid, sensitive methods to detect and identify *Y. pestis* in environmental and clinical samples, and an effective vaccine. The lack of readily available, rapid means of diagnosis is a major weak link in plague control. Currently, no plague vaccine is available. Previously used plague vaccines likely offered inadequate protection against pneumonic plague following aerosol challenge, and such protection is a high priority for any new vaccine.(102)

REFERENCES

1. Perry, R. D., and Fetherston, J. D., 1997, *Yersinia pestis* etiologic agent of plague, *Clin. Microbiol. Rev.* **10**:35–66.
2. Dennis, D., Gage, K. L., Gratz, N., *et al.*, 1999, *Plague Manual: Epidemiology, Distribution, Surveillance and Control,* World Health Organization, Geneva.
3. Inglesby, T. V., Dennis, D. T., Henderson, D. A., *et al.*, 2000, Plague as a biological weapon: medical and public health management. Working Group on Civilian Biodefense, *JAMA.* **283**:2281–2290.
4. Galimand, M., Guiyoule, A., Gerbaud, G., Rasoamanana, B., Chanteau, S., Carniel, E., and Courvalin, P., 1997, Multidrug Resistance in *Yersinia pestis* Mediated by a Transferable Plasmid. *New Engl. J. Med.* **337**:677–680.
5. Pollitzer, R., 1954, *Plague,* World Health Organization, Geneva.
6. Meyer, K. F., 1961, Pneumonic plague, *Bacteriol. Rev.* **25**:249–261.
7. Franz, D. R., Jahrling, P. B., Friedlander, A. M., *et al.*, 1997, Clinical recognition and management of patients exposed to biological warfare agents, *JAMA.* **278**:399–411.
8. Butler, T., 1983, *Plague and Other Yersinia Infections,* Plenum Medical Books, New York.
9. Wu, L.-T., Chun, J. W. H., Pollitzer, R. *et al.*, 1936, *Plague: A Manual for Medical and Public Health Workers,* Mercury Press, Shanghai.
10. Dennis, D. T., and Meier, F. A., 1997, Plague, in: *Pathology of Emerging Infections* (C. R. Horsburgh Jr., and A. M. Nelson, eds.), American Society for Microbiology, Washington, DC.
11. Wu, L.-T., 1926, *A Treatise on Pneumonic Plague,* League of Nations Health Organization, Geneva.
12. Alsofrom, D. J., Mettler, F. A., Jr., and Mann, J. M., 1981, Radiographic manifestations of plague in New Mexico (1975–1980). A review of 42 proved cases, *Radiology.* **139**:561–565.

13. Strong, R. P., Cromwell, B. C., and Teague, O., 1912, Studies on pneumonic plague and plague immunization. VII. Pathology, *Philippine J. Tropical Med.* **7**:203–221.
14. Finegold, M. J., Petery, J. J., Berendt, R. F., and Adams, H. R., 1968, Studies on the pathogenesis of plague. Blood coagulation and tissue responses of *Macaca mulatta* following exposure to aerosols of *Pasteurella pestis, Am. J. Pathol.* **53**:99–114.
15. Smith, L. J., 1998, Aerosols, in: *Textbook of Pulmonary Diseases,* 6th ed. (G. L. Baum, B. R. Celli, J. D. Crapo and J. B. Barlinsky, eds.) Lippincott-Raven Publishers, Philadelphia, pp. 313–320.
16. Strong, R. P., and Teague, O., 1912, Studies on pneumonic plague and plague immunization. VI. Bacteriology, *Philippine J. Sci.* **7**:187–202.
17. Chanteau, S., Rahalison, L., Ratsitorahina, M., Mahafaly, Rasolomaharo, M., Boisier, P., O'Brien, T., Aldrich, J., Keleher, A., Morgan, C. and Burans, J., 2000, Early diagnosis of bubonic plague using F1 antigen capture ELISA assay and rapid immunogold dipstick, *Int. J. Med. Microbiol.* **290**:279–283.
18. Davis, K. J., Fritz, D. L., Pitt, M. L., Welkos, S. L., Worsham, P. L. and Friedlander, A. M., 1996, Pathology of experimental pneumonic plague produced by fraction 1-positive and fraction 1-negative *Yersinia pestis* in African green monkeys (*Cercopithecus aethiops*), *Arch. Pathol. Lab. Med.* **120**:156–63.
19. Loiez, C., Herwegh, S., Wallet, F., Armand, S., Guinet, F. and Courcol, R. J., 2003, Detection of *Yersinia pestis* in sputum by real-time PCR, *J. Clin. Microbiol.* **41**:4873–4875.
20. Bercovier, H. and Mollaret, H. H., 1984, Genus XIV Yersinia, in: *Bergey's Manual of Systematic bacteriology* (Kreig, N. R. and Holt, J. G., eds.), Wiliams & Wilkins, Baltimore, pp. 498–506.
21. Hinnebusch, J., and Schwan, T. G., 1993, New method for plague surveillance using polymerase chain reaction to detect *Yersinia pestis* in fleas, *J. Clin. Microbiol.* **31**:1511–1514.
22. Leal, N. C., Abath, F. G., Alves, L. C., and de Almeida, A. M., 1996, A simple PCR-based procedure for plague diagnosis, *Rev. Inst. Med. Trop. Sao. Paulo.* **38**:371–373.
23. Higgins, J. A., Ezzell, J., Hinnebusch, B. J., Shipley, M., Henchal, E. A., and Ibrahim, M. S., 1998, 5′ nuclease PCR assay to detect *Yersinia pestis, J. Clin. Microbiol.* **36**:2284–2288.
24. Cao, L. K., Anderson, G. P., Ligler, F. S., and Ezzell, J., 1995, Detection of *Yersinia pestis* fraction 1 antigen with a fiber optic biosensor, *J. Clin. Microbiol.* **33**:336–341.
25. Feodorova, V. A., and Devdariani, Z. L., 2000, Development, characterisation and diagnostic application of monoclonal antibodies against *Yersinia pestis* fibrinolysin and coagulase. *J. Med. Microbiol.* **49**:261–269.
26. Byrne, W. R., Welkos, S. L., Pitt, M. L., Davis, K. J., Brueckner, R. P., Ezzell, J. W., Nelson, G. O., Vaccaro, J. R., Battersby, L. C., and Friedlander, A. M., 1998, Antibiotic treatment of experimental pneumonic plague in mice, *Antimicrob. Agents Chemother.* **42**:675–681.
27. Druett, H. A., Robinson, J. M., Henderson, D. W., Packman, L., and Peacock, S., 1956, Studies on respiratory infection. II. The influence of aerosol particle size on infection of the guinea pig with *Pasturella pestis. J. Hyg.* **54**:37–48.
28. Samoilova, S. V., Samoilova, L. V., Yezhov, I. N., Drozdov, I. G., and Anisimov, A. P., 1996, Virulence of pPst+ and pPst− strains of *Yersinia pestis* for guinea-pigs, *J. Med. Microbiol.* **45**:440–444.
29. Meyer, K. F., Quan, S. F., and Larson, A., 1948, Prophylactic immunization and specific therapy of experimental pneumonic plague. *Am. Rev. Tuberc.* **57**:312–321.
30. Smith, P. N., McCamish, J., Seely, J., and Cooke, G. M., 1957, The development of pneumonic plague in mice and the effect of paralysis of respiratory cilia upon the course of infection. *J. Infect. Dis.* **100**:215–222.
31. Smith, P. N., 1959, Pneumonic plague in mice: gross and histopathology in untreated and passively immunized animals, *J. Infect. Dis.* **104**:78–84.
32. Williams, J. E., and Cavanaugh, D. C., 1979, Measuring the efficacy of vaccination in affording protection against plague. *Bull World Health Organ.* **57**:309–313.
33. Wu, L.-T., and Eberson, F., 1917, Transmission of pulmonary and septicaemic plague among marmots. *J. Hyg.* **16**:1–11.

34. Wu, L.-T., and Jettmar, H. M., 1925–1926, A systematic experimental study of the pathology of pneumonic plague in the tarabagan and sisel (suslik), *North Manchurian Plague Prevention Service Reports 1925–1926* **5**:1–26.
35. Batzaroff, A., 1899, La pneumonie pesteuse experimentale, *Ann. Inst. Pasteur.* **13**:385–405.
36. Watson, R. P., Blanchard, T. W., Mense, M. G., and Gasper, P. W., 2001, Histopathology of experimental plague in cats, *Vet. Pathol.* **38**:165–172.
37. Ehrenkranz, N. J., and White, L. P., 1954, Hepatic function and other physiologic studies in monkeys with experimental pneumonic plague, *J. Infect. Dis.* **95**:226–231.
38. Ehrenkranz, N. J., and Meyer, K. F., 1955, Studies on immunization against plague. VIII. Study of three immunizing preparations in protecting primates against pneumonic plague, *J. Infect. Dis.* **96**:138–144.
39. Speck, R. S., and Wolochow, H., 1957, Studies on the experimental epidemiology of respiratory infections. VIII. Experimental pneumonic plague in *Macacus rhesus*, *J. Infect. Dis.* **100**:58–69.
40. Finegold, M. J., 1969, Pneumonic plague in monkeys. An electron microscopic study, *Am. J. Pathol.* **54**:167–185.
41. Chen, T. H., Elberg, S. S. and Eisler, D. M., 1977, Immunity in plague: protection of the vervet (*Cercopithecus aethips*) against pneumonic plague by the oral administration of live attenuated *Yersinia pestis*, *J. Infect. Dis.* **135**:289–293.
42. Welkos, S., Pitt, M. L., Martinez, M., Friedlander, A., Vogel, P. and Tammariello, R., 2002, Determination of the virulence of the pigmentation-deficient and pigmentation-/plasminogen activator-deficient strains of *Yersinia pestis* in non-human primate and mouse models of pneumonic plague, *Vaccine* **20**:2206–2214.
43. Martini, E., 1902, Ueber die Wirkung des Pestserums bei experimenteller Pestpneumonie an Ratten, Mausen, Katzen, Meerschweinchen und Kaninchen, *Klin Jahrb* **10**:137–176.
44. McCrumb, F. R., Jr., Larson, A., and Meyer, K. F., 1953, The chemotherapy of experimental plague in the primate host, *J. Infect. Dis.* **92**:273–287.
45. Gage, K. L., Dennis, D. T., Orloski, K. A., Ettestad, P., Brown, T. L., Reynolds, P. J., Pape, W. J., Fritz, C. L., Carter, L. G. and Stein, J. D., 2000, Cases of cat-associated human plague in the Western US, 1977–1998. *Clin. Infect. Dis.* **30**:893–900.
46. Cornelis, G. R., 2002, Yersinia type III secretion: send in the effectors, *J. Cell. Biol.* **158**:401–408.
47. Monack, D. M., Mecsas, J., Bouley, D. and Falkow, S., 1998, Yersinia-induced apoptosis in vivo aids in the establishment of a systemic infection of mice, *J. Exp. Med.* **188**:2127–2137.
48. Schesser, K., Spiik, A. K., Dukuzumuremyi, J. M., Neurath, M. F., Pettersson, S. and Wolf-Watz, H., 1998, The yopJ locus is required for Yersinia-mediated inhibition of NF-kappa B activation and cytokine expression: YopJ contains a eukaryotic SH2-like domain that is essential for its repressive activity, *Mol. Microbiol.* **28**:1067–1079.
49. Kerschen, E. J., Cohen, D. A., Kaplan, A. M., and Straley, S. C., 2004, The plague virulence protein YopM targets the innate immune response by causing a global depletion of NK cells, *Infect. Immun.* **72**:4589–4602.
50. Price, S. B., Cowan, C., Perry, R. D., and Straley, S. C., 1991, The *Yersinia pestis* V antigen is a regulatory protein necessary for $Ca^{2(+)}$-dependent growth and maximal expression of low-Ca^{2+} response virulence genes, *J. Bacteriol.* **173**:2649–2657.
51. Brubaker, R. R., 2003, Interleukin-10 and inhibition of innate immunity to Yersiniae: roles of Yops and LcrV (V antigen). *Infect. Immun.* **71**:3673–3681.
52. Cornelis, G. R., 2000, Molecular and cell biology aspects of plague, *Proc. Natl. Acad. Sci. USA* **97**:8778–8783.
53. Parkhill, J., Wren, B. W., Thomson, N. R., Titball, R. W., Holden, M. T., Prentice, M. B., Sebaihia, M., James, K. D., Churcher, C., Mungall, K. L., Baker, S., Basham, D., Bentley, S. D., Brooks, K., Cerdeno-Tarraga, A. M., Chillingworth, T., Cronin, A., Davies, R. M., Davis, P., Dougan, G., Feltwell, T., Hamlin, N., Holroyd, S., Jagels, K., Karlyshev, A. V., Leather, S.,

Moule, S., Oyston, P. C., Quail, M., Rutherford, K., Simmonds, M., Skelton, J., Stevens, K., Whitehead, S., and Barrell, B. G., 2001, Genome sequence of *Yersinia pestis*, the causative agent of plague, *Nature* **413**:523–527.

54. Du, Y., Rosqvist, R., and Forsberg, A., 2002, Role of fraction 1 antigen of *Yersinia pestis* in inhibition of phagocytosis. *Infect. Immun.* **70**:1453–1460.
55. Drozdov, I. G., Anisimov, A. P., Samoilova, S. V., Yezhov, I. N., Yeremin, S. A., Karlyshev, A. V., Krasilnikova, V. M. and Kravchenko, V. I., 1995, Virulent non-capsulate *Yersinia pestis* variants constructed by insertion mutagenesis, *J. Med. Microbiol.* **42**:264–268.
56. Welkos, S. L., Davis, K. M., Pitt, L. M., Worsham, P. L., and Freidlander, A. M., 1995, Studies on the contribution of the F1 capsule-associated plasmid pFra to the virulence of *Yersinia pestis*, *Contrib. Microbiol. Immunol.* **13**:299–305.
57. Meka-Mechenko, T. V., 2003, F1-negative natural *Y. pestis* strains, *Adv. Exp. Med. Biol.* **529**:379–381.
58. Sodeinde, O. A., Subrahmanyam, Y. V., Stark, K., Quan, T., Bao, Y., and Goguen, J. D., 1992, A surface protease and the invasive character of plague, *Science* **258**:1004–1007.
59. Lahteenmaki, K., Virkola, R., Saren, A., Emody, L., and Korhonen, T. K., 1998, Expression of plasminogen activator *pla* of *Yersinia pestis* enhances bacterial attachment to the mammalian extracellular matrix, *Infect. Immun.* **66**:5755–5762.
60. Bearden, S. W., Fetherston, J. D., and Perry, R. D., 1997, Genetic organization of the yersiniabactin biosynthetic region and construction of avirulent mutants in *Yersinia pestis*, *Infect. Immun.* **65**:1659–1668.
61. Bearden, S. W., and Perry, R. D., 1999, The Yfe system of *Yersinia pestis* transports iron and manganese and is required for full virulence of plague, *Mol. Microbiol.* **32**:403–414.
62. Gong, S., Bearden, S. W., Geoffroy, V. A., Fetherston, J. D. and Perry, R. D., 2001, Characterization of the *Yersinia pestis* Yfu ABC inorganic iron transport system, *Infect. Immun.* **69**:2829–2837.
63. Lindler, L. E., Klempner, M. S., and Straley, S. C., 1990, *Yersinia pestis* pH 6 antigen: genetic, biochemical, and virulence characterization of a protein involved in the pathogenesis of bubonic plague, *Infect. Immun.* **58**:2569–2577.
64. Payne, D., Tatham, D., Williamson, E. D. and Titball, R. W., 1998, The pH 6 antigen of *Yersinia pestis* binds to beta 1-linked galactosyl residues in glycosphingolipids, *Infect. Immun.* **66**:4545–4548.
65. Worsham, P. L., and Roy, C., 2003, Pestoides F, a *Yersinia pestis* strain lacking plasminogen activator, is virulent by the aerosol route, *Adv. Exp. Med. Biol.* **529**:129–131.
66. Benner, G. E., Andrews, G. P., Byrne, W. R., Strachan, S. D., Sample, A. K., Heath, D. G. and Friedlander, A. M., 1999, Immune response to Yersinia outer proteins and other *Yersinia pestis* antigens after experimental plague infection in mice, *Infect. Immun.* **67**:1922–1928.
67. Erickson, D. L., Endersby, R., Kirkham, A., Stuber, K., Vollman, D. D., Rabin, H. R., Mitchell, I., and Storey, D. G., 2002, *Pseudomonas aeruginosa* quorum-sensing systems may control virulence factor expression in the lungs of patients with cystic fibrosis, *Infect. Immun.* **70**:1783–1790.
68. Sokol, P. A., Sajjan, U., Visser, M. B., Gingues, S., Forstner, J. and Kooi, C., 2003, The CepIR quorum-sensing system contributes to the virulence of *Burkholderia cenocepacia* respiratory infections, *Microbiology* **149**:3649–3658.
69. Ulrich, R. L., Deshazer, D., Hines, H. B., *et al.*, and Jeddeloh, J. A., 2004, Quorum sensing: a transcriptional regulatory system involved in the pathogenicity of *Burkholderia mallei*, *Infect. Immun.* **72**:6589–6596.
70. Ulrich, R. L., Deshazer, D., Brueggemann, E. E., Hines, H. B., Oyston, P. C. and Jeddeloh, J. A., 2004, Role of quorum sensing in the pathogenicity of *Burkholderia pseudomalle*, *J. Med. Microbiol.* **53**:1053–1064.
71. Jarrett, C. O., Deak, E., Isherwood, K. E., Oyston, P. C., Fischer, E. R., Whitney, A. R., Kobayashi, S. D., DeLeo, F. R., and Hinnebusch, B. J., 2004, Transmission of *Yersinia pestis* from an infectious biofilm in the flea vector, *J. Infect. Dis.* **190**:783–792.

72. Locht, C., Bertin, P., Menozzi, F. D., and Renauld, G., 1993, The filamentous haemagglutinin, a multifaceted adhesion produced by virulent *Bordetella* spp, *Mol. Microbiol.* **9**:653–660.
73. Chain, P. S., Carniel, E., Larimer, F. W., Lamerdin, J., Stoutland, P. O., Regala, W. M., Georgescu, A. M., Vergez, L. M., Land, M. L., Motin, V. L., Brubaker, R. R., Fowler, J., Hinnebusch, J., Marceau, M., Medigue, C., Simonet, M., Chenal-Francisque, V., Souza, B., Dacheux, D., Elliott, J. M., Derbise, A., Hauser, L. J., and Garcia, E., 2004, Insights into the evolution of *Yersinia pestis* through whole-genome comparison with *Yersinia pseudotuberculosis, Proc. Natl. Acad. Sci. U S A* **101**:13826–13831.
74. Achtman, M., Zurth, K., Morelli, G., *et al.*, G., Torrea, G., Guiyoule, A., and Carniel, E., 1999, *Yersinia pestis,* the cause of plague, is a recently emerged clone of *Yersinia pseudotuberculosis, Proc. Natl. Acad. Sci. U S A* **96**:14043–14048.
75. Van Zonneveld, M., Droogh, J. M., Fieren, *et al.*, M. W., Gyssens, I. C., Van Gelder, T. and Weimar, W., 2002, *Yersinia pseudotuberculosis* bacteraemia in a kidney transplant patient, *Nephrol. Dial. Transplant.* **17**:2252–2254.
76. Devignat, R., 1951, Varietes de l'espece *Pasturella pestis*: nouvelle hyphotheses, *Bull WHO.* **4**:247–263.
77. Zhou, D., Han, Y., Song, Y., Huang, P., and Yang, R., 2004, Comparative and evolutionary genomics of *Yersinia pestis, Microbes. Infection.* **6**:1226–1234.
78. Anisimov, A. P., Lindler, L. E., and Pier., G. B., 2004, Intraspecific diversity of *Yersinia pestis, Clin. Microbiol. Rev.* **17**:434–464.
79. Deng, W., Burland, V., Plunkett, G., 3rd, Boutin, A., Mayhew, G. F., Liss, P., Perna, N. T., Rose, D. J., Mau, B., Zhou, S., Schwartz, D. C., Fetherston, J. D., Lindler, L. E., Brubaker, R. R., Plano, G. V., Straley, S. C., McDonough, K. A., Nilles, M. L., Matson, J. S., Blattner, F. R. and Perry, R. D., 2002, Genome sequence of *Yersinia pestis* KIM, *J. Bacteriol.* **184**:4601–4611.
80. Wren, B. W., 2003, The yersiniae- a model genus to study the rapid evolution of bacterial pathogens, *Nat. Rev. Microbiol.* **1**:55–64.
81. Guiyoule, A., Grimont, F., Iteman, I., Grimont, P. A., Lefevre, M., and Carniel, E., 1994, Plague pandemics investigated by ribotyping of *Yersinia pestis* strains, *J. Clin. Microbiol.* **32**:634–641.
82. Motin, V. L., Georgescu, A. M., Elliott, J. M., Hu, P., Worsham, P. L., Ott, L. L., Slezak, T. R., Sokhansanj, B. A., Regala, W. M., Brubaker, R. R., and Garcia, E., 2002, Genetic variability of *Yersinia pestis* isolates as predicted by PCR-based IS100 genotyping and analysis of structural genes encoding glycerol-3-phosphate dehydrogenase (*glpD*), *J. Bacteriol.* **184**:1019–1027.
83. Klevytska, A. M., Price, L. B., Schupp, J. M., Worsham, P. L., Wong, J., and Keim, P., 2001, Identification and characterization of variable-number tandem repeats in the *Yersinia pestis* genome, *J. Clin. Microbiol.* **39**:3179–3185.
84. Orent, W., 2004, *Plague,* Free Press, New York.
85. Gani, R., and Leach, S., 2004, Epidemiologic determinants for modeling pneumonic plague outbreaks, *Emerg. Infect. Dis.* **10**:608–614.
86. Won, W. D., and Ross, H., 1966, Effect of diluent and relative humidity on apparent viability of airborne *Pasteurella pestis, Appl. Microbiol.* **14**:742–745.
87. Rose, L. J., Donlan, R., Banerjee, S. N., and Arduino, M. J., 2003, Survival of *Yersinia pestis* on environmental surfaces, *Appl. Environ. Microbiol.* **69**:2166–2171.
88. Wu, L.-T., Chun, J. W. H. and Pollitzer, R., 1923, Plague in Manchuria. I. Observations made during and after the second Manchurian plague epidemic of 1920–21. II. The role of the tarabagan in the epidemiology of plague, *J. Hyg.* **21**:307–328.
89. bk Hirst, L. F., 1953, *The Conquest of Plague,* Clarendon Press, Oxford.
90. Gray, G. D., 1911, The septicemic and pneumonic plague outbreak in Manchuria and North China, *Lancet* **29**:1152–1163.
91. Farrar, R., 1912, Plague in Manchuria, *Proc. R. Soc. Med.* **5**:1–24.
92. Wu, L.-T., 1923, The second pneumonic plague epidemic in Manchuria, 1920–21, *J. Hyg.* **21**:262–288.
93. Dennis, D. T., 1994, Plague in India, *Br. Med. J.* **309**:893–894.

94. Ramalingaswami, V., 1995, Plague in India, *Nat. Med.* **1**:1237–1239.
95. Gabastou, J. M., Proano, J., Vimos, A., Jaramillo, G., Hayes, E., Gage, K., Chu, M., Guarner, J., Zaki, S., Bowers, J., Guillemard, C., Tamayo, H., and Ruiz, A., 2000, An outbreak of plague including cases with probable pneumonic infection, Ecuador, 1998, *Trans. R. Soc. Trop. Med. Hyg.* **94**:387–391.
96. Ratsitorahina, M., Chanteau, S., Rahalison, L., Ratsifasoamanana, L., and Boisier, P., 2000, Epidemiological and diagnostic aspects of the outbreak of pneumonic plague in Madagascar, *Lancet* **355**:111–113.
97. Altman, L. K., 1994, Was there or wasn't there a pneumonic plague epidemic? *New York Times*, Nov. 15, p. C3.
98. Nandan, G., 1994, Plague spreads in India but is "under control", *Br. Med. J.* **309**:897.
99. Court, C., 1994, Plague prompts worldwide action, *Br. Med. J.* **309**:897–898.
100. Rotz, L. D., Khan, A. S., Lillibridge, S. R., Ostroff, S. M., and Hughes, J. M., 2002, Public health assessment of potential biological terrorism agents, *Emerg. Infect. Dis.* **8**:225–230.
101. CDC Working Group, 2000, Biological and chemical terrorism: strategic plan for preparedness and response. Recommendations of the CDC Strategic Planning Workgroup, *MMWR Recomm. Rep.* **49**:1–14.
102. Titball, R. W., and Williamson, E. D., 2001, Vaccination against bubonic and pneumonic plague, *Vaccine.* **19**:4175–4184.

10

Coxiella burnetii, Q Fever, and Bioterrorism

J. D. MILLER, E. I. SHAW, and H. A. THOMPSON

1. INTRODUCTION

Coxiella burnetii is a gram-negative obligate intracellular bacterium that must survive and replicate in an acidified phagosome of an infected host cell (Fig. 1). Most closely related to *Legionella pneumophila*, it is a member of the γ-proteobacteria class in the Legionellales order. First discovered in the 1930s in Australia and the United States, it is an organism with a worldwide distribution and is the causative agent of query (Q) fever in humans. *C. burnetii*, like *Chlamydia*, has a complex life cycle with at least two forms, a metabolically active large cell variant (LCV) and a spore-like small cell variant (SCV). Unlike *Chlamydia*, both the cell forms are infectious. Moreover, the SCV (or another resilient form, see *C. burnetii* Lifecycle Stages section) can survive in the environment outside a host for years while remaining infectious and is small enough to be carried and dispersed for miles by wind. These properties have renewed interest in *C. burnetii*, classified as a category B critical biologic agent, as a biological weapon and terror agent.

2. A BRIEF HISTORY

C. burnetii was discovered in 1935 independently by two research groups in Australia and the United States. That year in Brisbane, Queensland, Australia, there was an outbreak of undiagnosed febrile illness in abattoir

J. D. MILLER, E. I. SHAW, and H. A. THOMPSON • Viral and Rickettsial Zoonoses Branch, Division of Viral and Rickettsial Diseases, Center for Disease Control, Atlanta, Georgia.

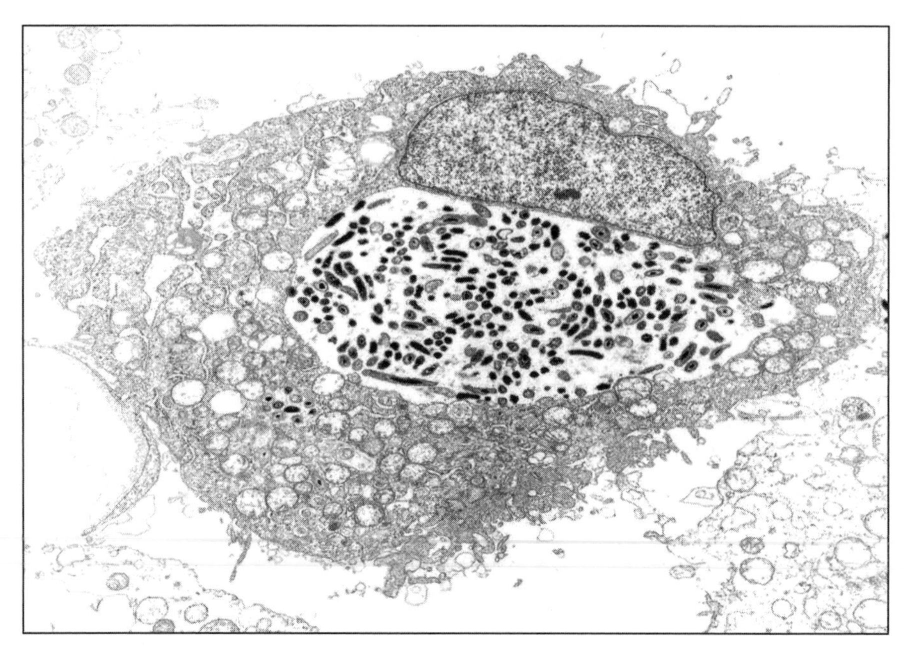

FIGURE 1. *Coxiella burnetii* M44 strain organisms in a rabbit kidney (RK-13) cell.

(slaughterhouse) workers. Edward Derrick described the illness (named query or Q fever), and was able to experimentally transmit the infection to guinea pigs.[1] Another researcher, Macfarlane Burnet, also reproduced the illness and detected the organisms in mice by using spleens of infected guinea pigs provided by Derrick. Machiavelli-stained spleen sections taken from infected animals displayed intracellular vacuoles filled with small rod-shaped organisms that appeared rickettsial in nature.[1,2] The researchers named this organism *Rickettsia burnetii.* Also in 1935, researchers at the Rocky Mountain Laboratory in Montana were investigating Rocky Mountain spotted fever (RMSF) when they established a febrile illness in guinea pigs fed upon by ticks (*Dermacentor andersoni*) obtained at Nine Mile creek in western Montana.[3] The symptoms of this illness were unlike RMSF, and it could be passed on to uninfected guinea pigs by infected blood. Attempts to grow the etiologic agent in axenic media failed. In 1936, researchers showed that the pathogen could pass through a filter much like a virus, and thus named the organism as *Rickettsia diaporica* because of this ability.[3] Herald Cox was the first to grow the organism in embryonated eggs.[4] In 1938 a laboratory worker was infected with the "Nine-Mile" agent in a laboratory accident, and it was found that his blood could transmit the febrile illness to guinea pigs.[5] Also, the connection between the Australian illness and the organism from ticks in Montana was made by experiments where animals inoculated with the Australian agent were protected from challenge by the patient's blood[5] The agent was renamed *Coxiella burnetii* to honor the contribution of Cox and Burnet.[6]

3. Q FEVER

C. burnetii and Q fever infections have been found throughout the world. From 1978 to 2002 in the United States, 538 cases of Q fever were reported by the state health departments to the Centers for Disease Control and Prevention (J.H. McQuiston, Epidemiology of Human Q Fever in the United States, 1978–2004, unpublished data). In the Unites States the estimated average annual incidence of reported Q fever in humans is 0.09 cases per million population (J.H. McQuiston, refered above, unpublished data). Of the reported cases, 76.2% are male and 44.4% of the people exposed had contact with livestock (J.H. McQuiston, refered above, unpublished data). The actual number of cases in the United States is unknown because Q fever has not historically been a reportable disease; it was made a nationally reportable disease in 1999 and 45 states considered it reportable in 2003. Diagnosis of Q fever is complicated because many acute infections of Q fever are either asymptomatic, mild, or resemble influenza symptoms.[2] Because of the lack of overt or unique symptomatic features, the actual incidence of Q fever is severely underreported worldwide and is probably more common than we know. The incubation period following exposure can range from 1 to 3 weeks.[7] Symptoms of acute Q fever are generally nonspecific and could include fever, frontal headache, myalgias, and unproductive cough.[7] Other symptoms may include light sensitivity, fatigue, rigors, night sweats, nausea, and vomiting.[2] Roughly 40% of patients reported with granulomatous hepatitis and over 60% with increased liver enzyme activity.[7,8] Approximately one third of patients have pronounced respiratory signs such as cough and radiographic changes in the lungs.[9] Q fever has also been linked to complications and abortions during pregnancy.[10,11] The organism has also been isolated from human placenta tissue.[12] Acute Q fever is usually self-limiting; however, doxycycline has been found effective for treatment.[13] Real-time polymerase chain reaction (PCR) has also shown that *C. burnetii* is susceptible to tetracycline, rifampicin, and ampicillin.[14] Some patients that are persistently infected with *C. burnetii* will develop chronic Q fever (1–16% of reported patients worldwide[15]), which is most commonly endocarditis, but can also manifest as an infection of vascular grafts or aneurysms, osteomyelitis, or hepatitis. Splenomegaly is a common clinical finding. Endocarditis may be accompanied by clubbing of fingers and prolonged fever.[9,13] Chronic Q fever is verified in the laboratory by the prevalence of phase I antibodies in the patient's serum. Some cases of chronic fatigue syndrome have also been associated with persistent infections of *C. burnetii,* with the organism being identified in bone marrow aspirates and liver tissue.[16–18] Treatments of chronic Q fever include surgical removal and replacement of infected heart valves, and prolonged therapy (up to 2 years) with doxycycline and hydroxychloroquine.[19,20]

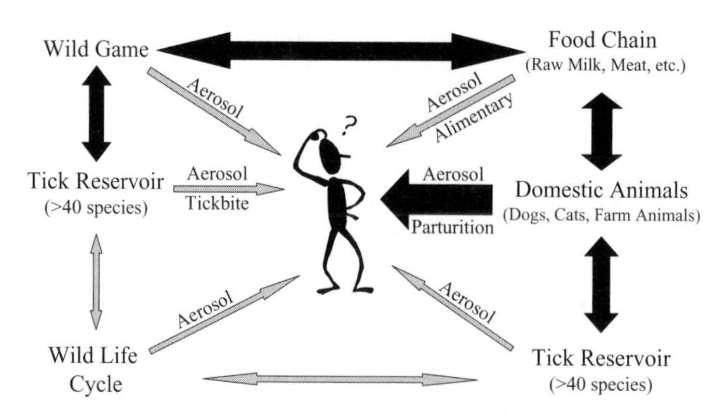

FIGURE 2. Transmission cycle of *Coxiella burnetii.* Potential mechanisms of human infection by *C. burnetii* are shown. Heavy black arrows signify major infection pathways, and thin gray arrows represent minor infection pathways. Humans are considered a dead-end host and human-to-human infections are very rare.

4. EPIDEMIOLOGY

Generally, *C. burnetii* is an organism that infects ruminants, and human infections are incidental (Fig. 2). Seroprevalence studies of animals in the United States show that goats have more frequent exposure to *C. burnetii* (41.6%) than either sheep (16.5%) or cattle (3.4%).[21] Generally, human infections are initiated from inhaling fine particulate matter generated by infected animals during parturition[22] or from desiccated urine or fecal matter.[21] It has been estimated that 1 g of infected placental tissue can contain as many as one billion organisms.[23] Humans have also been infected by eating contaminated milk products.[24] Occasionally, other infected animals, such as cats, dogs, pigeons, and wild animals, have been reported to cause Q fever infections in people.[21,25–27] There have also been reports of sexual transmission of *C. burnetii*, but this is extremely rare considering other modes of infection.[28] Although *C. burnetii* was originally isolated from ticks and has been identified in at least 40 species of ticks,[2,29] the role ticks play in transmission of the agent to animals is not completely defined and tick transmission to humans is rare.

While some information is available on incidence of infection of animals, most information comes from outbreaks where significant numbers of animals or humans were infected by the organism. The actual incidence of *C. burnetii* in nature in most regions is unknown. Without this information, clinicians could not distinguish a natural outbreak from an organized, intentional release of a biological agent. More serological and epidemiological studies must be performed to better understand how common *C. burnetii* infections are both worldwide and at regional levels.

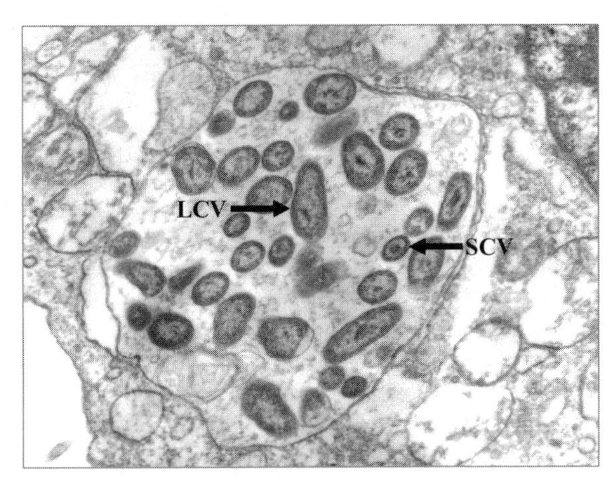

FIGURE 3. *Coxiella burnetii* Nine Mile organisms in an acidified phagosome within L929 cells. Large cell variants (LCV) and small cell variants (SCV) are identified by arrows.

5. *Coxiella burnetii* LIFECYCLE STAGES

The *C. burnetii* organism has a complex life cycle involving at least two cell types (Fig. 3); the spore-like small cell variant (SCV) and the metabolically active large cell variant (LCV).[30,31] Two other variant forms, the small dense cells (SDC) and the spore-like particle (SLP), are thought to exist, but their relationship to the SCV or the life cycle of the organism is poorly defined.[31,32] The SCV is 0.2–0.5 nm in diameter, rod-shaped, and is believed to be formed by asymmetrical division from its mother cell, the LCV, although the factors that induce differentiation are unknown.[31,32] It has been speculated that either nutritional or pH variation in the host cell triggers the conversion; however, there may be other signals that cause the differentiation that have not been identified.[31] It has been shown that the Q fever agent can pass through 40-μm filters[33]; this is now attributed to the SCV, SDC, and SLP forms (J.C. Williams, unpublished data). The SCV is resistant to physical and chemical stress, such as elevated temperature, desiccation, osmotic shock, ultraviolet light, and chemical disinfectants.[31,34–36] The ability of the SCV to survive outside of a host cell can be attributed to the higher concentration of peptidoglycan (2.7-fold more than the LCV) in its outer membrane,[30,37] appearing more dense than the LCV membrane with a smaller periplasmic space.[31] The organism protects its DNA in the SCV form by using histone-like proteins (Hq1 and ScvA) that bind to and condense the chromosomal DNA.[38,39] SCV particles are introduced into the environment by infected animals shedding the organisms through urine, fecal matter, or parturition and birthing fluids. Once the infectious matter has desiccated, the aerosolized particles can be inhaled by animals or humans thus initiating infection.[40] We presently believe that the SCV is the form responsible for infections contracted from an environmental, rather than an animal, source.

The LCV is a gram-negative pleomorphic rod with a distinguishable outer membrane, periplasmic space, and inner membrane.[31] The LCV differs from the SCV by its larger size (more than 1-μm long[32]), lack of ScvA protein,[39] appearance of a 29.5 kDa major outer membrane protein,[41] presence of protein translation factors,[42] susceptibility to physical and chemical conditions, and lower concentration of peptidoglycan in its outer membrane.[31,43] It is also the metabolically active variant in acidic conditions (pH 4.0–5.5).[44] While the SCV form is generally linked to Q fever exposure in the environment, in the laboratory it has been observed that both the LCV and the SCV forms can invade host cells.[2,45]

6. *Coxiella burnetii* GENOME

The *Coxiella burnetii* genome is defined by a single circular chromosome and, in all strains except the Scurry strain, by a resident plasmid that is present in low copy numbers (perhaps 2–5 copies per cell).[46,47] The origin of replication of the single circular chromosome was implied by the G-C skew of the chromosomal strands, and also by the location of the *dnaA*, which encodes the DnaA protein for replication initiation.[46] In earlier work, carried out by an “origin search and rescue” technique employing strains of *Escherichia coli*, an autonomous replication sequence (*ars*) was selected, isolated, and characterized.[48] Although the *ars* supports the replication of recombinant plasmids in both *E. coli* and *C. burnetii* and contains several DnaA protein binding sites, no evidence was found to suggest that it functioned as an origin during *C. burnetii* growth in tissue cultures.[48,49] The *ars* is located approximately 260 kb pairs to the left of the designated origin.

The *C. burnetii* Nine Mile strain genome consists of a chromosome of 1,995,275 base pairs and a resident plasmid (QpH1) of 37,393 base pairs.[46] Other *C. burnetii* strains are known to vary in the type of plasmid carried (QpH1, QpRS, QpDG, QpDV, or no plasmid).[47,50–54] A plasmidless strain (Scurry Q217), in which much of the plasmid information has been found inserted within the main chromosome, has also been characterized.[55] The genome and QpH1 plasmid contain 2,134 coding regions (possible genes), with 33.7% of these being hypothetical.[46] The chromosome also contains 29 insertion sequences, with 21 being the unique transposon IS1111.[46,56] All insertion sequences have more than 99% DNA identity, suggesting a recent introduction to the organism.[46] No insertion sequences have been found in the plasmid (QpH1). The Nine Mile strain genome has a G-C content of 42.6%, with a lower G-C content in the resident plasmid (39.3%).[46]

Unlike free-living bacteria, *C. burnetii* has been able to survive the potentially lethal loss of some of its genes by random changes (like point mutations, deletions, or frame shifts) because it can transport lost metabolic precursors from the host cell (much like *Rickettsia prowazekii*).[46,57–59] These 83 pseudogenes, while a significant number, represent a lower percentage of the total chromosome in *C. burnetii* than do pseudogenes of *R. prowazekii* (10.9% and ~24%, respectively).[46,57] This difference in pseudogene number could mean that *C. burnetii* became an intracellular parasite at a later time than *R. prowazekii*. *C. burnetii* has also shown a plasticity in its chromosome, with strains able to further modify their chromosome to form various avirulent phenotypes when grown in controlled laboratory conditions (see Phase Transition section).

7. LIPOPOLYSACCHARIDE

The *C. burnetii* lipopolysaccharide (LPS) is a macromolecule in the outer membrane that protects the organism from the external environment, shields it from complement binding,[60] and hides the charge of its outer membrane proteins and transporters.[61] The *C. burnetii* LPS is composed of 74% hydrophilic and 26% hydrophobic regions, with unique Lipid A and outer antigen sugar components.[62] The *C. burnetii* Nine Mile wild-type LPS is more than 1,000 times less toxic to egg embryos and mice than the LPS of either *E. coli* or *S. typhimurium*.[62,63] The lipid A moiety of the *C. burnetii* LPS is highly conserved between strains examined.[64,65] The structure consists of a diphosphorylated D-glucosamine disaccharide backbone with four acyl fatty acid chains.[64] The acyl chain lengths vary in *C. burnetii* strains, and mass differences indicate a microheterogeneity of substitutions of fatty acids of different lengths in the lipid A.[64,65] Like other endotoxins, the lipid A has a nonlamellar cubic aggregate structure that corresponds to a strictly conical shape of the lipid A moiety.[64] However, the inclination angle of the acyl chains to the D-glucosamine disaccharide backbone is much less (around 40°) than for other enterobacteria.[64]

The inner core region of the *C. burnetii* LPS has been examined in detail by studying the phase II variant LPS and resembles other enterobacterial LPS molecules.[66] This region is composed of D-mannose, D-glycero-D-manno-heptose, and 3-deoxy-D-manno-2-octulosonic acid (Kdo) in a ratio of 2:2:3.[66–68] The inner core attaches to the lipid A region at a Kdo molecule, and the outer antigen region attaches at the outer core from D-mannose and D-glycero-D-manno-heptose sugars.[68]

The outer antigen (O-antigen) region of the *C. burnetii* LPS is a complex structure of repeating components that is poorly defined. Molecules such as D-mannose, galactosaminuronic acid, glucosamine, L-rhamnose, 6-deoxy-3-*C*-methyl-D-gulose (virenose), 3-*C*-(hydroxymethyl)-L-lyxose (dihydo-hydroxystreptose), and other unknown components constitute the branches of the O-antigen region, but the exact sequence of the structure is unknown.[67,69,70] Virenose and dihydro-hydroxystreptose are rare sugars that have not been observed in other bacterial lipopolysaccharides, and their closest related sugars are found in antibiotics (virenomycin and hydroxystreptomycin, respectively).[67] This information is interesting because it may, when taken with the facts that (i) nearly one quarter of the transport genes are drug efflux systems and (ii) there are pseudogenes with homology to antimicrobial lantibiotics and β-lactamases,[46] suggest that *C. burnetii* was originally a free-living acidophilic organism competing with other bacteria in the environment. It is suggested, on the basis of phase transition studies of Nine Mile RSA514, that virenose and dihydro-hydroxystreptose are terminal sugars.[62,67] The O-antigen region is important to protect *C. burnetii* from compliment-mediated killing, because only the avirulent phase II organism is susceptible to complement.[60] A complete LPS is also considered one of the molecules important to immunity to *C. burnetii*, because vaccinating with phase II whole-cell vaccines does not confer lasting immunity.[71]

Most of the genes for the LPS synthesis, including synthesis of the outer repeating units of the O-antigen and the LPS inner core, are clustered into two regions of the genome.[46] These gene clusters also include large numbers of genes related to nucleotide sugar metabolism involved in the polymerization of the sugar branches in the O-antigen.[46] Unlike the rest of the LPS biosynthesis genes, the genes for the synthesis of Lipid A and Kdo components of the LPS are scattered throughout the genome.[46]

8. PHASE TRANSITION

When virulent wild-type *C. burnetii* (phase I) strains are passaged for many replication cycles in either embryonated eggs or tissue culture systems, the organisms (when plaque-purified) will undergo a permanent phase transition into avirulent variants (phase II).[58] Our knowledge of the events of phase transition from a wild-type *C. burnetii* to an avirulent variant is limited. In many strains of *C. burnetii* (Australian, M44 Grita, Priscilla, and Nine Mile) an avirulent phase variant with an LPS truncation occurs, but the genetic modifications of phase transition appear to be variable.[59] In the Nine Mile strain, two types of chromosomal deletions have been characterized. The Nine Mile phase II clone 4 organisms have a chromosomal deletion (roughly 26 kb pair region that consists primarily of O-antigen synthesis genes[58]), exhibit a more severely truncated LPS that lack both rare sugars (virenose and dihydrohydroxystreptose[72]) and are avirulent.[60,62,73,74] Despite the chromosomal deletion and modification, the phase II organism is still able to form an SCV cell form (R.A. Heinzen and E.I.Shaw, unpublished data). The Nine Mile RSA514 variant displays a larger chromosomal deletion (31.6 kb) in the same region as the phase II deletion, possesses an intermediate LPS, and is partially virulent.[58,75,76] The varied phenotypes from the Nine Mile phase variants suggest that phase II organisms may have other chromosomal deletions or mutations which have not been characterized, yet play a role in loss of the complete wild-type phenotype. It is also unknown if the Nine Mile phase transition is consistent, causing the same chromosomal mutations each time a laboratory-passed *C. burnetii* organism adapts. Phase transition in the Priscilla strain occurs by undergoing a loss of LPS sugars in three stages, eventually leading to a truncated LPS(77); however, there is no genetic information available on the avirulent phenotype. Little information is available on the M44 or Australian phase II strain (QD), variants that also display the truncated LPS phenotype and are avirulent.(78,79) PCR analysis has determined that both M44 and Australian phase II variants retain many of the genes lost in the deletion of Nine Mile strain phase II organisms, but the fidelity of these genes is unknown.(59)

The events of phase transition not only affect the phenotype and genome of the affected organism, but in the Nine Mile strain it appears to alter the invasion of the host cells. Phase II organisms can invade host cells faster and are more successful at establishing persistent infections in laboratory cell lines than phase I organisms.(80,81) *C. burnetii* Nine Mile phase II invasion studies suggest that the uptake process is mediated by both the leukocyte response integrin, $\alpha_v\beta_3$ (LRI) and CR3.(82) This is in agreement with earlier observations that showed that complement components (such as C3) can attach to and kill phase II organisms.(60) Once the phase II organism has attached to the host cell the membrane does not visibly ruffle and the organism cannot delay lysosome fusion to its endosome compartment,(83,84) suggesting that the phase II organism lacks a factor to modify the behavior of the host cell. The final vesicle of the *C. burnetii* phase II organism has characteristics of a mature phagolysosome, including the marker proteins Rab5, Rab7, and LAMP-1,(85–87) but it is unknown if the wild-type *C. burnetii* phagosome is different from the phase II phagosome. Because of the differences in invasion and survival between the wild-type and the phase-variant *C. burnetii* organisms, we cannot assume that the organisms react to or modify the host cell in an identical manner.

The phase transition event(s) also seem to modify the replication of Nine Mile strain organisms in certain growth systems; phase II organisms grown in embryonated eggs display consistently lower yields when compared to phase I organisms.(88,89) In some growth experiments using Baby Hamster Kidney (BHK-21) cells (J.D. Miller and H. Thompson, unpublished observations), we have observed the number of intracellular phase I organisms to be higher than that of phase II organisms cultured in identical conditions. This growth can be significantly improved with the addition of cytidine supplemented to the tissue culture media, but the reason for this improvement is unknown at this time. Because of our poor understanding of the phase transition and the genetic modifications the chromosome undergoes, it is probable that other modifications exist that are relieved by cytidine or a cytidine product, allowing for the increased phase II growth.

9. INVASION OF HOST CELLS

The *C. burnetii* bacterium has an advantage in that, once it has contaminated an environment, it can infect an animal or a person with an extremely low infectious dose (1–10 organisms).[40] Once the *C. burnetii* organism has been introduced into the host, either by inhalation, consumption, or other route, the organism can bind to and invade a host cell. In laboratory studies, *C. burnetii* has exhibited an ability to infect a wide variety of cell lines, including, but not limited to, human monocyte cells (THP-1), fibroblast cells, Vero cells, chick endodermal (primary culture) cells, Chinese hamster ovary cells (CHO cells), baby hamster kidney (BHK-21) cells, and mouse macrophage cell lines.[2,44,80,84,90–94]

Wild-type *C. burnetii* (phase I) organisms are thought to initiate infection of host cells by binding to leukocyte response integrin $\alpha_v\beta_3$ (LRI) and integrin-associated protein (IAP) complex, while not engaging the CR3.[82,95] During the attachment and invasion process, wild-type organisms induce activation of protein tyrosine kinases (PTK) and tyrosine phosphorylation of several substrates thought to interfere with cytoskeleton organization, causing the host cell surface to ruffle and change morphology.[84,96] The *C. burnetii* genome also contains 13 ankyrin repeat-containing proteins that could mediate interaction of the plasmid membrane with the membrane skeleton, aiding the invasion process.[46,97] Other genes such as homologues of the *L. pneumophila* EnhA, EnhB, and EnhC may be involved in cell entry, but their role in *C. burnetii* invasion needs to be investigated.[46] The organism is then internalized and shuttled through an endosomal pathway, where vacuole ATPases lower the pH to 5.5.[83] In laboratory experiments, the organism has been shown to transport amino acids in acidified media; however, it is unknown at what stage in the invasion process the organism becomes fully active (that is, able to replicate DNA and synthesize proteins).[83] It has also been shown to transport guanosine at neutral pH, but the reason for this ability is unknown.[88] In J774A.1 mouse macrophage cells, wild-type *C. burnetii* can delay the binding of lysosomes to its vesicle;[83] however, it is unknown if this event occurs in other cell lines. We also do not know how the organism modifies its vesicle to promote survival and replication. Current research is focusing on the signaling mechanisms used by *C. burnetii* to influence the host cell.

Once *C. burnetii* has begun replication within the acidified phagosome, it has a replication time ranging between 12 and 20 hours.[98] The organism has only one rRNA operon, which is consistent with organisms that have a slow growth rate.[99] This slow growth adaptation may have evolved to allow for the accumulation of the maximum bacterial load in a host cell, while gaining as much metabolic precursors from the host as possible before killing it. This would allow for the greatest number of *C. burnetii* organisms in the blood, increasing the opportunity for the organisms to find new host cells and continue the cycle. This is thought to be the reason for slow growth in other intracellular parasites as well, such as *R. prowazekii*.[100] In the growth compartment, both LCV and SCV forms are present,[101] perhaps ensuring survival for the organism once the host cell is lysed. *C. burnetii* organisms are thought to kill their host cells by replicating within the acidified phagosome compartment, enlarging the vacuole until it exerts enough internal pressure to lyse the host, thereby releasing the organisms.

10. ENVIRONMENT OF ACIDIFIED PHAGOSOME

Once the organism has established itself in the acidified phagosome, *C. burnetii* has to survive a harsh environment designed to destroy pathogens. In recent years, our understanding of the phagolysosome compartment has improved, but it is still incomplete. The membrane of a mature phagosome is a complex organelle composed of over 600 different proteins, some of which are unidentified.[102,103] Some of these proteins are H^+-translocating ATPases, which lower the pH of the compartment. The *C. burnetii* organism has adapted to the acidic pH and apparently can only replicate in this environment. Attempts to reproduce the conditions required for replication by axenic media have failed, showing that more than an acidic environment, glucose and a supplement of 20 amino acids and thymidine are required for the replication.[44,104,105] The organism has also adapted by having a much higher percentage of basic proteins when compared to other bacteria. This may be a backup mechanism for the organism to titrate or regulate the H^+ ions, along with predicted sodium ion/proton exchangers, from the cytoplasm if it cannot maintain its adenylate energy charge for cytoplasmic pH maintenance.[46]

It is unknown at this time how closely a mature phagolysosome resembles the acidified phagosome containing *C. burnetii* organisms. While *C. burnetii* occupies this acidified phagosome, it grows until it becomes the largest compartment in the host cell, dwarfing and pushing aside the cell nucleus.[31] This may be caused by the host cell fusing lysosomes and possibly endosomes to the *C. burnetii* phagosome in an effort to eliminate the organism. Conversely, the growth of the acidified phagosome compartment may be due to the organism somehow controlling the host cell and some of its functions. Within the compartment of a normal phagolysosome are a number of enzymes delivered by lysosomes, which are used to digest pathogens into peptides for MHC presentation as well as digesti DNA and protein material to nucleoside and amino acids for use by the cell. It is assumed that these enzymes are also present in the *C. burnetii* acidified phagosome. The cell also uses nitric oxide and oxygen radicals,[106] as well as cytokines such as TNF-α and IFN-γ, to induce cytotoxic or cell-mediated responses and clear the infection. *C. burnetii* is thought to protect itself from oxygen radicals and oxidative bursts within its compartment by using acid phosphatase, catalase, and superoxide dismutase (SOD) enzymes.[107–110] *C. burnetii* has been observed during *in vitro* acid activation experiments to release a number of unidentified proteins into the media.[111] The organism, at acidic pH, may also use a secretion system (possibly type I, II, or IV) to secrete a number of proteins into the environment.[46,112] At this time we are unsure of what specific proteins *C. burnetii* secretes, or what effect these proteins have on the host cell.

Despite the challenge the mature acidified phagosome poses to the organism and its survival, the compartment is also presumed to be rich in metabolic precursors the organism requires for metabolism and replication. It is known that phagolysosomes contain a large quantity of useful nutrients, including nucleosides, amino acids, phosphates, glucose, sulfates, and phosphates.[113,114] If these components are also available in the *C. burnetii* acidified phagosome, this wealth of resources would enable the parasite to become more energy efficient by discarding or inactivating portions of its genes.

11. METABOLIC PATHWAYS

Despite the elimination of 83 genes, sequence evidence suggests that *C. burnetii* has maintained most of its metabolic capabilities and retains the genes for glycolysis, the Entner–Doudoroff pathway, the electron transport chain, the Embden–Meyerhof–Parnas pathway, gluconeogenesis, the pentose phosphate pathway, and the tricarboxylic acid (TCA) cycle.[46] It also appears to have complete pathways for purines and pyrimidines, fatty acids, phospholipids, and cofactors.[46] The sequence also suggests that *C. burnetii* can utilize glucose, xylose, galactose, and glycerol.[46] The chromosome is missing the glyceraldehyde-3-phosphate (GAP) pyruvate pathway for IPP synthesis, as well as being auxotrophic for 11 amino acids (including leucine, isoleucine, phenylalanine, tryptophan, valine, histidine, lysine[46]).

12. TRANSPORT

Even though *C. burnetii* has many complete biochemical pathways for macromolecule synthesis, it also appears to transport many components such as amino acids, sugars, and nucleosides probably to conserve energy and its adenylate charge for other requirements. The genome sequence suggests it has 2 sugar transporters (xylose and glucose), 3 peptide transporters, and 15 amino acid transporters.[46] It has four predicted sodium ion/proton exchangers, presumably for maintenance of the cytoplasmic pH.[46] The genome also contains a number of sequences that share homology with various types of ABC (ATP binding cassette) transport proteins.[46] The genome sequence predicts that three mechanosensitive ion channels and three osmoprotectant transporters exist.[46] More than a quarter of the predicted transporters are predicted to be drug-efflux pumps, which may play a role in the protection from phagosome antimicrobial defensin proteins or resistance to antibiotics.[46] There is also a putative multidrug transporter protein found in a potential "pathogenicity island" flanked by insertion sequences.[46] There is no evidence, in the sequence or in biochemistry experiments, for an ATP/ADP exchanger to scavenge ATP from its host,[46,88] as is used by *R. prowazekii* or *Chlamydia trachomatis*.[115,116] This lack of an ATP/ADP exchanger makes sense because of the rarity of ATP within the environment of an acidified phagosome.[113]

Genome analysis has also suggested homology with type I, II, and IV secretion systems.[46] The proposed type IV secretion system, which has many similarities to the *L. pneumophila* icm-dot genes,[117–119] has components that can complement function with similar genes in *L. pneumophila* (*dotB, icmS, icmT*, and *icmW* [112,119]). This system may play a role in the subjugation of the host cell; however, more research into its function and the components that may be released by the secretion apparatus must be performed before any conclusion can be reached.

Much of our current knowledge of *C. burnetii* transport comes from research using acidified media. First developed in the early 1980s by Hackstadt and Williams,[44] it was found that a simple media at pH 4.0–5.5 would allow DNA and protein synthesis to occur so long as an energy source (glutamate) was available.[120] This media is incomplete in that it will not support replication of the organism. It is unknown if it is a metabolic precursor that is missing in the media, or if it is either a lack of communication with or a factor from the host cell that prevents *C. burnetii* growth. Using the acidified media, information on glucose and glutamate transport and usage as well as amino acid transport has been obtained.[92,105,120,121] DNA synthesis also occurs in the acidified medium, but evidence suggests that new chromosome initiations may not form during *in vitro* acid activation.[105] Inclusion of a complete amino acid supplement in acidified media resulted in a large increase in protein production,[92,105] a fact that can now be appreciated since it is known from genome studies that the organism must be auxotrophic for 11 amino acids.[46] It has also been shown that *C. burnetii* will secrete proteins in the acidified media, although these proteins have not yet been identified.[111]

We have found that, despite the fact that sequence information suggesting purine and pyrimidine biosynthesis pathways exist, nucleosides are transported and incorporated by *C. burnetii* organisms.[88] We have also found that purine nucleosides (inosine, adenosine, thymidine) appear to be actively transported and concentrated in the cytoplasm, while pyrimidine nucleoside transport likely relies on passive or facilitated diffusion (Fig. 4).[88] It may be that, even with the existing purine and pyrimidine biochemical pathways, *C. burnetii* prefers to transport available nucleosides from the host cell to conserve energy.

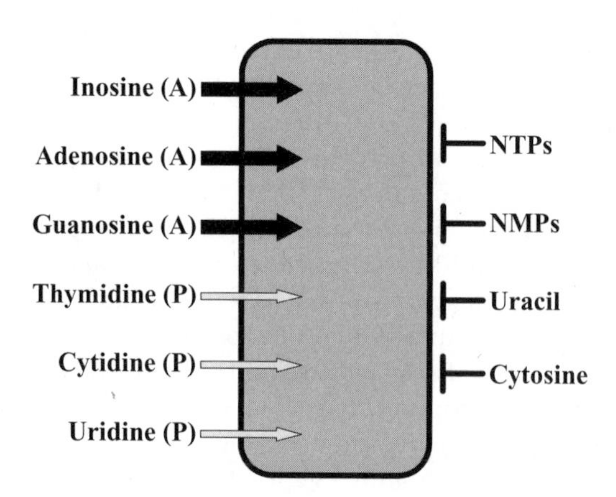

FIGURE 4. Diagram of *Coxiella burnetii* nucleoside transport. Active transport mechanisms are represented by (A) and heavy black arrows. Passive transport mechanisms are represented by (P) and thin gray arrows. Compounds not transported into the cytoplasm of *C. burnetii* Nine Mile strain are represented by blocked lines.

13. TRANSFORMATION STUDIES

The ability to stably introduce foreign DNA or manipulate the *C. burnetii* chromosome has proven to be a difficult pursuit. A step forward occurred when an autonomous replication sequence, a possible origin for *C. burnetii* DNA replication, was identified.(48) This sequence was utilized as an origin in a plasmid containing an ampicillin-resistance gene and transformed into *C. burnetii* Nine Mile wild-type organisms by electroporation, resulting in the first transformant of the organism.(49) Other research groups have transformed *C. burnetii* Nine Mile strain with a green fluorescence protein,(122) but the technology is still in its infancy. *C. burnetii* transformation experiments will remain difficult due to the limited antibiotics available for selection because of the number of antibiotics reserved for treatment of chronic Q fever. More work is required in this area before foreign DNA can be introduced reliably and remain stable in the organism.

14. *Coxiella burnetii* OUTBREAKS IN THE MILITARY

For an agent that was discovered in 1938, *C. burnetii* has played a surprising role in troop illnesses in the military. There have been thousands of identified cases of Q fever in the military across the world, ranging from World War II (WWII) to the first Persian Gulf War.(123) The first documented troop illnesses caused by Q fever were observed during WWII. It has been estimated that Q fever in Northern Italy may have accounted for roughly 75% of the cases of atypical pneumonia in certain army hospitals.(124) It must be stated that most Q fever cases during WWII were not identified by serology, but by symptoms.[123] Balkangrippe, an illness in German troops serving during WWII in Bulgaria, Italy, Crimea, Greece, Ukraine, Corsica, and Yugoslavia, had symptoms that are similar or identical to acute infections of Q fever with a mortality rate of less than 1%.[125] Over 1,000 cases of balkangrippe were documented by Dr. Dennig, a consultant physician to the German forces, during an epidemic in 1941.[125] In 1945, the causative agent for some cases of balkangrippe was positively identified as *C. burnetii*.[125] From December 1944 to June 1945, 1,700 troops, mostly stationed in Italy, were known to have Q fever.[123,125] In these cases contact with farm animals was not an absolute requirement.[123] From 1946 to 1956, Greek armed forces continued to have large numbers of troops hospitalized from Q fever infections.[123] Other smaller outbreaks documented in military personnel include 66 cases of Q fever occurring in 1948 at an army recruit training camp in Switzerland,[125] an epidemic in Algeria involving 175 French soldiers,[125] 78 cases of acute Q fever in British troops stationed in Cyprus between December 1974 through June 1975 (associated with an epidemic of spontaneous abortions in goats and sheep),[123] and 2 cases of Q fever during the Persian Gulf War.[123] In the documented Q fever epidemics involving the military, the loss of manpower in the involved units ranged from 23 to 77%, which could drastically affect unit efficiency and mission readiness.[125] While the significantly large outbreaks during WWII were probably caused by *C. burnetii* contaminated barns and bedding,[125] Q fever will continue to occur in military units because of the dust generated by the movement of heavy machinery and the worldwide distribution of *C. burnetii*.

15. *Coxiella burnetii* AS A BIOLOGICAL WEAPON

Biological warfare agents, as defined by the World Health Organization (WHO), are agents that display the following features: the ability to multiply in a host over time, infectivity (proportion of exposed population that becomes infected), virulence (relative severity of disease), lethality (ability to cause death in a population), pathogenicity (capacity to cause disease), and incubation period (time elapsed between exposure and first signs of infection).[126] *C. burnetii* is an attractive agent due to its high infectivity, its remarkable stability for storage, its stability and retention of infectivity for long periods of time once released in the environment, no requirement for complicated release mechanisms, and its capability to be spread by wind. First, *C. burnetii* is a naturally occurring organism with a worldwide distribution, and unlike anthrax spores, the SCV cell form can be carried by the wind for miles without further modification.[22] It also has the advantage of being able to grow large amounts of the organism in the proper cell system. It is highly infectious, requiring only 1–10 organisms to initiate an infection in an exposed host.[40] This infectious dose is lower than other developed biological weapons.[127] A commissioned WHO study suggested that if 50 kg of *C. burnetii* were released in a 2 km line upwind from a city of 500,000 people, the agent would be able to travel over 20 km. Further, there would be 125,000 incapacitated people and 150 deaths.[126] Once introduced into a host, it is able to not only replicate and cause disease, but in some cases can persistently infect the host and cause more serous diseases such as endocarditis.[13,18] *C. burnetii* has an incubation period of 1–3 weeks before classical acute Q fever symptoms appear.[2] *C. burnetii* can be lethal, but it will not transmit from person to person, and generally it kills less than 1% of those exposed (far lower than other agents such as anthrax).[125,126] *C. burnetii* is considered an incapacitating agent, and would be useful in spreading panic throughout an exposed population because of its very low infectious dose. Because so many infections would occur after exposure (despite the fact that roughly 50% of cases are asymptomatic), psychological consequences after an attack would probably includeacute stress disorder, posttraumatic stress disorder, depression, psychiatric disability, disruption of communal routine, and increased use of public health facilities.[128] The possibility of panic and psychological disorders would probably increase if *C. burnetii* were used in combination with a more lethal agent such as anthrax, causing high infections and a higher death rate.[127]

Several countries, including the United States, began biological weapon programs during or after WWII. The Japanese studied *C. burnetii* in their infamous germ warfare program during WWII.[129] *C. burnetii* was investigated and stockpiled by the US military as an incapacitating agent.[129] In the early 1950s, Project Whitecoat was initiated, which used animal models and volunteers from the Seventh Day Adventist Church to test the infectivity of aerosolized *C. burnetii.*[130] These tests included detonating biological munitions in a one million liter hollow metallic sphere (dubbed the "eight-ball") to determine the vulnerability of the volunteers to aerosolized pathogens and to test the efficacy of vaccines and antibiotic prophylaxis.[129,131] Fortunately, during these tests no deaths occurred in any of the volunteers.[130] In 1969 and 1970, President Nixon terminated the offensive biological weapons program by executive order, and all antipersonnel biological warfare stocks were destroyed between May 1971 and May 1972.[130,131]

Despite the destruction of our biological weapons, some extremist groups and at least 10 nations are believed to have continued to research and develop biological agents.[132] Some sources indicate that Russia continued to maintain a biological weapons program well after it signed the 1972 biological weapons convention.[127] When Russia finally disbanded its research groups on bioterrorism, Libya, Iran, Syria, and North Korea reportedly sought to employ the Russian scientists to improve their own programs.[132] It was found in 1995 that the Japanese cult Aum Shinrikyo not only used sarin gas to attack the Tokyo subway, but was developing *Bacillus anthracis, Clostridium botulinum,* and *C. burnetii* for potential release.[129,131] *C. burnetii* will continue to be a concern not only because of its availability in nature and its infectivity, but because so little is understood about its natural occurrence that it might be difficult to discern the difference between a natural epidemic and a premeditated release.

16. STERILIZATION/DISINFECTION

Historically, *C. burnetii* has been considered a difficult organism to eliminate. Many of the decontamination procedures used for other biological agents are unsuccessful in eliminating *C. burnetii*, presumably because of the SCV, SDC, and SLP forms. *C. burnetii* is also the second highest cause of laboratory infections. The SCV is susceptible to 70% ethanol, 5% chloroform, or 5% Micro-Chem Plus (National Chemical Laboratories, Philadelphia, PA) when applied for 30 minutes.(36) Soon after the discovery of *C. burnetii* in dairy herds and milk in Southern California, careful thermal inactivation studies were performed to modify the pasteurization methods and inactivate the organism.(34,133) Based upon these kill curves, small volumes (1.0 ml) containing liquid suspensions of the organism are routinely inactivated by heating at 80°C for 1 hour to ensure laboratory safety when necessary. Organisms distributed for use as antigens in serological applications are inactivated by exposure to cobalt (gamma) irradiation.(134) Humidified formaldehyde gas is effective if the contamination is contained within a room or an area that can be isolated and sealed.(36) Filtration sterilization does not work because of the small size of the SCV form. When an accidental release occurs, laboratory workers should use precautions such as HEPA-filtered respirators and disposable garments to prevent inhalation by aerosol. It is extremely difficult to contain or disinfect an environmental contamination because of factors such as wind and the difficulty of containing a spill.

17. DETECTION METHODS

Currently, Q fever is diagnosed and *C. burnetii* are detected by a number of methods. While some assays look for evidence of infection, others detect the presence of the *C. burnetii* organism directly. The type of samples available often dictates the methods used for analysis. In human and animal infections, blood is the usual sample taken. However, samples as diverse as heart valves, liver biopsy, and placental tissue are not uncommon, especially for suspected cases of chronic Q fever. A variety of assays have been developed in an attempt to diagnose and detect *C. burnetii* infections in a timely and sensitive manner.

The current sera diagnostic "gold standard" used for the detection of an infection caused by *C. burnetii* is indirect fluorescence antibody (IFA) microscopy analysis.(135,136) An example of a positive IFA is shown in Fig. 5. This assay is designed to detect a rise in serum antibodies specific to *C. burnetii* phase I and

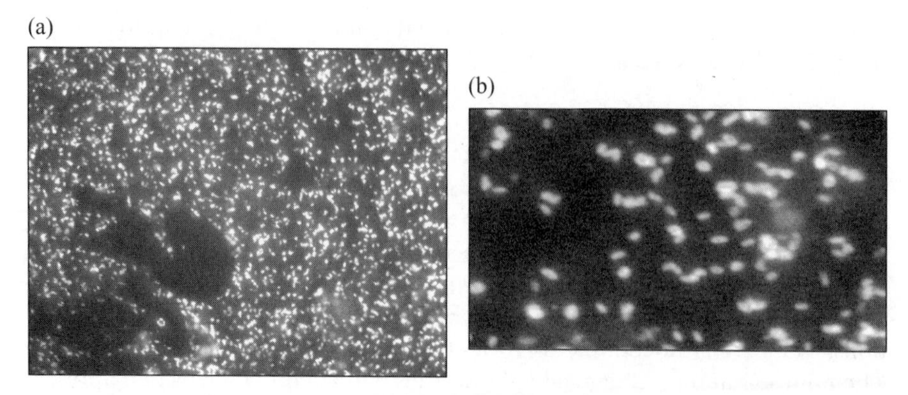

FIGURE 5. Indirect fluorescence antibody microscopy of *Coxiella burnetii* in 1% yolk sac. *C. burnetii* homogenized yolk sac was diluted to 1% in PBS pH 7.4 and spotted onto microscope slides. The slides were acetone-fixed, treated with human primary antibody and goat α-human FITC-labeled secondary antibody, and examined under (a) 400 and (b) 1,000 magnification using a UV light source.

phase II antigens. Ideally, in a case of acute Q fever, a rise in antibody titer of fourfold or more is required for a positive diagnosis. In acute Q fever cases in humans, the antibody titer against phase II antigen will usually (but not always) be higher than that against phase I. However, there are factors that must be taken into account when using IFA as a diagnostic tool for Q fever. These are (i) when the sample was taken in relation to the onset of symptoms and (ii) the possibility that the antibodies are from a previous infection with *C. burnetii.* As serum antibodies to *C. burnetii* do not appear until around 10 days postinfection, samples taken prior to that time having *C. burnetii* specific antibodies may be the result of a prior infection. This question is resolved by defining the class of antibody present in the sera. An initial infection will present sera containing IgM class antibodies early in an immune response (10 days after onset of symptoms), while sera from a previous infection would contain IgG antibodies.[135] While employed less frequently, other serodiagnostic assays are also used to diagnose acute Q fever. They include complement fixation, ELISA, and immunhistochemistry (IHC).[137–139] IHC is especially useful in confirming chronic Q fever cases from heart valve samples and in placental infections.[140]

Direct detection of the *C. burnetii* organism is often more difficult. The most sensitive, and most commonly used, is the PCR method designed to detect the IS1111 gene of *C. burnetii.*[141–143] IS1111 is a bacterial insertion sequence found on the *C. burnetii* Nine Mile genome in 21 copies[46] (see *Coxiella burnetii* Genome section). This high multiplicity makes the IS1111 DNA fragment an attractive target for the detection of *C. burnetii* DNA in a sample; however, there are some applications where using IS1111 probes do not work well (H.A. Thompson, unpublished data). Typically, DNA extractions from blood, tissue, or environmental samples are assayed using this method. It has the capacity to detect as few as five organisms in a reaction.[142] Traditionally, the PCR approach has been performed using gel-based analysis, but real-time PCR detection methods using fluorescent probes are being rapidly developed which have equal or greater sensitivity, while also being faster and less labor-intensive (H.A. Thompson, unpublished data). In the case of a suspected intentional release of *C. burnetii*, the initial samples would represent environmental material in which the ability to detect DNA or whole organism would be crucial. New approaches in processing environmental samples are being undertaken in a variety of settings and hold promise for heightening our ability to identify *C. burnetii* through detecting whole organism or specific DNA markers.

Another method for direct detection is to isolate and culture the organism. This is not normally done as a means of diagnosis and is usually reserved for confirmation and strain isolation. While clinical isolates have been successfully isolated by a tissue culture shell-vial method,[144] the difficulty and relatively poor success rate of the method makes it impractical for most laboratories. In addition, isolation of *C. burnetii* from environmental samples would pose significant problems as this method is acutely susceptible to contamination from environmental flora. An alternative isolation method for *C. burnetii* is the use of mice.[145] With its low ID_{50}in rodent models,[76,146] a sample containing as few as 1–10 *C. burnetii* will cause an infection in mice.[40] Subsequent amplification of the organism will occur within the mouse, and isolation of the bacterium from the spleen is then possible. An advantage in using this method is that possible nonpathogenic contaminating flora would be eliminated by the mouse immune system.

18. TREATMENTS OF Q FEVER

As stated previously, Q fever infections manifest as either acute or chronic disease (see Q Fever section). The treatments of the disease forms are quite different. Generally, acute Q fever is treated with 200 mg tetracycline (doxycycline) daily for 14 days.[2,147] Tetracyclines and fluoroquinolones are contraindicated for pregnant women and children younger than 8 years, and while erythromycin treatment has proven effective for some Q fever cases,[148,149] others have not responded well.[150] The current suggested antibiotic treatment for pregnant women is co-trimoxazole for the duration of the pregnancy.[151] Q fever endocarditis is the most common form of chronic Q fever. Treatment of these chronic cases of Q fever requires a prolonged antibiotic regimen. Doxycycline and hydroxychloroquine are used in combination for at least 18 months,[152] with 2–3 years a suggested regimen.[20] Cases where this regimen cannot be administered because of patient intolerance are treated for 3 years with a combination of doxycycline and ofloxacin.[153] For Q fever endocarditis patients, surgical replacement or repair of the effected valve is usually required, in which case it must be performed in conjunction with antibiotic therapy.[154,155] The concomitant antibiotic treatment is undertaken to prevent reinfection of the repaired valve.[2,156]

19. VACCINE

The best preventative measure against contracting Q fever is vaccination. The first vaccine, a formalin-killed and ether-extracted *C. burnetii* solution in 10% yolk sac, was developed by Smadel in 1948.[157] The current vaccine, Q-VAX® (CSL Limited), is a 30 μg dose of formalin-killed *C. burnetii* Henzerling strain phase I cellular vaccine licensed, used and produced in Australia.[16,158] Q fever vaccines are considered experimental in the United States and are not freely available to the public. The United States Army employs a whole-cell killed vaccine very similar to the Australian Q-VAX®. Generally administered to people in abattoir, laboratory, and agriculture-based occupations, *C. burnetii* vaccines could also play a role in protecting soldiers at risk when occupying regions with histories of high Q fever incidences or biological attacks. Q-VAX® is thought to give protection from Q fever for at least 5 years.[158] While nearly 100% effective,[158] the vaccine can have side effects that include subcutaneous abscesses or lipomata formations in the area of the vaccine injection if people have already been exposed to *C. burnetii* or been vaccinated before.[159,160] These side effects can mostly be avoided by eliminating patients who test positive to a skin test or a blood antibody titer. Another vaccine, a chloroform–methanol residue developed in the late 1970s by Rocky Mountain Labs, has been tested in human volunteers[161] and at one time was being developed as an alternative to the whole cell vaccine.[162,163] The current status of the chloroform–methanol residue vaccine is uncertain. While it may not have the local reactions that occur with administration of killed cellular vaccines such as Q-VAX®, it is unknown if it is as effective in preventing aerosolized *C. burnetii* infections in humans.

REFERENCES

1. Derrick, E. H., 1937, "Q" fever, new fever entity: Clinical features, diagnosis, and laboratory investigation, *Med. J. Aust.* **2**:281–299.
2. Maurin, M., and Raoult, D., 1999, Q fever, *Clin. Microbiol. Rev.* **12**:518–553.
3. Davis, G. E., and Cox, H. R., 1938, A filter-passing infectious agent isolated from ticks. I. Isolation from *Dermacentor andersoni*, reactions in animals, and filtration experiments, *Public Health Rep.* **53**:2259–2261.
4. Cox, H. R., and Bell, E. J., 1939, The cultivation of *Rickettsia diaporica* in tissue culture and in the tissues of developing chicken embryos, *Public Health Rep.* **54**:2171–2175.
5. Dyer, R. E., 1938, A filter-passing infectious agent isolated from ticks. Human infection, *Public Health Rep.* **53**:2277–2282.

6. Philip, C. B., 1948, Comments on the name of the Q fever organism, *Public Health Rep.* **63**:58.
7. Mcquiston, J. H., Childs, J. E., and Thompson, H. A., 2002, Q fever, *JAVMA* **221**:796–799.
8. Raoult, D., Tissot-Dupont, H., Foucault, C., Gouvernet, J., Fournier, P. E., Bernit, E., Stein, A., Nesri, M., Harle, J. R. and Weiller, P. J., 2000, Q fever 1985–1998. Clinical and epidemiologic features of 1,383 infections, *Medicine (Baltimore)* **79**:109–123.
9. Marrie, T. J., 1988, Q fever: Clinical signs, symptoms, and pathophysiology, in: *Biology of Rickettsial Diseases* (D. H. Walker, ed.), CRC Press, Boca Raton, FL.
10. Stein, A., and Raoult, D., 1998, Q fever during pregnancy: A public health problem in southern France, *Clin. Infect. Dis.* **27**:592–596.
11. Langley, J. M., Marrie, T. J., Leblanc, J. C., Almudevar, A., Resch, L., and Raoult, D., 2003, *Coxiella burnetii* seropositivity in parturient women is associated with adverse pregnancy outcomes, *Am. J. Obstet. Gynecol.* **189**:228–232.
12. Syrucek, L., Sobeslavsky, O., and Gutvirth, I., 1958, Isolation of *Coxiella burnetii* from human placentas, *J. Hyg. Epidemiol. Microbiol. Immunol.* **2**:29–35.
13. Marrie, T. J., and Raoult, D., 2002, Update on Q fever, including Q fever endocarditis, *Curr. Clin. Top. Infect. Dis.* **22**:97–124.
14. Brennan, R. E., and Samuel, J. E., 2003, Evaluation of *Coxiella burnetii* antibiotic susceptibilities by real-time PCR assay, *J. Clin. Microbiol.* **41**:1869–1874.
15. Siegman-Igra, Y., Kaufman, O., Keysary, A., Rzotkiewicz, S., and Shalit, I., 1997, Q fever endocarditis in Israel and a worldwide review, *Scand. J. Infect. Dis.* **29**:41–49.
16. Marmion, B. P., Ormsbee, R. A., Kyrkou, M., Wright, J., Worswick, D. A., Izzo, A. A., Esterman, A., Feery, B., and Shapiro, R. A., 1990, Vaccine prophylaxis of abattoir-associated Q fever: Eight years' experience in Australian abattoirs, *Epidemiol. Infect.* **104**:275–287.
17. Ayres, J. G., Flint, N., Smith, E. G., Tunnicliffe, W. S., Fletcher, T. J., Hammond, K., Ward, D., and Marmion, B. P., 1998, Post-infection fatigue syndrome following Q fever, *Q J Med.* **91**:105–123.
18. Harris, R. J., Storm, P. A., Lloyd, A., Arens, M., and Marmion, B. P., 2000, Long-term persistence of *Coxiella burnetii* in the host after primary Q fever, *Epidemiol. Infect.* **124**:543–549.
19. Raoult, D., Houpikian, P., Tissot Dupont, H., Riss, J. M., Arditi-Djiane, J., and Brouqui, P., 1999, Treatment of Q fever endocarditis: comparison of 2 regimens containing doxycycline and ofloxacin or hydroxychloroquine, *Arch. Intern. Med.* **159**:167–173.
20. Calza, L., Attard, L., Manfredi, R., and Chiodo, F., 2002, Doxycycline and chloroquine as treatment for chronic Q fever endocarditis, *J. Infect.* **45**:127–129.
21. Mcquiston, J. H., and Childs, J. E., 2002, Q fever in humans and animals in the United States, *Vector Borne Zoonotic Dis.* **2**:179–191.
22. Welsh, H. H., Lennette, E. H., Abinanti, F. R., and Winn, J. F., 1958, Air-borne transmission of Q fever: The role of parturition in the generation of infective aerosols, *Ann. N Y Acad. Sci.* **70**:528–540.
23. Welsh, H. H., Lennette, E. H., Abinanti, F. R., Winn, J. F., and Kaplan, W., 1959, Q fever studies. XXI. The recovery of *Coxiella burnetii* from the soil and surface water of premises harboring infected sheep, *Am. J. Hyg.* **70**:14–20.
24. Fishbein, D. B., and Raoult, D., 1992, A cluster of *Coxiella burnetii* infections associated with exposure to vaccinated goats and their unpasteurized dairy products, *Am. J. Trop. Med. Hyg.* **47**: 35–40.
25. Marrie, T. J., Durant, H., Williams, J. C., Mintz, E., and Waag, D. M., 1988, Exposure to parturient cats: A risk factor for acquisition of Q fever in Maritime Canada, *J. Infect. Dis.* **158**:101–108.
26. Stein, A., and Raoult, D., 1999, Pigeon pneumonia in provence: A bird-borne Q fever outbreak, *Clin. Infect. Dis.* **29**:617–620.
27. Buhariwalla, F., Cann, B., and Marrie, T. J., 1996, A dog-related outbreak of Q fever, *Clin. Infect. Dis.* **23**:753–755.

28. Milazzo, A., Hall, R., Storm, P. A., Harris, R. J., Winslow, W., and Marmion, B. P., 2001, Sexually transmitted Q fever, *Clin. Infect. Dis.* **33**:399–402.
29. Lee, J. H., Park, H. S., Jang, W. J., Koh, S. E., Park, T. K., Kang, S. S., Kim, B. J., Kook, Y. H., Park, K. H., and Lee, S. H., 2004, Identification of the *Coxiella* Sp. detected from *Haemaphysalis longicornis* ticks in Korea, *Microbiol. Immunol.* **48**:125–130.
30. Mccaul, T. F., and Williams, J. C., 1981, Developmental cycle of *Coxiella burnetii*: Structure and morphogenesis of vegetative and sporogenic differentiations, *J. Bacteriol.* **147**:1063–1076.
31. Mccaul, T. F., 1991, The developmental cycle of *Coxiella burnetii*, in: *Q Fever: The Biology of Coxiella burnetii* (J.C. Williams and H. A. Thompson, eds.), CRC Press, Boca Raton, FL.
32. Heinzen, R. A., Hackstadt, T., and Samuel, J. E., 1999, Developmental biology of *Coxiella burnettii*, *Trends Microbiol.* **7**:149–154.
33. Kordova, N., 1959, Filterable particles of *Coxiella burnetii*, *Acta Virol.* **3**:25–36.
34. Ransom, S. E., and Huebner, R. J., 1951, Studies on the resistance of *Coxiella burnetii* to physical and chemical agents, *Am. J. Hyg.* **53**:110–119.
35. Malloch, R. A. and Stoker, M. G., 1952, Studies on the susceptibility of *Rickettsia burneti* to chemical disinfectants, and on techniques for detecting small numbers of viable organisms, *J. Hyg. (Lond)* **50**:502–514.
36. Scott, G. H., and Williams, J. C., 1990, Susceptibility of *Coxiella burnetii* to chemical disinfectants, *Ann. N Y Acad. Sci.* **590**:291–296.
37. Amano, K., Williams, J. C., Mccaul, T. F., and Peacock, M. G., 1984, Biochemical and immunological properties of *Coxiella burnetii* cell wall and peptidoglycan-protein complex fractions, *J. Bacteriol.* **160**:982–988.
38. Heinzen, R. A., and Hackstadt, T., 1996, A developmental stage-specific histone H1 homolog of *Coxiella burnetii*, *J. Bacteriol.* **178**:5049–5052.
39. Heinzen, R. A., D. Howe, Mallavia, L. P., Rockey, D. D., and Hackstadt, T., 1996, Developmentally regulated synthesis of an unusually small, basic peptide by *Coxiella burnetii*, *Mol. Microbiol.* **22**:9–19.
40. Tigertt, W. D., Benenson, A. S., and Gochenour, W. S., 1961, Airborne Q fever, *Bacteriol. Rev.* **25**:285–293.
41. Mccaul, T. F., Banerjee-Bhatnagar, N., and Williams, J. C., 1991, Antigenic differences between *Coxiella burnetii* cells revealed by postembedding immunoelectron microscopy and immunoblotting, *Infect. Immun.* **59**:3243–3253.
42. Seshadri, R., Hendrix, L. R., and Samuel, J. E., 1999, Differential expression of translational elements by life cycle variants of *Coxiella burnetii*, *Infect. Immun.* **67**:6026–6033.
43. Amano, K., and Williams, J. C., 1984, Sensitivity of *Coxiella burnetii* peptidoglycan to lysozyme hydrolysis and correlation of sacculus rigidity with peptidoglycan-associated proteins, *J. Bacteriol.* **160**:989–993.
44. Hackstadt, T., and Williams, J. C., 1981, Biochemical stratagem for obligate parasitism of eukaryotic cells by *Coxiella burnetii*, *Proc. Natl. Acad. Sci. USA* **78**:3240–3244.
45. Wiebe, M. E., Burton, P. R., and Shankel, D. M., 1972, Isolation and characterization of two cell types of *Coxiella burnetii* phase I, *J. Bacteriol.* **110**:368–377.
46. Seshadri, R., Paulsen, I. T., Eisen, J. A., Read, T. D., Nelson, K. E., Nelson, W. C., Ward, N. L., Tettelin, H., Davidsen, T. M., Beanan, M. J., Deboy, R. T., Daugherty, S. C., Brinkac, L. M., R. Madupu, Dodson, R. J., Khouri, Lee, K. H., Carty, H. A., Scanlan, D., Heinzen, R. A., Thompson, H. A., Samuel, J. E., Fraser, C. M., and Heidelberg, J. F., 2003, Complete genome sequence of the Q-fever pathogen *Coxiella burnetii*, *Proc. Natl. Acad. Sci. USA* **100**: 5455–5460.
47. Samuel, J. E., Frazier, M. E., Kahn, M. L., Thomashow, L. S., and Mallavia, L. P., 1983, Isolation and characterization of a plasmid from phase I *Coxiella burnetii*, *Infect. Immun.* **41**:488–493.
48. Suhan, M., Chen, S. Y., Thompson, H. A., Hoover, T. A., Hill, A., and Williams, J. C., 1994, Cloning and characterization of an autonomous replication sequence from *Coxiella burnetii*, *J. Bacteriol.* **176**:5233–5243.

49. Suhan, M. L., Chen, S. Y., and Thompson, H. A., 1996, Transformation of *Coxiella burnetii* to ampicillin resistance, *J. Bacteriol.* **178**:2701–2708.
50. Lin, Z., and Mallavia, L. P., 1994, Identification of a partition region carried by the plasmid QpH1 of *Coxiella burnetii*, *Mol. Microbiol.* **13**:513–523.
51. Valkova, D., and Kazar, J., 1995, A new plasmid (QpDV) common to *Coxiella burnetii* isolates associated with acute and chronic Q fever, *FEMS Microbiol. Lett.* **125**:275–280.
52. Minnick, M. F., Heinzen, R. A., Douthart, R., Mallavia, L. P., and Frazier, M. E., 1990, Analysis of QpRS-specific sequences from *Coxiella burnetii*, *Ann. N Y Acad. Sci.* **590**:514–522.
53. Willems, H., Lautenschlager, S., Radomski, K.-U., Jager, C., and Baljer, G., *Coxiella burnetii* plasmid types, in: *Rickettsia and Rickettsial Diseases at the Turn of the Third Millenium*, (D. Raoult, and P. Brouqui, eds.), Elsevier Press, Paris.
54. Samuel, J. E., Frazier, M. E., and Mallavia, L. P., Correlation of plasmid type and disease caused by *Coxiella burnetii*, *Infect. Immun.* **49**:775–779.
55. Willems, H., Ritter, M., Jager, C., and Thiele, D., Plasmid-homologous sequences in the chromosome of plasmidless *Coxiella burnetii* Scurry Q217, *J. Bacteriol.* **179**:3293–3297.
56. Hoover, T. A., Vodkin, M. H., and Williams, J. C., 1992, A *Coxiella burnetti* repeated DNA element resembling a bacterial insertion sequence, *J. Bacteriol.* **174**:5540–5548.
57. Andersson, S. G., Zomorodipour, A., Andersson, J. O., Sicheritz-Ponten, T., Alsmark, U. C., Podowski, R. M., Naslund, A. K., Eriksson, A. S., Winkler, H. H., and Kurland, C. G., 1998, The genome sequence of *Rickettsia prowazekii* and the origin of mitochondria, *Nature* **396**: 133–140.
58. Hoover, T. A., Culp, D. W., Vodkin, M. H., Williams, J. C., and Thompson, H. A., 2002, Chromosomal DNA deletions explain phenotypic characteristics of two antigenic variants, phase II and RSA 514 (crazy), of the *Coxiella burnetii* nine mile strain, *Infect. Immun.* **70**:6726–6733.
59. Thompson, H. A., Hoover, T. A., Vodkin, M. H., and Shaw, E. I., 2003, Do chromosomal deletions in the lipopolysaccharide biosynthetic regions explain all cases of phase variation in *Coxiella burnetii* strains? An update, *Ann. N Y Acad. Sci.* **990**:664–670.
60. Vishwanath, S., and Hackstadt, T., 1988, Lipopolysaccharide phase variation determines the complement-mediated serum susceptibility of *Coxiella burnetii*, *Infect. Immun.* **56**:40–44.
61. Hackstadt, T., 1990, The role of lipopolysaccharides in the virulence of *Coxiella burnetii*, *Ann. N Y Acad. Sci.* **590**:27–32.
62. Amano, K., Williams, J. C., Missler, S. R., and Reinhold, V. N., 1987, Structure and biological relationships of *Coxiella burnetii* lipopolysaccharides, *J. Biol. Chem.* **262**:4740–4747.
63. Hackstadt, T., Peacock, M. G., Hitchcock, P. J., and Cole, R. L., 1985, Lipopolysaccharide variation in *Coxiella burnetti*: Intrastrain heterogeneity in structure and antigenicity, *Infect. Immun.* **48**:359–365.
64. Toman, R., Garidel, P., Andra, J., Slaba, K., Hussein, A., Koch, M. H., and Brandenburg, K., 2004, Physicochemical characterization of the endotoxins from *Coxiella burnetii* strain Priscilla in relation to their bioactivities, *BMC Biochem.* **5**:1.
65. Toman, R., Hussein, A., Palkovic, P., and Ftacek, P., 2003, Structural properties of lipopolysaccharides from *Coxiella burnetii* strains Henzerling and S, *Ann. N Y Acad. Sci.* **990**:563–567.
66. Toman, R., and Skultety, L., 1994, Analysis of the 3-deoxy-D-manno-2-octulosonic acid region in a lipopolysaccharide isolated from *Coxiella burnetii* strain Nine Mile in phase II, *Acta. Virol.* **38**:241–243.
67. Mayer, H., Radziejewska-Lebrecht, J., and Schramek, S., 1988, Chemical and immunochemical studies on lipopolysaccharides of *Coxiella burnetii* phase I and phase II, *Adv. Exp. Med. Biol.* **228**:577–591.
68. Toman, R., and Skultety, L., 1996, Structural study on a lipopolysaccharide from *Coxiella burnetii* strain Nine Mile in avirulent phase II, *Carbohydr. Res.* **283**:175–185.
69. Toman, R., and Kazar, J., 1991, Evidence for the structural heterogeneity of the polysaccharide component of *Coxiella burnetii* strain Nine Mile lipopolysaccharide, *Acta Virol.* **35**:531–537.

70. Toman, R., Skultety, L., Ftacek, P., and Hricovini, M., 1998, NMR study of virenose and dihydrohydroxystreptose isolated from *Coxiella burnetii* phase I lipopolysaccharide, *Carbohydr. Res.* **306**:291–296.
71. Ormsbee, R. A., Bell, E. J., Lackman, D. B., and Tallent, G., 1964, The influence of phase on the protective potency of Q fever vaccine, *J. Immunol.* **92**:404–412.
72. Schramek, S., Radziejewska-Lebrecht, J., and Mayer, H., 1985, 3-C-branched aldoses in lipopolysaccharide of phase I *Coxiella burnetii* and their role as immunodominant factors, *Eur. J. Biochem.* **148**:455–461.
73. Stoker, M. G., and Fiset, P., 1956, Phase variation of the Nine Mile and other strains of *Rickettsia burneti*, *Can. J. Microbiol.* **2**:310–321.
74. Amano, K., and Williams, J. C., 1984, Chemical and immunological characterization of lipopolysaccharides from phase I and phase II *Coxiella burnetii*, *J. Bacteriol.* **160**:994–1002.
75. Vodkin, M. H., and Williams, J. C., 1986, Overlapping deletion in two spontaneous phase variants of *Coxiella burnetii*, *J. Gen. Microbiol.* 132(Pt 9):2587–2594.
76. Moos, A., and Hackstadt, T., 1987, Comparative virulence of intra- and interstrain lipopolysaccharide variants of *Coxiella burnetii* in the guinea pig model, *Infect. Immun.* **55**:1144–1150.
77. Ftacek, P., Skultety, L., and Toman, R., 2000, Phase variation of *Coxiella burnetii* strain Priscilla: Influence of this phenomenon on biochemical features of its lipopolysaccharide, *J. Endotoxin. Res.* **6**:369–376.
78. Vodkin, M. H., Williams, J. C., and Stephenson, E. H., 1986, Genetic heterogeneity among isolates of *Coxiella burnetii*, *J. Gen. Microbiol.* 132(Pt 2):455–463.
79. Williams, J. C., Thomas, L. A., and Peacock, M. G., 1986, Identification of phase-specific antigenic fractions of *Coxiella burnetti* by enzyme-linked immunosorbent assay, *J. Clin. Microbiol.* **24**:929–934.
80. Baca, O. G., Akporiaye, E. T., Aragon, A. S., Martinez, I. L., Robles, M. V., and Warner, N. L., 1981, Fate of phase I and phase II *Coxiella burnetii* in several macrophage-like tumor cell lines, *Infect. Immun.* **33**:258–266.
81. Baca, O. G., and Paretsky, D., 1983, Q fever and *Coxiella burnetii*: A model for host-parasite interactions, *Microbiol. Rev.* **47**:127–149.
82. Capo, C., Lindberg, F. P., Meconi, S., Zaffran, Y., Tardei, G., Brown, E. J., Raoult, D., and Mege, J. L., 1999, Subversion of monocyte functions by *Coxiella burnetii*: impairment of the cross-talk between alphavbeta3 integrin and CR3, *J. Immunol.* **163**:6078–6085.
83. Howe, D., and Mallavia, L. P., 2000, *Coxiella burnetii* exhibits morphological change and delays phagolysosomal fusion after internalization by J774A.1 cells, *Infect. Immun.* **68**:3815–3821.
84. Meconi, S., Jacomo, V., Boquet, P., Raoult, D., Mege, J. L., and Capo, C., 1998, *Coxiella burnetii* induces reorganization of the actin cytoskeleton in human monocytes, *Infect. Immun.* **66**:5527–5533.
85. Beron, W., Gutierrez, M. G., Rabinovitch, M., and Colombo, M. I., 2002, *Coxiella burnetii* localizes in a Rab7-labeled compartment with autophagic characteristics, *Infect. Immun.* **70**:5816–5821.
86. Heinzen, R. A., Scidmore, M. A., Rockey, D. D., and Hackstadt, T., 1996, Differential interaction with endocytic and exocytic pathways distinguish parasitophorous vacuoles of *Coxiella burnetii* and *Chlamydia trachomatis*, *Infect. Immun.* **64**:796–809.
87. Lem, L., Riethof, D. A., Scidmore-Carlson, M., Griffiths, G. M., Hackstadt, T., and Brodsky, F. M., 1999, Enhanced interaction of HLA-DM with HLA-DR in enlarged vacuoles of hereditary and infectious lysosomal diseases, *J. Immunol.* **162**:523–532.
88. Miller, J. D., and Thompson, H. A., 2002, Permeability of *Coxiella burnetii* to ribonucleosides, *Microbiology* **148**:2393–2403.
89. Waag, D., Williams, J. C., Peacock, M. G. and Raoult, D., 1991, Methods of isolation, amplification, and purification of *Coxiella burnetii*, in: *Q Fever: The Biology of Coxiella burnetii* (J. C. Williams and H. A. Thompson, eds.), CRC Press, Boca Raton, FL.

90. Burton, P. R., Stueckemann, J., Welsh, R. M., and Paretsky, D., 1978, Some ultrastructural effects of persistent infections by the rickettsia *Coxiella burnetii* in mouse L cells and green monkey kidney (Vero) cells, *Infect. Immun.* **21**:556–566.
91. Khavkin, T., Sukhinin, V., and Amosenkova, N., 1981, Host-parasite interaction and development of infraforms in chicken embryos infected with *Coxiella burnetii* via the yolk sac, *Infect. Immun.* **32**:1281–1291.
92. Zuerner, R. L., and Thompson, H. A., 1983, Protein synthesis by intact *Coxiella burnetii* cells, *J. Bacteriol.* **156**:186–191.
93. Schneider, W., 1989, Titration of *Coxiella burnetii* in Buffalo green monkey (BGM) cell cultures, *Zentralbl Bakteriol.* **271**:77–84.
94. Veras, P. S., Moulia, C., Dauguet, C., Tunis, C. T., Thibon, M., and Rabinovitch, M., 1995, Entry and survival of *Leishmania amazonensis* amastigotes within phagolysosome-like vacuoles that shelter *Coxiella burnetii* in Chinese hamster ovary cells, *Infect. Immun.* **63**:3502–3506.
95. Mege, J. L., Maurin, M., Capo, C., and Raoult, D., 1997, *Coxiella burnetii*: The 'query' fever bacterium. A model of immune subversion by a strictly intracellular microorganism, *FEMS Microbiol. Rev.* **19**:209–217.
96. Meconi, S., Capo, C., Remacle-Bonnet, M., Pommier, G., Raoult, D., and Mege, J. L., 2001, Activation of protein tyrosine kinases by *Coxiella burnetii*: Role in actin cytoskeleton reorganization and bacterial phagocytosis, *Infect. Immun.* **69**:2520–2526.
97. Batrukova, M. A., Betin, V. L., Rubtsov, A. M., and Lopina, O. D., 2000, Ankyrin: structure, properties, and functions, *Biochemistry (Mosc)* **65**:395–408.
98. Zamboni, D. S., Mortara, R. A., and Rabinovitch, M., 2001, Infection of Vero cells with *Coxiella burnetii* phase II: Relative intracellular bacterial load and distribution estimated by confocal laser scanning microscopy and morphometry, *J. Microbiol. Methods* **43**:223–232.
99. Afseth, G., Mo, Y. Y., and Mallavia, L. P., 1995, Characterization of the 23S and 5S rRNA genes of *Coxiella burnetii* and identification of an intervening sequence within the 23S rRNA gene, *J. Bacteriol.* **177**:2946–2949.
100. Winkler, H. H., 1995, *Rickettsia prowazekii*, ribosomes and slow growth, *Trends Microbiol.* **3**: 196–198.
101. Mccaul, T. F., Williams, J. C., and Thompson, H. A., 1991, Electron microscopy of *Coxiella burnetii* in tissue culture. Induction of cell types as products of developmental cycle, *Acta Virol.* **35**:545–556.
102. Desjardins, M., 2003, ER-mediated phagocytosis: A new membrane for new functions, *Nat. Rev. Immunol.* **3**:280–291.
103. Garin, J., Diez, R., Kieffer, S., Dermine, J. F., Duclos, S., Gagnon, E., Sadoul, R., Rondeau, C., and Desjardins, M., 2001, The phagosome proteome: Insight into phagosome functions, *J. Cell Biol.* **152**:165–180.
104. Thompson, H. A., 1991, Metabolism in *Coxiella burnetii*, in: *Q Fever: The Biology of Coxiella burnetii* (J. C. Williams and H. A. Thompson, eds.), CRC Press, Boca Raton, FL.
105. Chen, S. Y., Vodkin, M., Thompson, H. A., and Williams, J. C., 1990, Isolated *Coxiella burnetii* synthesizes DNA during acid activation in the absence of host cells, *J. Gen. Microbiol.* **136**(Pt 1):89–96.
106. Howe, D., Barrows, L. F., Lindstrom, N. M., and Heinzen, R. A., 2002, Nitric oxide inhibits *Coxiella burnetii* replication and parasitophorous vacuole maturation, *Infect. Immun.* **70**:5140–5147.
107. Samuel, J. E., Kiss, K., and Varghees, S., 2003, Molecular pathogenesis of *Coxiella burnetii* in a genomics era, *Ann. N Y Acad. Sci.* **990**:653–663.
108. Baca, O. G., Li, Y. P., and Kumar, H., 1994, Survival of the Q fever agent *Coxiella burnetii* in the phagolysosome, *Trends Microbiol.* **2**:476–480.
109. Li, Y. P., Curley, G., Lopez, M., Chavez, M., Glew, R., Aragon, A., Kumar, H., and Baca, O. G., 1996, Protein-tyrosine phosphatase activity of *Coxiella burnetii* that inhibits human neutrophils, *Acta Virol.* **40**:263–272.

110. Baca, O. G., Roman, M. J., Glew, R. H., Christner, R. F., Buhler, J. E., and Aragon, A. S., 1993, Acid phosphatase activity in *Coxiella burnetii*: A possible virulence factor, *Infect. Immun.* **61**:4232–4239.
111. Redd, T., and Thompson, H. A., 1995, Secretion of proteins by *Coxiella burnetii*, *Microbiology* 141 (Pt 2):363–369.
112. Zamboni, D. S., Mcgrath, S., Rabinovitch, M., and Roy, C. R., 2003, *Coxiella burnetii* express type IV secretion system proteins that function similarly to components of the Legionella pneumophila Dot/Icm system, *Mol. Microbiol.* **49**:965–976.
113. Pisoni, R. L., and Thoene, J. G., 1991, The transport systems of mammalian lysosomes, *Biochim. Biophys. Acta* **1071**:351–373.
114. Thompson, H. A., 1988, Relationship of the physiology and composition of *Coxiella burnetii* to the *Coxiella*-host cell interaction, in: *Biology of Rickettsial Diseases* (D. H. Walker, ed.), CRC Press, Boca Raton, FL.
115. Winkler, H. H., 1976, Rickettsial permeability. An ADP-ATP transport system, *J. Biol. Chem.* **251**:389–396.
116. Tjaden, J., Winkler, H. H., Schwoppe, C., Van Der Laan, M., Mohlmann, T., and Neuhaus, H. E., 1999, Two nucleotide transport proteins in *Chlamydia trachomatis*, one for net nucleoside triphosphate uptake and the other for transport of energy, *J. Bacteriol.* **181**:1196–1202.
117. Segal, G., Russo, J. J., and Shuman, H. A., 1999, Relationships between a new type IV secretion system and the icm/dot virulence system of *Legionella pneumophila*, *Mol. Microbiol.* **34**:799–809.
118. Sexton, J. A., and Vogel, J. P., 2002, Type IVB secretion by intracellular pathogens, *Traffic* **3**:178–185.
119. Zusman, T., Yerushalmi, G., and Segal, G., 2003, Functional similarities between the icm/dot pathogenesis systems of *Coxiella burnetii* and *Legionella pneumophila*, *Infect. Immun.* **71**:3714–3723.
120. Hackstadt, T., and Williams, J. C., 1981, Incorporation of macromolecular precursors by *Coxiella burnetii* in an axenic medium, in: *Rickettsiae and Rickettsial Diseases* (W. Burgdorfer and R. L. Anacker, eds.), Academic Press, New York.
121. Hendrix, L., and Mallavia, L. P., 1984, Active transport of proline by *Coxiella burnetii*, *J. Gen. Microbiol.* 130(Pt 11):2857–2863.
122. Lukacova, M., Valkova, D., Quevedo Diaz, M., Perecko, D., and Barak, I., 1999, Green fluorescent protein as a detection marker for *Coxiella burnetii* transformation, *FEMS Microbiol. Lett.* **175**:255–260.
123. Byrne, W. R., 1997, Q fever, in: *Medical Aspects of Chemical and Biological Warfare* (F. R. Sidell, E. T. Takafuji, and D. R. Franz, eds.), TMM Publications, Washington, DC.
124. Moe, J. B., and Pedersen, Jr., C. E., 1980, The impact of rickettsial diseases on military operations, *Mil Med.* **145**:780–785.
125. Spicer, A. J., 1978, Military significance of Q fever: A review, *J. Roy. Soc. Med.* **71**:762–767.
126. Report of a WHO group of consultants 1970, Health aspects of chemical and biological weapons. WHO, Geneva.
127. Bellamy, R. J., and Freedman, A. R., 2001, Bioterrorism, *QJ Med.* **94**:227–234.
128. Holloway, H. C., and Engel, Jr., C. C., 1997, The threat of biological weapons: Prophylaxis and mitigation of psychological and social consequences, *JAMA* **278**:425–427.
129. Kagawa, F. T., Wehner, J. H., and Mohindra, V., 2003, Q fever as a biological weapon, *Semin. Respir. Infect.* **18**:183–195.
130. Noah, D. L., Huebner, K. D., Darling, R. G., and Waeckerle, J. F., 2002, The history and threat of biological warfare and terrorism, *Emerg. Med. Clin. North Am.* **20**:255–271.
131. Christopher, G. W., and Pavlin, J. A., 1997, Biological warfare: A historical perspective, *JAMA* **278**:412–417.
132. Henderson, D. A., 1999, The looming threat of bioterrorism, *Science* **283**:1279–1282.
133. Enright, J. B., Sadler, W. W., and Thomas, R. C., 1957, Thermal inactivation of *Coxiella burnetii* and its relation to pasteurization of milk, *Public Health Monogr.* **54**:1–30.

134. Scott, G. H., Mccaul, T. F., and Williams, J. C., 1989, Inactivation of *Coxiella burnetii* by gamma irradiation, *J. Gen. Microbiol.* **135**(Pt 12):3263–3270.
135. Field, P. R., Hunt, J. G., and Murphy, A. M., 1983, Detection and persistence of specific IgM antibody to *Coxiella burnetii* by enzyme-linked immunosorbent assay: A comparison with immunofluorescence and complement fixation tests, *J. Infect. Dis.* **148**:477–487.
136. Peter, O., Dupuis, G., Burgdorfer, W., and Peacock, M., 1985, Evaluation of the complement fixation and indirect immunofluorescence tests in the early diagnosis of primary Q fever, *Eur. J. Clin. Microbiol.* **4**:394–396.
137. Muhlemann, K., Matter, L., Meyer, B., and Schopfer, K., 1995, Isolation of *Coxiella burnetii* from heart valves of patients treated for Q fever endocarditis, *J. Clin. Microbiol.* **33**:428–431.
138. Mccaul, T. F., and Williams, J. C., 1990, Localization of DNA in *Coxiella burnetii* by post-embedding immunoelectron microscopy, *Ann. N Y Acad. Sci.* **590**:136–147.
139. Thiele, D., Karo, M., and Krauss, H., 1992, Monoclonal antibody based capture ELISA/ELIFA for detection of *Coxiella burnetii* in clinical specimens, *Eur. J. Epidemiol.* **8**:568–574.
140. Brouqui, P., Dumler, J. S., and Raoult, D., 1994, Immunohistologic demonstration of *Coxiella burnetii* in the valves of patients with Q fever endocarditis, *Am. J. Med.* **97**:451–458.
141. Stein, A., and Raoult, D., 1992, Detection of *Coxiella burnetti* by DNA amplification using polymerase chain reaction, *J. Clin. Microbiol.* **30**:2462–2466.
142. Willems, H., Thiele, D., Frolich-Ritter, R., and Krauss, H., 1994, Detection of *Coxiella burnetii* in cow's milk using the polymerase chain reaction (PCR), *Zentralbl Veterinarmed B* **41**:580–587.
143. Musso, D., and Raoult, D., 1995, *Coxiella burnetii* blood cultures from acute and chronic Q-fever patients, *J. Clin. Microbiol.* **33**:3129–3132.
144. Raoult, D., Vestris, G., and Enea, M., 1990, Isolation of 16 strains of *Coxiella burnetii* from patients by using a sensitive centrifugation cell culture system and establishment of the strains in HEL cells, *J. Clin. Microbiol.* **28**:2482–2484.
145. Perrin, T. K., and Bengston, I. A., 1942, The histopathology of experimental Q fever in mice, *Public Health Rep.* **57**:790–794.
146. Scott, G. H., Williams, J. C., and Stephenson, E. H., 1987, Animal models in Q fever: Pathological responses of inbred mice to phase I *Coxiella burnetii*, *J. Gen. Microbiol.* 133 (Pt 3):691–700.
147. Spelman, D. W., 1982, Q fever: A study of 111 consecutive cases, *Med. J. Aust.* **1**:547–548, 551, 553.
148. D'angelo, L. J., and Hetherington, R., 1979, Q fever treated with erythromycin, *Br. Med. J.* **2**:305–306.
149. Perez-Del-Molino, A., Aguado, J. M., Riancho, J. A., Sampedro, I., Matorras, P., and Gonzalez-Macias, J., 1991, Erythromycin and the treatment of *Coxiella burnetii* pneumonia, *J. Antimicrob. Chemother.* **28**:455–459.
150. Marrie, T. J., 1990, *Coxiella burnetii* (Q fever), in: *Principles and Practice of Infectious Diseases* (G. L. Mandell and J. E. Bennett, eds.), Churchill Livingtone, New York.
151. Raoult, D., Fenollar, F., and Stein, A., 2002, Q fever during pregnancy: Diagnosis, treatment, and follow-up, *Arch. Intern. Med.* **162**:701–704.
152. Raoult, D., 1993, Treatment of Q fever, *Antimicrob. Agents Chemother.* **37**:1733–1736.
153. Levy, P. Y., Drancourt, M., Etienne, J., Auvergnat, J. C., Beytout, J., Sainty, J. M., Goldstein, F., and Raoult, D., 1991, Comparison of different antibiotic regimens for therapy of 32 cases of Q fever endocarditis, *Antimicrob. Agents Chemother.* **35**:533–537.
154. Krisstinsson, A., and Bentall, H. H., 1967, Medical and surgical treatment of Q fever endocarditis, *Lancet* **2**:693–695.
155. Fernandez-Guerrero, M. L., Muelas, J. M., Aguado, J. M., Renedo, G., Fraile, J., Soriano, F., and De Villalobos, E., 1988, Q fever endocarditis on porcine bioprosthetic valves. Clinicopathologic features and microbiologic findings in three patients treated with doxycycline, cotrimoxazole, and valve replacement, *Ann. Intern. Med.* **108**:209–213.
156. Pedoe, H. D., 1970, Apparent recurrence of Q fever endocarditis following homograft replacement of aortic valve, *Br. Heart J.* **32**:568–570.

157. Smadel, J. E., Snyder, M. J., and Robbins, F. C., 1948, Vaccination against Q fever, *Am. J. Hyg.* **47**:71–81.
158. Ackland, J. R., Worswick, D. A., and Marmion, B. P., 1994, Vaccine prophylaxis of Q fever. A follow-up study of the efficacy of Q-Vax (CSL) 1985–1990, *Med. J. Aust.* **160**:704–708.
159. Marmion, B. P., Ormsbee, R. A., Kyrkou, M., Wright, J., Worswick, D., Cameron, S., Esterman, A., Feery, B., and Collins, W., 1984, Vaccine prophylaxis of abattoir-associated Q fever, *Lancet* **2**:1411–1414.
160. Mills, A. E., Murdolo, V., and Webb, S. P., 2003, A rare local granulomatous complication of Q fever vaccination, *Med. J. Aust.* **179**:166.
161. Fries, L. F., Waag, D. M., and Williams, J. C., 1993, Safety and immunogenicity in human volunteers of a chloroform-methanol residue vaccine for Q fever, *Infect. Immun.* **61**:1251–1258.
162. Waag, D. M., England, M. J., and Pitt, M. L., 1997, Comparative efficacy of a *Coxiella burnetii* chloroform:methanol residue (CMR) vaccine and a licensed cellular vaccine (Q-Vax) in rodents challenged by aerosol, *Vaccine* **15**:1779–1783.
163. Waag, D. M., England, M. J., Tammariello, R. F., Byrne, W. R., Gibbs, P., Banfield, C. M., and Pitt, M. L., 2002, Comparative efficacy and immunogenicity of Q fever chloroform:methanol residue (CMR) and phase I cellular (Q-Vax) vaccines in cynomolgus monkeys challenged by aerosol, *Vaccine* **20**:2623–2634.

11

Genomic and Proteomic Approaches Against Q Fever

JAMES E. SAMUEL, LAURA R. HENDRIX, KASI RUSSELL, and GUOQUAN ZHANG

1. INTRODUCTION

Coxiella burnetii is an obligate intracellular bacterium that causes a world-wide zoonotic disease, Q fever. The organism is an occupational hazard and can develop as an acute self-limiting illness or occasionally manifest as a chronic infection, with poor prognosis of resolution. The organism passively enters host cells via actin-based cytoskeletal reorganization and replicates in a parasitophorus vacuole with markers similar to typical phagolysosomes. Recent progress with genomic and proteomic technologies provide a variety of new research opportunities for understanding the molecular pathogenesis of the agent as well as developing diagnostic and vaccine strategies to respond to disease outbreaks. Review of host parasite interactions, insights provided by genomic analysis of the prototype isolate, Nine Mile, and immune response to infection and protective vaccination are included in this chapter. Additionally, new opportunities for research are highlighted.

2. DISEASE AND THREAT

Coxiella burnetii, the etiological agent of "Q-fever", is a category-B bioterrorism agent that is highly infective to both humans and livestock. The first description of the disease was the result of efforts to understand a mysterious disease outbreak in Australia.[1] Cattle, sheep, and goats are the primary

JAMES E. SAMUEL, LAURA R. HENDRIX, KASI RUSSELL, and GUOQUAN ZHANG • Texas A&M University System Health Science Center, College Station, TX 77843, USA.

reservoirs, but isolates have been obtained from a wide variety of wild vertebrates and arthropods.[2,3] Human infection arises primarily from aerosol transmission, and *C. burnetii* can withstand desiccation and remain infectious in contaminated soils for several years.[4] Ease of dissemination via aerosol, environmental persistence, and high infectivity ($ID_{50} = 1$) make *C. burnetii* a serious threat for military personnel and civilians. This agent has been weaponized and mass-produced under various biological warfare programs.[5–7] The disease manifestations of *C. burnetii* infection in humans can be separated into acute and chronic illnesses. Acute disease commonly presents as flu-like illness with hallmark cyclic fever and periorbital headache. Pneumonitis and hepatitis are common complications, but acute disease is almost always self-limiting. Various antibiotics, including tetracycline, are effective for abrogating acute disease. Clinically, the illness falls within the group of FUO (fever of unknown origin) syndromes and is not commonly recognized or diagnosed. In many areas a high percentage of the population (10–20%) has serological evidence of previous infection. In contrast, chronic infection is much less common and has a grim prognosis. Chronic disease most frequently manifests as endocarditis and hepatitis with recognition of these infections increasing worldwide. Chronic infections appear to be associated with a suppression of the cell-mediated immune system. Antigen-driven lymphoproliferation and interferon-γ (IFN-γ) synthesis are down-regulated.[8–10] Elevated expression of IL-10 is associated with poor disease outcome.[11] Chronic infections have not responded well to a variety of antibiotic regimens,[12,13] with patients receiving a combination of doxycycline plus chloroquine administered over 1–2 years showing the best outcome.[14]

3. HOST–PARASITE INTERACTION

Coxiella burnetii replicate in vacuoles in a variety of fibroblast, epithelial and macrophage-like cell lines with minimal effects on the host cell. *C. burnetii*-infected cells show an up-regulation of transferrin receptor synthesis resulting in an increase in intracellular iron. The inhibition of *C. burnetii* replication by desferrioxamine, an intracellular iron chelator, indicates the pathogen requires iron for growth.[15] Adherence and entry into host cells appears to be a passive process, as inactivated *C. burnetii* are endocytosed in a microfilament-dependent manner at a rate equal to that of viable bacteria.[16] Other than an increase in the rate of bacterial uptake in the presence of specific antibody,[17] there is little information available on host cell receptors involved in entry. However, several studies in THP-1 human monocyte/macrophage cell lines have described a strategy by virulent *C. burnetii* to subvert the phagocytic process by altering the localization of host receptors. Virulent phase I *C. burnetii*, having an LPS with a complete O-side chain, were found to bind to THP-1 cells via a complex of leukocyte response integrin (LRI) $\alpha_v\beta_3$ and integrin-associated protein

(IAP). However, avirulent phase II organisms, having a truncated LPS, entered THP-1 cells via complement CR3 receptors.[18] A transient reorganization of the host cell actin cytoskeleton by virulent but not avirulent *C. burnetii* stimulated morphological changes, seen as protrusions, in human monocytes.[19] CR3 was excluded from the actin- and *C. burnetii*-containing protrusions.[20] In this system, virulent organisms were taken up poorly, but survived, while avirulent *C. burnetii* were readily taken up and were killed.[18] In earlier studies, purified phase I LPS was able to reduce entry of both phase I and phase II *C. burnetii* into L cells, possibly through a similar mechanism.[16]

After engulfment, early phagosomes undergo maturation to form an endocytic compartment with a pH of approximately 4.8.[21] *Coxiella burnetii* require an acidic pH to activate growth and metabolism inside the host cell and for metabolic activation *in vitro*.[22–24] They maintain this acidic pH in persistently infected cells.[25] Many studies, using primarily avirulent phase II cells, have shown that vacuoles in which *C. burnetii* replicate fuse with lysosomes and contain lysosomal markers, including acid phosphatase,[21,26,27,28] cathepsin D,[27] 5′ nucleotidase,[28] vacuolar H^+-ATPase,[27] Rab 7,[29] and the lysosomal glycoproteins LAMP1 and LAMP2.[27] In THP-1 cells, phase I *C. burnetii* were taken up poorly but survived in a vacuole that did not contain cathepsin D. This same study found that phase II *C. burnetii* were readily taken up but were killed in phagolysosomes that were cathepsin D-positive. This is the first report to indicate a difference in the survival of phase I and phase II *C. burnetii* in any cell type *in vitro*.[18]

Coxiella burnetii exist as small and large cell variants (LCV) inside host cells.[30] Small cell variants (SCV) have a condensed chromatin and a thick cell wall. LCV resemble typical gram-negative bacteria and are thought to be the metabolically active form. Both forms are believed to be infectious.[30] Howe and Mallavia[15] used antibodies to ScvA, a protein present only in SCV[31] to show that the number of SCV decrease in the first hour following infection in a J774 mouse macrophage cell line. In an *in vitro* assay, numbers of SCV decreased when *C. burnetii* were incubated at pH 5.5, but not at pH 7 or 4.5. They also found that viable phase I *C. burnetii* appeared to fuse with lysosomes to a lesser extent in the first 6 hours after infection than did dead *C. burnetii* or latex beads, as demonstrated by the number of vacuoles colocalizing with thorium dioxide or acid phosphatase over time. They hypothesized a delay or lack of fusion allowed the development of the metabolically active LCV.

Coxiella burnetii protein synthesis was shown to be required for the fusion of early phagosomes to form large parasitophorus vacuoles (LPV), but not for the acquisition of lysosomal markers. Replication of *C. burnetii* was also not required for LPV formation.[32] Rab7, a GTPase that controls transport and fusion of late endosomes and lysosomes, was required for the formation of LPV in *C. burnetii*-infected HeLa cells.[29] Strains of inbred mice have been shown to differ in their ability to control the development of LPV, and this ability correlates with the susceptibility of the mice to infection with *C. burnetii.*

Strains of mice that were able to control the development of LPV were able to restrict *C. burnetii* multiplication.[33] These results may correlate with earlier findings that showed activated macrophages and secretion of cytokines, including INF-γ, are required to control *C. burnetii* replication in vitro,[17,34] as several of the susceptible mouse strains have defects in activation of macrophages. Howe *et al.*[35] have gone on to show that nitric oxide controls the production of LPV in infected L929 mouse fibroblasts following treatment with INF-γ and tumor necrosis factor-α. Similar results were obtained in primary mouse macrophages.[36] Brennan *et al.* demonstrated a role for both oxygen radical (H_2O_2 and O-) as well as nitric oxide radicals in control of *C. burnetii* replication in INF-γ-treated mouse monocytes/macrophages.[37] The authors also demonstrated that both radicals are expressed at low levels by unstimulated infected cells and dramatically affect the replication rate of the pathogen. Finally, this study showed the role of both stress responsive radical products *in vivo* using knockout mice to show enhanced disease in animals lacking genes required to express these radicals.

4. SECRETION OF VIRULENCE FACTORS

Genomic sequencing has revealed that *C. burnetii* possess a type IV secretion system similar to its closest relative, *Legionella pneumophila.*[38] For *Legionella*, type IV secretion is required for proper trafficking of the bacterium in the host cell.[39,40] Several of the genes from the *C. burnetii* type IV operon have been shown to complement deletions of homologues in *L. pneumophila.*[41] Transcriptional data suggest *C. burnetii* expresses the type IV operon in infected cells. It is likely that *C. burnetii* also use this system to secrete molecules into the host cell to modify its environment; however, such effector molecules have not yet been identified.

Three enzymes have been suggested to play a role in intracellular survival based on their activities in other pathogens. Macrophage infectivity potentiator (Mip), a peptidylprolyl isomerase, was first isolated in *Legionella pneumophila* through discovery of a mutant attenuated for infection and survival in macrophages.[42,43] *C. burnetii* Mip is expressed in three forms from a single mRNA species using alternate translational initiation sites. Two smaller forms (15 and 15.5 kDa) remain in the cytoplasm, while a larger product (23.5 kDa) is exported to the periplasm and outer membrane via a signal sequence.[44,45]

Com-1 is a periplasmic and outer membrane protein found to have homology to the active site of disulfide oxidoreductase enzymes, such as *E. coli* DsbA.[46] These enzymes are required by several pathogens for the proper folding of virulence determinants.[47] The *C. burnetii com-1* gene was able to complement a *dsbA* mutant and the purified recombinant protein was enzymatically active.

Several pathogens, including *Legionella micdadei*, have been shown to block the oxidative burst of phagocytic cells by expressing an acid phosphatase enzyme.(48) Baca *et al.*(49) have shown that a partially purified acid phosphatase from *C. burnetii* sonic extract blocks production of superoxide anion from fMetLeuPhe-stimulated neutrophils. Using a series of heteromolybdate complex inhibitors, they showed the *C. burnetii* phosphatase had a characteristic pattern of inhibition different from the host cell acid phosphatase.(50) An inhibitor of *C. burnetii* acid phosphatase greatly reduced the percentage of infected L929 cells in a persistently infected cell line.(49)

5. PATHOGENESIS

The obligate intracellular nature of *C. burnetii* has hindered a definition of its virulence factors due to the difficulty in generating and testing defined mutants. Other than the requirement for an LPS with a complete O-side chain (phase I), which is only weakly pyrogenic, *C. burnetii* has no obvious virulence factors. It is a persistent but relatively passive parasite that confers no detectable disturbance to the growth or viability of the host cell and appears to reside in an unmodified phagolysosomal vacuole in most cell types. Studies using THP-1 cells and circulating human monocytes, however, have indicated a possible difference between virulent and avirulent *C. burnetii* in uptake, trafficking, and survival. Virulent *C. burnetii* survive through altered phagosome maturation, but do not replicate well in THP-1 cells and human monocytes. Exogenous IFN-γ added to THP-1 cells induced the killing of virulent *C. burnetii*, through the restoration of vacuole maturation, as shown by the acquisition of cathepsin D.(51) Increased replication of virulent *C. burnetii* in monocytes occurred with the addition of IL-10, a macrophage-deactivating cytokine. Infected monocytes from patients having chronic Q fever endocarditis were shown to overproduce IL-10, allowing for the continued replication of *C. burnetii* in the bloodstream.(52)

Acute Q fever has been modeled in animals as fever development in guinea pigs(53) and as splenomegaly(54) or lethality(55) in mice. As few as 10 virulent organisms of the Nine Mile strain administered intraperitoneally in guinea pigs caused fever within 5 days. Moos and Hackstadt(56) found that 10 inclusion-forming units (IFU) of an acute disease isolate induced fever in guinea pigs, but even 10^6 IFU of a chronic disease isolate could not cause fever, although infection could be confirmed in both cases by isolation of organisms from their spleens. Immunocompetent animal hosts have not yet been able to model chronic disease or endocarditis following *C. burnetii* infection. Several mouse models have employed an immunodeficient host to show enhancement of the severity of disease using acute disease isolates. Athymic mice were unable to clear *C. burnetii* from the blood or spleen and developed chronic disease.(57) Severe combined immunodeficient (SCID) mice infected with the Nine Mile phase I acute disease isolate showed persistent symptoms, severe chronic lesions, and death.[58] Heart lesions were similar to those in humans with chronic Q fever endocarditis[59] and included focal calcifications and *C. burnetii*-containing macrophages.[58] Other experiments to produce chronic disease in animals relied on methods to create an immunodeficient host, including the use of steroids,[60] whole-body irradiation,[60] or cyclophosphamide treatment.[61] Others have modeled endocarditis using intracardiac catheters[62] or electrocoaggulation-induced valvular lesions.[63] While these studies point to the role of the immune system and/or undamaged heart valves in the prevention of chronic disease due to *C. burnetii* infection, they do not answer the question of whether certain isolates have differing virulence potentials.

6. ISOLATE DIVERSITY AND VIRULENCE

Coxiella burnetii has been isolated from various sources including ticks, milk and human cases of acute and chronic Q fever worldwide. Previous studies have shown that *C. burnetii* isolates originating from ticks, milk, and human cases of acute Q fever differ in plasmid type,[64] lipopolysaccharide (LPS) profiles,[65] and chromosomal DNA restriction endonuclease fragment patterns[66] from many isolates originating from chronic Q fever patients. The differences at the phenotypic and molecular levels between acute and chronic disease isolates suggested that acute or chronic Q fever might be caused by different isolates of *C. burnetii.* Studies on several *C. burnetii* isolates from Europe detected either the QpH1 plasmid specific sequences[67,68] or a plasmid type (QpDV)[69] in both acute and chronic disease isolates, suggesting there was no specific gene(s) on plasmids responsible for a specific virulence phenotype. This data supported the notion that chronic disease could result from isolates associated with acute disease and that host-specific factors may be more important in determining disease presentation. Antigen structures including LPS and membrane proteins are considered to play an important role in the development of protective immunity to infection by *C. burnetii.* Hackstadt[65] reported the antigenic variation in LPS of *C. burnetii* isolates from various sources. To *et al.* reported antigenic differences among 18 isolates originating from various sources by immunoblotting. The genetic and antigenic heterogeneity among *C. burnetii* strains suggests Q fever vaccines may need to be developed to protect against challenge from different antigenic groups. However, Ormsbee *et al.* demonstrated phase I whole cell vaccine generated cross-protection against challenge by various *C. burnetii* strains in a guinea pig fever model.[71] Recently, study from our laboratory also indicated that there was complete cross-protection between Nine Mile phase I and Scurry strain (a chronic disease prototype isolate) (K. Russell *et al.*, manuscript in preparation). These studies suggested that a vaccine containing a single killed phase I organism can provide full protection against different isolates.

Coxiella burnetii undergoes a phase variation phenomenon in which virulent phase I (smooth-LPS) convert to an avirulent phase II (rough-LPS) upon serial passage in a non-immunologically competent host. An early study suggested that phase I whole cell vaccine (WCV-PI) was more protective than phase II whole cell vaccine (WCV-PII) against virulent phase I challenge in a guinea pig model.[71] Hackstadt *et al.*[65] demonstrated that the LPS were structurally and antigenically varied between phase I and phase II cells, but the protein components were shared. Since there has been no direct evidence to demonstrate either phase I LPS or phase I unique protein antigens are involved in the development of protective immunity, it remains unclear why the protective efficacy is different between WCV-PI and WCV-PII. The only characterized difference between Nine Mile phase I and phase II organisms is LPS core polysaccharide or O-side chain expression, supporting the hypothesis that phase I LPS plays a critical role in the development of protective immunity. Further comparison of immunogenicity between WCV-PI and WCV-PII may provide important evidence for understanding the fundamental mechanism of protective immunity.

Since *Coxiella* infection may cause reproductive disorder, infertility, or abortion in ruminants, prevention of *Coxiella* infection in domestic animals has a high economic impact. But, more importantly, since cattle, sheep, and goats are considered the main reservoirs for *Coxiella* infection, controlling *Coxiella* infection in these animals could decrease the transmission of the infection to other domestic animals and humans. Although acute Q fever is a treatable disease in humans, infection can lead to severe chronic and occasionally fatal disease. Previous studies indicated that vaccination provided the best way to control Q fever and *Coxiella* infection in humans and animals.[73–77] Several vaccines, including formalin-inactivated phase I or phase II whole cell, attenuated *C. burnetii* and chloroform:methanol residue (CMR) subunit of Nine Mile phase I vaccine have been tested in animals and humans.[73,78–80] Formalin-inactivated WCV-PI was an effective vaccine in protecting against the disease in humans and animals. However, vaccination with WCV-PI can induce severe local or occasional systemic reactions in previously sensitized individuals. Vaccination with WCV-PI required pre-screening of potential vaccinees by skin tests, serological tests, or *in vitro* lymphocyte proliferation assays. However, pre-screening of vaccinees is time-consuming, costly, and may not prevent the development of adverse vaccination reactions. In addition, vaccines have not completely prevented shedding of *C. burnetii* into the environment in animals.

7. ACQUIRED IMMUNITY

Several reports have described the characterization of acquired immunity to *Coxiella* infections in animals and humans.[8,81–85] These studies suggested that both humoral and cell-mediated immune responses are important for protection against *Coxiella* infection, with cell-mediated immunity probably playing the critical role in eliminating *C. burnetii* in experimental animals, while specific antibodies accelerate the process. Early studies indicated that infection and vaccination with *C. burnetii* in animals and humans induced significant antibody responses against *C. burnetii* antigens, suggesting humoral immunity plays a role in the protection of Q fever. Behymer *et al.*[74] reported the long-term persistence of agglutinating antibodies in dairy cattle after vaccination with WCV-PI. Ackland *et al.*[73] showed that patients with acute Q fever developed IgM-specific antibodies to phase I and IgM, IgG, IgA and CF antibodies to phase II antigen, while patients with chronic Q fever have undetectable IgM-specific antibody to phase I or II and induced high level IgG, IgA and CF antibodies to both phase I and II antigens. They also indicated that seronegative volunteers vaccinated with WCV-PI generated dominant IgM antibody response to phase I and lower IgM and IgG and CF antibodies to phase II antigen, but seropositive subjects developed IgA- and IgG-specific antibody response to phase I and CF and IgG class response to phase II. These studies suggested that subclass antibodies may play an important role in the host immune defense and that measurement of subclass antibody responses in Q fever patients may be useful for differential diagnosis of acute and chronic Q fever. Previous studies also demonstrated that antibody plays a direct role in resistance to *C. burnetii* infections. Abinanti and Marmion[86] first reported that mixtures of antibody and *C. burnetii* organisms were not infectious in experimental animals, suggesting that antibody plays a role in the control of *Coxiella* infection. Peacock *et al.* also demonstrated that anti-phase I IgM antibody suppressed the growth of *C. burnetii* in mouse spleen when mixed with the suspension of organisms prior to inoculation. Studies on the efficacy of formalin-killed phase I and II vaccines in humans and experimental animals demonstrated that antibodies were involved in the resistance that developed against *C. burnetii* antigens.[71,87] Several *in vitro* studies also indicated that treatment of *C. burnetii* with immune serum made the organisms more susceptible to phagocytosis and to destruction by normal polymorphonuclear leukocytes or macrophages in culture. These studies provided strong support for the hypothesis that humoral immunity was important in the development of acquired resistance. Humphres and Hinrichs[81] found that immune serum could alter the degree of infection within infected mice and enhance clearance of the rickettsia by the macrophage population of the host. However, treatment of athymic mice with immune serum 24 hours before challenge with *C. burnetii* had no effect on rickettsial multiplication within the spleens of these T-cell-deficient animals. This study suggested that specific antibodies were able to accelerate the initial interactions of the inductive phase of the cellular immune response and promote a more rapid development of activated macrophages to a level that could control *C. burnetii* replication. On the other hand, this study provided evidence supporting the notion that cell-mediated immunity plays a critical role in controlling *C. burnetii* replication.

The role of cell-mediated immunity in *C. burnetii* infection and vaccination has been studied in animals and humans. Kishimoto *et al.* showed that peritoneal macrophages from guinea pigs previously immunized with phase I antigen were capable of killing phase I *C. burnetii* in the absence of homologous immune serum.[84,85] Subsequently, they demonstrated that both infection and vaccination with *C. burnetii* developed cell-mediated immune responses in guinea pigs as determined by the inhibition of macrophage migration and lymphocyte transformation assays.[57,88] Several studies also indicated that infection and vaccination with *C. burnetii* in humans induces the long-lived ability of peripheral blood lymphocytes to proliferate when cultured with *C. burnetii* antigens. These studies demonstrated that cell-mediated immunity was required for protection against *C. burnetii* infection. As in other infections with obligate intracellular pathogens, host defense in Q fever appears to be dependent on cell-mediated immunity in which specifically activated T cells enhance the microbicidal machinery of macrophages. Izzo *et al.* also reported that T lymphocytes are the major contributor to the cellular immune response to *C. burnetii*, as measured by the proliferation of circulating blood lymphocytes on antigen challenge.[82] Recent studies have shown that activation of guinea pig monocytes, THP-1 monocytes, L929 murine fibroblasts, and primary mouse macrophages with IFN-γ resulted in the inhibition of *C. burnetii* replication.[17,34–36,89] *In vitro* and *in vivo* studies also demonstrated that IFN-γ and TNF-α play important roles in the host defense processes leading to the elimination of *C. burnetii*.[34,89] IFN-γmediated killing of *C. burnetii* and death of infected monocytes were dependent on TNF-α. Thus, cell-mediated immune response is required for clearance of a *C. burnetii* infection, and IFN-γ plays a key role in controlling *C. burnetii* replication.

8. WHOLE CELL *C. burnetii* VACCINES

Experimental vaccines prepared from phase I or phase II organisms have been developed for several decades to prevent the disease in animals and humans.[73–77] However, one early study demonstrated that formalin-inactivated phase I *C. burnetii* was 100 to 300 times more effective than phase II organisms in guinea pigs in eliciting antibody and protection against challenge with virulent phase I organisms, with cross-protection among various *C. burnetii* strains in vaccinated guinea pigs.[71] WCV-PI has been the most extensively tested vaccine for prevention of Q fever in animals and humans. Previous studies have demonstrated that WCV-PI was able to induce both humoral and cell-mediated immune responses to *C. burnetii* antigens and was effective in preventing Q fever in animals and humans. Vaccination of domestic animals with WCV-PI has shown a high level of protection against low fetal weight and chronic infertility.[76] Biberstein *et al.* also indicated that vaccination of dairy cattle with WCV-PI significantly reduced the shedding of *C. burnetii* in the milk,[90] suggesting that vaccination may be a good long-term strategy to reduce the spread of the organisms in this species. In Europe, a combination vaccine containing phase II *C. burnetii* and *Chlamydia psittaci* has been marketed to protect cattle and goats against reproductive disorders caused by these two organisms. However, a report indicated that a Q fever outbreak was associated with exposure to vaccinated goats and their unpasteurized dairy products.[91] In addition, this vaccine was suspected of increasing the shedding of *C. burnetii* in milk for several months when administered to previously infected animals.[76] These data suggested that phase II vaccine was not able to effectively prevent the natural infection in animals. Since vaccination of domestic animals only provided protection in *C. burnetii*-negative animals and WCV-PI was not able to completely control the shedding of the organisms in fluids and birth products, Q fever vaccines are currently not widely used in domestic animals in most countries.

Vaccine prophylaxis in humans has been restricted to occupationally at-risk individuals including livestock workers, veterinarians, research laboratory workers, and personnel of research animal facilities. Several vaccines consisting of either formalin-inactivated phase I or phase II whole cell, attenuated phase II *C. burnetii,* or a chloroform:methanol residue (CMR) fraction of Nine Mile phase I have been developed and tested in humans.[73,78–80] There is no vaccine licensed for use in the United States. However, individuals at risk can be vaccinated with an investigational phase I cellular vaccine. Smadel *et al.* first reported that a vaccine prepared from the Henzerling strain (isolated from a soldier in Italy) was highly immunogenic in both guinea pigs and humans.[92] Subsequently, this vaccine was widely used for laboratory workers. A formalin-inactivated whole cell *C. burnetii* vaccine (Q-Vax, Commonwealth Serum Laboratories) produced from phase I Henzerling strain *C. burnetii* was the only licensed vaccine for use in Australia in March 1989. Evaluation of the efficacy of Q-Vax in abattoir workers demonstrated that this vaccine was very effective in preventing clinical Q fever.[73] Marmion, *et al.*,[93] reported that among 924 nonimmune abattoir workers in Australia, no Q fever case was diagnosed within 18 months of vaccination, whereas 34 cases were recorded among 1349 unvaccinated employees. In a later investigation, the same group compared the incidence of Q fever among vaccinated and unvaccinated abattoir workers.[73] The results indicated that two cases of Q fever were diagnosed among 2555 employees who received a single subcutaneous dose (30 μg) of Q-Vax, while 55 cases were identified among 1365 unvaccinated employees. Since the two Q fever cases in vaccinated employees were within a few days of vaccination and may have represented coincidental natural infection and vaccination, the protective efficacy of Q-Vax was considered 100%. This study also demonstrated that the duration of protection was greater than 5 years. These investigations suggested that Q-Vax provided extraordinary protection against the disease. However, adverse effects such as erythema or tenderness at the site of vaccine inoculation was observed in most of the vaccinated individuals, while more severe adverse reactions including transient headache, shivering, and flu-like symptoms occurred in 10% to 18% of vaccinated subjects.[93] Several early studies also demonstrated that vaccination with whole-cell phase I *C. burnetii* could result in severe local or systemic adverse reactions, especially when administered to previously infected populations, and repeat vaccination could develop severe persistent reactions. Screening for prior immunity using a skin test can reduce adverse reactions in vaccinated humans.

An attenuated vaccine that was prepared from M-44 *C. burnetii* strain in phase II has been successfully tested on Soviet volunteers and has been proposed for use in humans.[78] However, previous studies with this vaccine in guinea pigs have indicated that the organisms persisted in the animals for a long time and caused mild lesions in the heart, spleen, and liver, suggesting a potential for reactivation of the infection in the vaccinated host. Therefore this live vaccine was not considered safe for humans.

Subsequently, several different chemical extraction procedures have been applied for treatment of *C. burnetii* whole cells vaccines in attempts to develop a vaccine with good immunogenicity and reduced adverse effects. Kazar *et al.* developed a chemovaccine using trichloroacetic acid-extracted antigen from *C. burnetii* Nine Mile phase I strain and tested this vaccine in persons who had a high risk of *C. burnetii* exposure in Czechoslovakia.[80] The results of this study indicated that there was a significant antibody response to phase I antigen in vaccinated persons, but more severe local and systemic reactions occurred in previously infected individuals. A chloroform:methanol residue (CMR) Q fever vaccine from phase I Henzerling strain *C. burnetii* was developed jointly at the Rocky Mountain Laboratories and US Army Research Institute for Infectious Disease in the late 1970s. Initial testing indicated that CMR did not cause adverse reactions in mice at doses several times larger than doses of WCV that caused severe adverse effects.[94] Several studies have demonstrated that CMR was nontoxic, immunogenic, and protective in mice, guinea pigs, and sheep and could provoke a delayed hypersensitivity response without persistent granulomata in previously sensitized guinea pigs. Recent study also showed that CMR vaccine was immunogenic and gave protection equivalent to Q-Vax in a nonhuman primate aerosol challenge model.[95] CMR significantly reduced adverse reactions in the animals tested. However, a safety test of CMR vaccine in 35 volunteers indicated that CMR at 30 and 60 μg caused a minimal reaction, while higher doses caused reactions that were qualitatively similar to the common reactions to 30 μg doses of Q-Vax observed by Marnion and colleagues in Australia.[96] Although chemically inactivated *C. burnetii* vaccines, either "chemovaccine" or CMR, have shown reduced adverse reactions in vaccinated animals, there are still no vaccines that completely eliminate the adverse reaction. Therefore, efforts are underway to create a safe and effective vaccine that can be administered to a population without pre-screening.

9. NEW OPPORTUNITIES WITH GENOMIC AND PROTEOMIC APPROACHES

9.1. Genomic Comparison of Isolate Groups

The breadth of knowledge concerning this organism was fundamentally changed with the elucidation of the Nine Mile genome. Because of its obligate intracellular nature, the generation of specific mutations has not been possible. Understanding the role in pathogenesis or other processes for specific genetic loci has been very limited. The most productive approach, prior to genomics, was to characterize specific genes by cloning into a surrogate expression system (*Escherichia coli*) and evaluating the phenotype. But these results are generally indirect and based on the homologous gene function in genetically tractable organism(s). Genomics provides a systematic approach via *in silico* screening. One of the immediate applications of this approach will be the elucidation of selected isolate genomes to compare with the Nine Mile prototype. Studies comparing various isolates suggest significant phylogenetic diversity among isolates. Some phenotypic differences have been reported for isolates. These isolates currently represent an immediately available genetic diversity with which to understand important issues such as virulence determinants. Therefore, it is likely that comparative genomic sequences will identify loci that may be responsible for selected phenotypes. Additionally, comparative genomics will also allow the development of a model for evolutionary diversity for this pathogen. *C. burnetii* are likely confined to an intracellular replication niche that does not allow co-mingling with DNA from other organisms, except perhaps the host. Therefore, genetic diversity to allow for evolutionary adaptation may be provided primarily from the host. An example of such acquisition may be the histone-like protein (Hq1) described by Heinzen *et al.*[97] This gene has no close homologues among eubacterium and may have originated through horizontal transfer from a eukaryotic host. Comparative genomics may also provide clues to the requirement for open reading frames (orfs) for which the annotation has not identified a functional homologue. Comparison between isolates for this large group of predicted orfs might identify genes that have been lost or frame-shifted, implying that they do not encode important functions.

9.2. Development of New-Generation Vaccines

A new generation vaccine for Q fever will be required to confer protection against infection and also have the ability to be administered without prior screening for immunity. To overcome the problem of current vaccines, a few efforts have attempted to develop a subunit protein vaccine. Williams and co-workers demonstrated that a 29 kDa protein, P1, purified from phase I Nine Mile strain *C. burnetii*, could confer protection from a lethal challenge in mice.[98] Zhang *et al.* demonstrated that a partially purified 67 kDa antigen from *C. burnetii* could confer full protection in both guinea pigs and mice.(70) These studies suggested that subunit protein vaccines could provide protection against *C. burnetii* infection. However, since these two proteins were not cloned and are not well characterized, no single protective protein has been confirmed to deliver protection. Our studies have initially focused on identification and characterization of immunogenic proteins as defined by strong reactivity with infection-derived serum. Four candidate antigens, including Com1, P1, Cb-Mip, and P28, have been previously cloned and characterized by our group.(44,46,54,99) The protective efficacies of these recombinant proteins were tested in a sublethal challenge BALB/c mouse model using protection from the development of severe splenomegally as an indicator of vaccinogenic activity.(100) However, the results indicated that the selected recombinant proteins did not individually confer significant protection against infection in this model. These results suggested that other unidentified antigens or multiple recombinant proteins and/or an appropriate delivery system would be required for development of protective immunity. New approaches facilitated by proteomics may be important in characterizing the critical protective epitopes encoded by *C. burnetii.* Libraries of the complete set of orfs cloned into expression vectors will allow expression of each orf and subsequent evaluation of vaccinogenic activity. Alternately, generation of a clone of each orf in a DNA vaccine expression vector will allow testing in either pooled groups of orf clones or single clones for protection in new and more sensitive experimental animal infection models.

ACKNOWLEDGEMENT

This work was supported by Public Health Service Grants AI37744 and AI448191, from the National Institute of Allergy and Infectious Diseases.

REFERENCES

1. Derrick, E. H., 1937, "Q" fever, a new fever entity: clinical features, diagnosis, and laboratory investigation, *Med. J. Aust.* **2**:281–299.
2. Babudieri, C., 1959, Q fever: a zoonosis, *Adv. Vet. Sci.* **5**:81–84.
3. Lang, G. H., 1990, Coxiellosis (Q fever) in animals, in: *Q Fever. Volume 1: The Disease* (T. J. Marrie, ed.), CRC Press, Boca Raton, FL, pp. 23–49.
4. Tigertt, W. D., Benenson, A. S., and Gochenour W. S., 1961, Airborne Q fever, *Bacteriol. Rev.* **25**:285–293.
5. Cekanac, R., Lukac, V., and Coveljiâc, M., 2002, Epidemija Q groznice u jednoj jedinici Vojske Jugoslavije u ratnim uslovimà. Vojnosanitetski pregled, *Mil. Med. Pharm. Rev.* **59**:157–160.
6. Regis, E., 1999, *The Biology of Doom: The History of America's Secret Germ Warfare Project,* Henry Holt and Associates, New York.
7. Spicer, A. J., 1978, Military significance of Q fever: a review, *J. Royal Soc.. Med.* **71**:762–767.

8. Izzo, A. A., and Marmion, B. A., 1993, Variation in interferon-γ responses to *Coxiella burnetii* antigens with lymphocytes from vaccinated and naturally infected subjects, *Clin. Exp. Immunol.* **94**:507–514.
9. Koster, F. T., Williams, J. C., and Goodwin, J. S., 1985, Cellular immunity in Q fever: modulation of responsiveness by a suppressor T cell-monocyte circuit, *J. Immunol.* **135**:1067–1072.
10. Koster, F. T., Williams, J. C., and Goodwin, J. S., 1985, Cellular immunity in Q fever: specific unresponsiveness in Q fever endocarditis, *J. Infect. Dis.* **152**:1283–1288.
11. Capo, C., Zaffran, Y., Zugan, F., Houpikian, P., Raoult, D., and Mege, J. L., 1996, Production of interleukin-10 and transforming growth factor β by peripheral blood mononuclear cells in Q fever endocarditis, *Infect. Immun.* **64**:4143–4150.
12. Peacock, M. G., Fiset, P., Ormsbee, R. A., and Wisseman, C. L. Jr., 1979, Antibody response in man following a small intradermal inoculation with *Coxiella burnetii* phase I vaccine, *Acta Virol.* **23**:73–81.
13. Raoult, D., 1993, Treatment of Q fever. *Antimicro. Agents Chemother.* **37**:1733–1736.
14. Maurin, M., and Raoult, D., 1996, Optimum treatment of intracellular infection, *Drugs* **2**:45–55.
15. Howe, D., and Mallavia, L. P., 1999, *Coxiella burnetii* infection increases transferrin receptors on J774A.1 cells, *Infect. Immun.* **67**:3236–3241.
16. Baca, O. G., Klassen, D. A., and Aragon, A. S., 1993, Entry of *Coxiella burnetii* into host cells. *Acta Virologica* **37**:143–155.
17. Hinricks, D. J., and Jerrells, T. R., 1976, In vitro evaluation of immunity to *Coxiella burnetii, J. Immunol.* **117**:996–1103.
18. Capo, C., Lindberg, F. P., Meconi, S., Zaffran, Y., Tardei, G., Brown, E. J., Raoult, D., and Mege, J., 1999, Subversion of monocyte functions by *Coxiella burnetii*: impairment of the cross-talk between αvβ2 integrin and CR3, *J. Immun.* **163**:6078–6085.
19. To, H., Hotta, A., Zhang, G. Q., Nguyen, S., Ogawa, M., Yamaguchi, T., Fukushi, H., Avnano, K., Hirai, K., 1998, Antigenic of polypeptides of *Coxiella burnetii* isolates, *Microbiol. Immunol.* **42**:81–85.
20. Capo, C., Moynault, A., Collette, Y., Olive, D., Brown, E. J., Raoult, D., and Mege, J. L., Unitâe des Rickettsies, 2003, *Coxiella burnetii* avoids macrophage phagocytosis by interfering with spatial distribution of complement receptor 3, *J. immunol.* **170**:4217–4225.
21. Akporiaye, E. T., Rowatt, J. D., Aragon, A. A., and Baca, O. G., 1983, Lysosomal response of a murine macrophage-like cell line persistently infected with *Coxiella burnetii, Infect. Immun.* **40**:1155–1162.
22. Hackstadt, T., and Williams, J. C., 1981, Biochemical stratagem for obligate parasitism of eukaryotic cells by *Coxiella burnetii, Proc. Nat. Acad. Sci. U S A* **78**:3240–3244.
23. Hackstadt, T., and Williams, J. C., 1983, pH dependence of the *Coxiella burnetii* glutamate transport system, *J. Bacteriol.* **154**:598–603.
24. Hendrix, L., and Mallavia, L. P., 1984, Active transport of proline by *Coxiella burnetii, J. Gen. Microbiol.* **130**:2857–2863.
25. Maurin, M., Benoliel, A. M., Bongrand, P., and Raoult, D., 1992, Phagolysosomes of *Coxiella burnetii*-infected cell lines maintain an acidic pH during persistent infection, *Infect. Immun.* **60**:5013–5016.
26. Burton, P. R., Stueckemann, J., Welsh, R. M., and Paretsky, D., 1978, Some ultrastructural effects of persistent infections by the rickettsia *Coxiella burnetii* in mouse L cells and green monkey kidney (Vero) cells, *Infect. Immun.* **21**:556–566.
27. Heinzen, R. A., Scidmore, M. A., Rockey, D. D., and Hackstadt, T., 1996, Differential interaction with endocytic and exocytic pathways distinguish parasitophorous vacuoles of *Coxiella burnetti* and *Chlamydia trachomatis, Infect. Immun.* **64**:796–809.
28. Burton, P. R., Kordova, N., and Paretsky, D., 1971, Electron microscopic studies of the rickettsia *Coxiella burnetii*: entry, lysosomal response, and fate of rickettsial DNA in L-cells. *Can. J. Microbiol.* **17**:143–150.

29. Berâon, W., Gutierrez, M. G., Rabinovitch, M., and Colombo, M. I., Instituto de Histologâia y Embriologâia, 2002, Coxiella burnetii localizes in a Rab7-labeled compartment with autophagic characteristics, *Infect. Immun.* **70**:5816–5821.
30. Wiebe, M. E., Burton, P. R., and Shankel, D. M., 1972, Isolation and characterization of two cell types of *Coxiella burnetii, J. Bacteriol.* **110**:368–377.
31. Heinzen, R. A., Howe, D., Mallavia, L. P., Rockey, D. D., and Hackstadt, T., 1996, Developmentally regulated synthesis of an unusually small, basic peptide by *Coxiella burnetii, Mol. Microbiol.* **22**:9–19.
32. Howe, D., Melnicâakovâa, J., Barâak, I., and Heinzen, R. A., 2003, Maturation of the *Coxiella burnetii* parasitophorous vacuole requires bacterial protein synthesis but not replication, *Cell. Microbial.* **5**:469–80.
33. Zamboni, D. S., 2004, Genetic control of natural resistance of mouse macrophages to Coxiella burnetii infection in vitro: macrophages from restrictive strains control parasitophorous vacuole maturation. *Infect. Immun.* **72**:2395–2399.
34. Turco, J., Thompson, H. A., and Winkler, H., 1984, Interferon-γ inhibits growth of *Coxiella burnetii* in mouse fibroblasts, *Infect. Immun.* **45**:781–783.
35. Howe, D., Barrows, L. F., Lindstrom, N. M., and Heinzen, R. A., 2002, Nitric oxide inhibits*Coxiella burnetii* replication and parasitophorous vacuole maturation. *Infect. Immun.* **70**:5140–5147.
36. Zamboni, D. S., and Rabinovitch, M., 2003, Nitric oxide partially controls *Coxiella burnetii* phase II infection in mouse primary macrophages. *Infect. Immun.* **71**:1225–1233.
37. Brennan, R. E., Russell, K., Zhang, G. Q., and Samuel, J. E., 2004, Both inducible nitric oxide synthase and NADPH oxidase contribute to the control of virulent phase I *Coxiella burnetii* infections, *Infect. Immun.* **72**:6666–6675.
38. Seshadri, R., Heidelberg, J. F., Paulsen, I. T., Eisen, J. A., Read, T. D., Nelson, K. E., Ward, N., Nelson, W. C., Tettelin, H., Davidsen, T. M., Beanan, M. J., Deboy, R. T., Daugherty, S. C., Brinkac, L. M., Madupu, R., Dodson, R. J., Lee, K. H., Carty, H. A., Scanlan, D., Thompson, H. A., Heinzen, R. A., Samuel, J. E., and Fraser, C. M., 2003, Complete genome sequence of the Q-fever pathogen, *Coxiella burnetii, Proc. Natl. Acad. Sci. U S A* **100**:5455–5460.
39. Segal, G., and Shuman, H. A., 1998, Intracellular multiplication and human macrophage killing by *Legionella pneumophila* are inhibited by conjugal components on IncQ plasmid RSF1010. *Mol. Microbiol.* **30**:197–208.
40. Vogel, J. P., and Isberg, R. R., 1999, Cell biology of Legionella pneumophila, *Curr. Opin. Microb.* **2**:30–34.
41. Zamboni, D. S., McGrath, S., Rabinovitch, M., and Roy, C. R., 2003, Coxiella burnetii express type IV secretion system proteins that function similarly to components of the Legionella pneumophila Dot/Icm system, *Mol. Microbiol.* **49**:965–976.
42. Cianciotto, N. P., Bangsborg, J. M., Eisenstein, B. I., and Engleberg, N. C., 1990, Identification of mip-like genes in the genus Legionella, *Infect. Immun.* **58**:2912–2918.
43. Cianciotto, N. P., Eisenstein, B. I., Mody, C. H., Toews, G. B., and Engleberg, N. C., 1989, A Legionella pneumophila gene encoding a species-specific surface protein potentiates initiation of intracellular infection, *Infect. Immun.* **57**:1255–1262.
44. Mo, Y. Y., Cianciotto, N. P., and Mallavia, L. P., 1995, Molecular cloning of a Coxiella burnetii gene encoding a macrophage infectivity potentiator (Mip) analogue, *Microbiology* **141**:2861–2871.
45. Mo, Y. Y., Seshu, J., Wang, D., and Mallavia, L. P., 1998, Synthesis in Escherichia coli of two smaller enzymically active analogues of Coxiella burnetii macrophage infectivity potentiator (CbMip) protein utilizing a single open reading frame from the cbmip gene, *Biochem. J.* **335**:67–77.
46. Hendrix, L. R., Mallavia, L. P., and Samuel, J. E., 1993, Cloning and sequencing of *Coxiella burnetii* outer membrane protein gene *com1, Infect. Immun.* **61**:470–477.

47. Yu, J., McLaughlin, S., Freedman, R. B., and Hirst, T. R., 1993, A homologue of the *Escherichia coli* DsbA protein involved in disulfide bond formation is required form enterotoxin biogenesis in *Vibrio cholerae, Mol. Microbiol.* **6**:1949–1958.
48. Saha, A. K., Dowling, J. N., LaMarco, K. I., Das, S., Remaley, A. T., Olomu, N., Pope, M., and Glew, R. H., 1985, Properties of an acid phosphatase from *Legionella micdadei* which blocks superoxide anion production by human neutrophils, *Arch. Biochem. Biophys.* **243**:150–160.
49. Baca, O. G., Roman, M. J., Glew, R. H., Christner, R. F., Buhler, J. E., and Aragon, A. S., 1993, Acid phosphatase activity in *Coxiella burnetii*: a possible virulence factor, *Infect. Immun.* **61**:4232–4239.
50. Li, Y. P., Curley, G., Lopez, M., Chavez, M., Glew, R., Aragon, A., Kumar, H., and Baca, O. G., 1996, Protein-tyrosine phosphatase activity of *Coxiella burnetii* that inhibits human neutrophils. *Acta Virolog.* **40**:163–272.
51. Ghigo, E., Capo, C., Tung, C.-H., Raoult, D., Gorvel, J., and Mege, J., 2002, *Coxiella burnetii* survival in THP-1 monocytes involves the impairment of phagosome maturation: IFN-g mediates its restoration and bacterial killing, *J. Immunol.* **168**:4488–4495.
52. Ghigo, E., Capo, C., Raoult, D., and Mege, J. L., 2001, Interleukin-10 stimulates *Coxiella burnetii* replication in human monocytes through tumor necrosis factor down-modulation: role in microbicidal defect of Q fever, *Infect. Immun.* **69**:2345–2352.
53. Heggers, J. P., Billups, L. H., Hinrichs, D. J., and Mallavia, L. P., 1975, Pathophysiologic features of Q fever-infected guinea pigs. *Am. J. Vet. Res.* **36**:1047–1052.
54. Zhang, G. Q., Kiss, K., Seshadri, R., Hendrix, L. R., and Samuel, J. E., 2004, Identification and cloning of immunodominant antigens of Coxiella burnetii, *Infect. Immun.* **72**:844–852.
55. Scott, G. H., Williams, J. C., and Stephenson, E. H., 1987, Animal models in Q fever: pathological responses of inbred mice to phase I *Coxiella burnetii, J. Gen. Microbiol.* **133**:691–700.
56. Moos, A., and Hackstadt, T., 1987, Comparative virulence of intra-and interstrain lipopolysaccharide variants of *Coxiella burnetii* in the guinea pig model, *Infect. Immun.* **55**:1144–1150.
57. Kishimoto, R. A., Rozmiarek, H., and Larson, R. W., 1978, Experimental Q fever infection in congenitally athymic nude mice, *Infect. Immun.* **22**:69–71.
58. Andoh, M., T. Naganawa, T., Hotta, A., Yamaguchi, T., Fukushi, H., Masegi, T., and Hirai, K., 2003, SCID mouse model for lethal Q fever, *Infect. Immun.* **71**:4717–4723.
59. Lepidi, H., Houpikian, P., Liang, Z., and Raoult, D., 2003, Cardiac valves in patients with Q fever endocarditis: microbiological, molecular, and histologic studies, *J. Infect. Dis.* **187**:1097–1106.
60. Sidwell, R. W., Thorpe, B. D., and Gebhardt, L. P., 1964, Studies of latent Q fever infections II. Effects of multiple cortisone injections, *Am. J Hyg.* **79**:320–327.
61. Atzpodien, E., Baumgartner, W., Artelt, A., and Thiele, D., 1994, Valvular endocarditis occurs as a part of a disseminated *Coxiella burnetii* infection in immunocompromised Balb/cj ($H-2^d$) mice infected with the Nine Mile isolate of *C. burnetii, J. Infect. Dis.* **170**:223–226.
62. Moos, A., Vishwanath, S., and Hackstadt, T., 1988, Experimental Q fever endocarditis in rabbits. In *Abstracts of the Seventh National Meeting of the American Society for Rickettsiology and Rickettsial Diseases 1988.* American Society for Rickettsiology and Rickettsial Diseases, Santa Fe, NM.
63. La Scola, B., Lepidi, H., Maurin, M., and Raoult, D., 1998, A guinea pig model for Q fever endocarditis. *J. Infect. Dis.* **178**:278–281.
64. Samuel, J. E., Frazier, M. E., and Mallavia, L. P., 1985, Correlation of plasmid type and disease caused by *Coxiella burnetii. Infect. Immun.* **49**:775–779.
65. Hackstadt, T., 1986, Antigenic variation in the phase I lipopolysaccaride of *Coxiella burnetii* isolates, *Infect. Immun.* **52**:337–340.
66. Hendrix, L. R., Samuel, J. E., and Mallavia, L. P., 1991, Differentiation of *Coxiella burnetii* isolates by analysis of restriction-endonuclease-digested DNA separated by SDS-PAGE, *J. Gen. Microbiol.* **137**:269–276.

67. Stein, A., and Raoult, D., 1993, Lack of pathotype specific gene in human *Coxiella burnetii* isolates, *Microb. Pathog.* **15**:177–185.
68. Thiele, D., and Willems, H., 1994, Is plasmid based differentiation of *Coxiella burnetii* in 'acute' and 'chronic' isolates still valid? *Eur. J. Epidemiol.* **10**:427–434.
69. Valkova, D., and J. Kazar, J., 1995, A new plasmid (QpDV) common to *Coxiella burnetii* isolates associated with acute and chronic fever, *FEMS Microbiol. Lett.* **125**:275–280.
70. Zhang, Y. X., Zhi, N., Yu, S. R., Li, Q. R., Yu, G. C., and Zhang, X., 1994, Protective immunity induced by 67 K outer membrane protein of phase I *Coxiella burnetii* in mice and guinea pigs, *Acta Virolog.* **38**:327–332.
71. Ormsbee, R. A., Bell, E. J., Lackman, D. B., and Tallent, G., 1964, The influence of phase on the protective potency of Q fever vaccine, *J. Immunol.* **92**:404–411.
72. Amano, K., and Williams J. C., 1984, Sensitivity of *Coxiella burnetii* peptidoglycan to lysosome hydrolysis and correlation of sacculus rigidity with peptidoglycan-associated proteins, *J. Bacteriol.* **160**:989–993.
73. Ackland, J. R., Worswick, D. A., and Marmion, B. P., 1994, Vaccine prophylaxis of Q fever. A follow-up study of the efficacy of Q- Vax (CSL) 1985–1990, *Med. J. Aust.* **160**:704–708.
74. Behymer, D. E., Biberstein, E. L., Riemann, H. P., Franti, C. E., Sawyer, M., Ruppanner, R., and Crenshaw, G. L., 1976, Q fever (*Coxiella burnetii*) investigation in dairy cattle:challenge of immunity after vaccination, *Am. J. Vet. Res.* **37**:631–634.
75. Marmion, B. P., Ormsbee, R. A., Kyrkou, M., Wright, J., Worswick, D. A., Izzo, A. A., Esterman, A., Feery, B., and Shapiro, R. A., 1990, Vaccine prophylaxis of abattoir-associated Q fever: eight years' experience in Australian abattoirs. *Epidemiol. Infect.* **104**:275–287.
76. Schmeer, N., Muller, P., Langel, J., Krauss, H., Frost, J. W., and Wieda, J., 1987, Q fever vaccines for animals. *Zentralbl. Bakteriol. Microbiol. Hyg. [A].* **267**:79–88.
77. Waag, D. M., 1990, Acute Q fever, in: *Q Fever. Volume 1: The Disease* (T. J. Marrie,ed.), CRC Press, Boca Raton, FL, pp. 107–123.
78. Genig, V. A., 1968, A live vaccine 1/M-44 against Q-fever for oral use, *J. Hyg., Epidemiol., Microbiol. Immunol.* **12**:265–273.
79. Hoover, T. A., M. H., V., and Williams, J. C., 1992, A *Coxiella burnetii* repeated DNA element resembling a bacterial insertion sequence, *J. Bacteriol.* **174**:5540–5548.
80. Kazar, J., Brezina, R., Palanova, A., Turda, B., and Schramek, S., 1982, Immunogenicity and reactogenicity of Q fever chemovaccine in persons professionally exposed to Q fever in Czechoslovakia, *Bull. World Health Organ.* **60**:389–394.
81. Humphres, R. C., and Hinrichs, D. J., 1981, Role for antibody in *Coxiella burnetii* infection, *Infect. Immun.* **31**:641–645.
82. Izzo, A. A., Marmion, B. P., and Hackstadt, T., 1991, Analysis of the cells involved in the lymphoproliferative response to *Coxiella burnetii* antigens. *Clinic. Exp. Immunol.* **85**:98–108.
83. Izzo, A. A., Marmion, B. P., and Worsick, D. A., 1988, Markers of cell-mediated immunity after vaccination with an inactivated, whole-cell Q fever vaccine, *J. Infect. Dis.* **157**:781–789.
84. Kishimoto, R. A., 1977, Appearance of cellular and humoral immunity in gûinea pigs after infection with *Coxiella burnetii* administration in small particle aerosol, *Infect. Immun.* **16**:518–521.
85. Kishimoto, R. A. J. W. J., Kenyon, R. H., Ascher, M. S., Larson, E. W., and Pederson, C. E., Jr., 1978, Cell mediated immune response of guinea pigs to an inactivated phase I *Coxiella burnetii* vaccine, *Infect. Immun.* **19**:194–198.
86. Abinanti, F. R., and Marmion, B. P., 1957, Protective or neutralizing antibody in Q fever, *Am. J. Hyg.* **66**:173–195.
87. Anacker, R. L., Lackman, D. B., Pickens, E. G., and Ribi, E., 1962, Antigenic and skin-reactive properties of fractions of *Coxiella burnetii*, *J. Immunol.* **89**:145–153.
88. Kishimoto, R. A., Veltri, B. J., Shirley, F. G., Canonico, P. G., and Walker, J. S., 1977, Fate of *Coxiella burnetii* in macrophages from immune guinea pigs, *Infect. Immun.* **15**:601–607.

89. Dellacasagrande, J., Capo, C., Raoult, D., and Mege, J.-L., 1999, IFN-γ-mediated control of *Coxiella burnetii* survival in monocytes: the role of cell apoptosis and TNF, *J. Immunol.* **176**:2259–2265.
90. Biberstain, E. L., H. P. Reiemann, H. P., and Franti, C. E., 1977, Vaccination of dairy cattle against Q fever (*Coxiella burnetii*): results of field trials, *Am. J. Vet. Res.* **38**:189–193.
91. Fishbein, D. B., and Raoult, D., 1992, A cluster of *Coxiella burnetii* infections associated with exposure to vaccinated goats and their unpasteurized dairy products, *Am. J. Tropic. Med. Hyg.* **47**:35–40.
92. Smadel, J. E., Snyder, M. J., and Robbins, F. C., 1948, Vaccine Against Q Fever, *Am. J. Hyg.* **47**:71–78.
93. Marmion, B. P., Ormsbee, R. A., Kyrkou, M., Wright, J., Worswick, D., Cameron, S., Esterman, A., Feey, B., and Collins, W., 1984, Vaccine prophylaxis of abattoir-associated Q fever, *Lancet* **324**:1411–1414.
94. Williams, J. C., and Cantrell, J. L., 1982, Biological and Immunological properties of *Coxiella burnetii* vaccines in C57BL/10ScN endotoxin-nonresponder mice, *Infect. Immun.* **35**:1091–1102.
95. Waag, D. M., England, M. J., Tammariello, R. F., Byrne, W.R., Gibbs, P., Banfield, C. M., and M. Pitt, M. L., 2002, Comparative efficacy and immunogenicity of Q fever chloroform: methanol residue (CMR) and phase I cellular (Q-Vax) vaccines in cynomolgus monkeys challenged by aerosol, *Vaccine* **20**:2623–2634.
96. Fries, L. F., Waag, D. M., and Williams, J., 1993, Safety and immunogenicity in human volunteers of a chloroform-methanol residue vaccine for Q fever, *Infect. Immun.* **61**:1251–1258.
97. Heinzen, R. A., and T. Hackstadt, T., 1996, A developmental stage-specific histone H1 homolog of *Coxiella burnetii, J. Bacteriol.* **178**:5049–5052.
98. Williams, J. C., Peacock, M. G., Waag, D. M., Kent, G., England, M. J., Nelson, G., and Stephenson, E. H., 1990, Vaccines against Coxiellosis and Q fever, *Annal. N Y Acad. Sci.* **590**:88–111.
99. Varghees, S., Kiss, K., Frans, G., Braha, O., and Samuel, J. E., 2002, Cloning and porin activity of the major outer membrane protein P1 from *Coxiella burnetii, Infect. Immun.* **70**:6741–6750.
100. Zhang, G. Q., and Samuel, J. E., 2003, Identification and cloning potentially protective antigens of *Coxiella burnetii* using sera from mice experimentally infected with Nine Mile phase I. *Rickettsiology: Present and Future Directions* **990**:510–520.

12

Rickettsia rickettsii and Other Members of the Spotted Fever Group as Potential Bioweapons

DONALD H. BOUYER and DAVID H. WALKER

1. INTRODUCTION

Since the anthrax attacks in the fall of 2001, there has been an increase in the level of anxiety felt throughout society in general about the use of biological agents as weapons. Although much of the nation's and the world's attention has been focused upon anthrax, botulinum toxin, and Ebola as weapons,[1] there are other microbes that also pose a significant threat due to their potential as bioweapons. Characteristics that are shared by each of these organisms are that they are easily obtainable in nature, difficult to detect, and highly infectious at a low dose, can be easily transmitted by aerosol, and have a short incubation period. One frequently overlooked genus of bacteria that fulfills these criteria and also poses a significant threat is the *Rickettsia.* Members of this genus, such as *Rickettsia rickettsii,* a spotted fever group (SFG) rickettsia, have long been recognized as inherently dangerous with many reports of accidental infections and even deaths because of inhalational transmission to scientists who have worked with these organisms.[2–5] *Rickettsia prowazekii,* a member of the typhus group, not only occurs in periodic outbreaks,[6,7] but also has the infamous history of having been developed as a bioweapon by both the Japanese Army during World War II and the former Soviet Union.[8,9] Because of its history and reputation, many scientists suspect that *R. prowazekii* is the most probable

DONALD H. BOUYER and DAVID H. WALKER • Department of Pathology and Center for Biodefense and Emerging Infectious Diseases, University of Texas Medical Branch at Galveston, Galveston TX 77555-0609 USA.

Rickettsia to be utilized as a bioweapon. There is also a substantial possibility that *R. rickettsii* or another member of the SFG would be utilized and that we should also prepare for that scenario. *R. rickettsii* is the most pathogenic rickettsia, and other pathogenic members of the SFG are formidable in their own right and could have a potential devastating effect if loosed upon an unsuspecting society and an unprepared medical and public health system.

2. SFG RICKETTSIAE WITH BIOWEAPON POTENTIAL

One of the misconceptions accepted by the general public when considering the "bioweapon potential" is that the microbial agent will cause mass numbers of deaths. While that is a real concern, one should not overlook the "terror impact" of a potential agent. The effectiveness of a bioweapon can be also measured in its impact upon altering or inhibiting the daily routines of a community. For example, if clusters of patients with general symptoms of fever, headache, and myalgia begin to appear in clinics or hospitals a few days after the aerosol release from a small airplane, not only would emergency services be overwhelmed, but media coverage would also launch the populace into panic. If the disease is difficult to diagnose rapidly and it causes a range of clinical severity from life threatening to temporarily incapacitating, one could easily imagine the chaos that would ensue. The above scenario could easily be caused through the utilization of any of the pathogenic members of the SFG (Table I). The pathogenic members of the SFG are *R. rickettsii, R. conorii, R. australis, R. akari, R. africae/R. parkeri, R. japonica, R. honei*, and *R. sibirica.*[10] The members of this group have world-wide distribution with continuous reporting of new endemic areas.[11] The most devastating disease caused by a *Rickettsia* is Rocky Mountain spotted fever (RMSF). In the pre-antibiotic era, case fatality rates were 23%, with 80% fatalities in some regions and outbreaks.[10] Its case fatality rate is similar to that of bubonic plague and tularemia and greater than that of Lassa fever and Rift Valley fever, which are Category A agents. In nature it is transmitted through the bite of *Dermacentor varabilis, D. andersoni, Rhipicephalus sanguineus*, or *Amblyomma cajennense* ticks. *Rickettsia conorii* causes boutonneuse or Mediterranean spotted fever (MSF) and is transmitted by *Rhipicephalus sanguineus* ticks and has a case-fatality rate of 1–5%.[10] Rickettsial diseases caused by *R. australis* (Queensland tick typhus), *R. honei* (Flinders Island spotted fever), *R. akari* (rickettsialpox), *R. africae* (African tick bite fever), and *R. japonica* (Japanese spotted fever) are less severe than that of RMSF.[10] As stated previously, these less severe rickettsial diseases are still of importance because an outbreak would temporarily overwhelm the health care system and inhibit the normal daily activities of society.

TABLE I
Pathogenic Members of the Spotted Fever Group of Rickettsiae

Organism	Vector(s)	Distribution
Rickettsia rickettsii	*Dermacentor andersoni, Dermacentor varabilis, Rhipicephalus sanguineus, Amblyomma cajennense*	The Americas
R. africae/R. parkeri	*Amblyomma* species	Africa, the Americas
R. akari	*Liponysossides sanguineus*	Northern Hemisphere temperate zones
R. australis	*Ixodes holocyclus*	Australia
R. conorii	*Rhipicephalus sanguineus*	Mediterranean, sub-Saharan Africa, India
R. japonica	Presumably *Haemaphysalis flava, Dermacentor taiwanensis*	Asia
R. honei	*Aponomma hydrosauri*	Australia, Asia
R. sibirica	*Dermacentor, Haemaphysalis,* and *Hyalomma* species	Asia, Eastern Russia

3. FEASIBILITY OF OBTAINING, PROPAGATING, STABLIZING, AND WEAPONIZING SFG RICKETTSIAE

Scientists tend to believe that in order to create a weapon, the newest and most recent technology is required. Among legitimate scientists, select agent research is regulated by myriad rules and governing bodies. However, in order to protect society against future threats, we will need to understand the mindset of terrorists, evaluate all possibilities, and recognize what work could be accomplished in a basement, garage, or warehouse laboratory.

If one takes a critical analysis of the requirements as set forth by CDC and NIH for an organism to be considered as a potential bioterror weapon, it is easy to see how *R. rickettsii* could be converted into a weapon. First, let us examine the feasibility of obtaining the etiologic rickettsia. One method to obtain a SFG rickettsia would be to criminally obtain it from an existing source. However, most laboratories that conduct research on SFG rickettsiae are heavily regulated, and their stocks are secured and constantly monitored by authorities. Another potential source is economically desperate scientists engaged in bioweapons research in the former Soviet Union. Biopreparat was engaged in research on *R. prowazekii,* and joint efforts are underway to secure these agents.[12] Another scenario would involve collecting a large number of ticks from an endemic region and isolating rickettsiae in guinea pigs, cell culture, or embryonated eggs. All the potential terrorist would need for this is a tub to hold the animals or an incubator for eggs or flasks of cells. This level of laboratory safety is in stark contrast to what is done in laboratories in developed countries. The cumulative result of incidents involving laboratory exposure is that all present-day rickettsial research must be conducted in a biosafety level 3 laboratory that adheres to strict safety procedures to prevent aerosol-transmitted infections.[1] In order to combat self-infections by aerosolized rickettsiae, the terrorist could self-medicate with tetracycline.

The drawback to such an approach is that because highly virulent strains of RMSF kill their tick hosts, the overall tick infection rate is less than 0.1%, making collection of thousands of ticks likely to be necessary for isolation of *R. rickettsii.*[13–15] Niebylski and others reported that 94% of infected *D. andersoni* larvae died before molting into the nymphal stage of development.[15] Nymphs that acquired *R. rickettsii* during feeding fared slightly better with only 35% of the ticks dying before molting into the adult stage. Vertical transmission of rickettsiae from the mother tick was only observed in 39% of the offspring.[15] Although the tick lethality property may limit the incidence of RMSF in nature, *R. rickettsii* isolations from ticks are regularly achievable.

One aspect of potential RMSF and SFG rickettsial ecology that has been neglected is whether SFG rickettsiae have the ability to exist in a stable, extracellular, dormant form in arthropod feces. *R. prowazekii* has such a dormant form and has been observed in louse feces, and *R. typhi* in flea feces,[16] and it stands to reason that a similar ability could be shared by rickettsiae of the SFG. This hypothesis needs to be further investigated.

Once the starting material is obtained, the next step would be to propagate sufficient amounts of the organism to be utilized in an attack. The primary methods to multiply the rickettsial agent would be to use either embryonated eggs or cell culture. All that the potential bioterrorist would need to successfully propagate a rickettsial agent is a cell culture hood to prevent contamination of their cultures and an incubator. For growth in embryonated eggs, only an incubator would suffice.

Since all *Rickettsia* species are obligately intracellular organisms, the bioterrorist would need a method to stabilize the agent.[1] A stable form can be accomplished by lyophilization of the cultures. This material could then be milled to particles between 1–5 μm diameter. Particles of such a small size travel deep into the pulmonary alveoli and are retained in the lungs.[17] To complete the weaponization process, the agent could be treated chemically to prevent clumping by electrostatic forces.

4. METHODS OF DISPERSAL

The weaponized *R. rickettsii* could now be placed in a mechanism for aerosol dispersal, such as a crop duster sprayer, and dispersed over a location by plane or dispersed in a semi-enclosed space, such as a subway or in the air conditioning ducts of a building, where it could be spread by the environmental circulation of air.

Once in aerosol form, a person would only need to inhale a few organisms for the disease to occur.[3,4] Aerosol studies where target animals were exposed

TABLE II
Comparison of the Infectivity of SFG Rickettsiae and Other Bacterial Select Agents

Bacteria	Infectious dose by aerosol	Incubation period	Transmission routes
Bacillus anthracis	8,000–50,000 organisms	2–45 days	Aerosol
Burkholderia mallei	Unknown (apparently low)	1–14 days	Aerosol, direct contact with nasal secretion of infected equines
Coxiella burnetii	10 organisms	2–3 weeks	Aerosol, direct contact infected animals
Francisella tularensis	5–10 organisms (10^6–10^8 by ingestion)	1–14 days depending on route of transmission	Aerosol, handling of infected animals
Yersinia pestis	100 organisms	2–6 days	Aerosol
Rickettsia rickettsii and pathogenic members of the SFG	<10 organisms	3–14 days	Aerosol

to various doses of *R. rickettsii* show that as few as 0.8 rickettsia are required to cause disease in guinea pigs(18) and 1.5 yolk sac LD_{50} can infect rhesus and cynomolgus monkeys.(19,20) Male guinea pigs were exposed via different routes (aerosol, nasal, conjunctival, gastric, and subcutaneous) to dilutions of *R. rickettsii* and monitored for clinical signs of infection. All animals that inhaled at least 80 organisms by aerosol became ill with a mortality rate of 75%, and 25% of animals that received a dose that was calculated as 0.8 rickettsia became ill.(18) Saslaw and Carlise observed in their nonhuman primate aerosol-model that 93% (56 of 60) of the animals developed clinical signs of RMSF and 75% of the monkeys that had clinical infection died within 7–24 days post-exposure.(19) The animals became febrile 5–7 days after exposure with the appearance of other symptoms such as rash, lethargy, and anorexia 1–2 days after onset of fever.(19) The quantity of rickettsiae needed for infection by aerosol is comparable to that of other potential biological weapons (Table II).

5. PATHOGENESIS OF AEROSOL TRANSMISSION

Numerous studies of accidental laboratory infections and experimental animal studies provide a detailed analysis of the pathology of RMSF transmitted by aerosol.(2–4,18–20) All of the studies found that RMSF infections acquired by aerosol could not be distinguished from naturally occurring infections. Laboratory workers who were infected by aerosol had fever, chills, myalgia, headache, and rash on their extremities.(2,3) In the study by Pike,(4) the number of laboratory infections with *R. rickettsii* attributed to aerosols (217) was greater than that of parenteral (45 cases) or animal/ectoparasite exposure (66).

Aerosol-infected monkeys developed[19,20] typical lesions with perivascular infiltration of lymphocytes.

6. AVAILABLE METHODS FOR DIAGNOSIS, TREATMENT, AND PREVENTION

It is easy to envision difficulties in diagnosis arising from the utilization of SFG rickettsiae in a bioterrorist attack. Rickettsioses are notoriously difficult to diagnose and are under- and misdiagnosed even under normal clinical situations.[21–23] It is not difficult to imagine that during an attack out of the tick season with patients presenting nonspecific systemic symptoms, a rickettsial disease would very likely not be considered even by physicians convinced of a bioterror attack. Anthrax or one of the hemorrhagic fever viruses would be more likely to be included in the differential diagnosis.

Another problem in the diagnosis of rickettsial diseases is that there is currently no widely available test that is reliably diagnostic during the acute stage of the diseases. For the diagnosis of rickettsial diseases, most clinical laboratories use the immunofluorescence assay (IFA), which is considered the "gold standard" for the detection of antibodies to rickettsiae.[24] The advantage of the IFA is that the antigen contains all the rickettsial conformational proteins and group-shared lipopolysaccharide antigens. Current recommendations are that a diagnostic titer for RMSF is ≥ 64, or for a more confident laboratory confirmation a four-fold change between paired acute and convalescent serum specimens is required.[24,25] Other serologic tests that could be used for diagnosis are the indirect immunoperoxidase assay, Western immunoblotting, and enzyme immunoassays (EIA).[26–29] The problem with these serological assays is that antibodies to most SFG rickettsiae are not detected until the second week of illness, considerably after appropriate therapeutic decisions should have been made.[24] Currently, the best method for diagnosis of acute rickettsial infections is PCR amplification of selected rickettsial genes such as the 17-kDa lipoprotein gene, citrate synthase, 16S rRNA (*rrs*), *omp*A, and *omp*B.[24] Unfortunately, there are only a few laboratories in the United States that perform molecular diagnosis of rickettsioses on a regular basis. There is also a lack of commercially available kits for rickettsial molecular diagnostics.

In the case of rickettsial diseases, doxycycline or another tetracycline is the recommended treatment of choice, with chloramphenicol as the alternative drug. However, there are reports of rickettsiae being rendered resistant to antibiotics by experimental or naturally occurring means.[9,30] With the real possibility of an antibiotic-resistant rickettsia being used in a bioterrorist attack, it is imperative that the development and evaluation of new antirickettsial drugs be emphasized by the scientific community.

There have been previous attempts to develop effective vaccines against RMSF.[20,31,32] Although the vaccines developed had some success in experimental settings, all were discontinued because of the occurrence of laboratory-acquired infections in vaccinated individuals.[3,34]

7. NEEDED COUNTERMEASURES

It is our opinion that we are woefully underprepared for a bioterrorist attack using *R. rickettsii* or any other member of the spotted fever group. We lack a widely available diagnostic test for rickettsioses that is effective during the acute stage of illness. There has been little development of therapeutics for tetracycline- and/or chloramphenicol-resistant rickettsiae; and no new vaccine has replaced the old vaccine that was withdrawn from the market.

Although there is reason for concern, we should acknowledge the recent advances in the field of rickettsiology that would allow us to rapidly address the biothreat of SFG rickettsiae. First, there have been two significant advances utilizing proteomics and real-time PCR to improve point-of care (POC) diagnostics for acute-stage rickettsial diseases. La Scola and Raoult[34] developed an antigen capture assay for detection of *R. conorii* in circulating endothelial cells. In this method, infected endothelial cells are "captured" using magnetic beads coated with monoclonal antibodies against a human endothelial cell surface antigen and stained for the presence of intracellular rickettsiae using immunofluorescence. This method has a sensitivity of 50% and a specificity of 94%.[34] Labruna and others recently developed a real-time PCR assay for the quantification of *Rickettsia* species in ticks that can detect 1 copy of *R. rickettsii.*[35] This assay has been shown to detect both spotted fever and the typhus group rickettsiae. There are also ongoing efforts to develop proteomic assays to detect and enhance the signal from rickettsia-specific protein antigens and to define the "biosignature" of the human innate immune response to rickettsial infections. All of the aforementioned tests focus on the acute stage of the disease in order to allow clinicians to begin effective treatment in a more expedited fashion than normally occurs.

There has also been an increased effort to develop improved vaccines by focusing on subunit vaccines instead of the whole organism approach of yesterday.[36,37] Experimental vaccines were developed using epitopes of outer membrane protein A and outer membrane protein B to stimulate protective immunity against *R. conorii.* A multivalent vaccine combination of DNA encoding the protective epitopes and booster immunization with its corresponding recombinant proteins provided protection against a lethal challenge with *R. conorii.* These antigens could potentially be components of an improved human vaccine. There are also efforts underway to utilize microarray technology and the ever-growing number of sequenced *Rickettsia* species genomes not only to identify new vaccine candidates, but also to evaluate the mechanisms of rickettsial pathogenesis and to identify targets for therapeutic intervention.

In the current geopolitical climate, there is an increased possibility of attacks upon society using biological agents. It would be foolhardy to prepare only for 2–3 agents due to their notoriety especially when there are reports that many organisms have been weaponized throughout history. *R. rickettsii* and the other members of the spotted fever group are some of the organisms that fall in the true threat category. They are easy to obtain, weaponize, and use. It will be only through intense and continued research efforts to develop improved diagnostics, treatments, and vaccines that society can brace itself against attack.

REFERENCES

1. Walker, D. H., 2003, Principles of malicious use of infectious agents to create terror: reasons for concern for organisms of the genus *Rickettsia*, *Ann. NY Acad. Sci.* **990**:739–742.
2. Johnson, J. E., III and Kadull, P. J., 1967, Rocky Mountain spotted fever acquired in a laboratory, *N. Eng. J. Med.* **277**:842–847.
3. Calia, F. M., Bartelloni, P. J., and McKinney, R. W., 1970, Rocky Mountain spotted fever, laboratory infection in a vaccinated individual, *JAMA.* **211**: 2012–2014.
4. Pike, R. M., 1976, Laboratory-associated infections: summary and analysis of 3921 cases, *HLS* **13**:105–114.
5. Weiss, E., 1988, History of Rickettsiology, in: *Biology of Rickettsial Diseases*, vol. 1, (D. H. Walker ed.),CRC Press, Boca Rata, FL, pp. 16–28.
6. Raoult, D., Ndihokubwayo, J. B., Tissot-Dupont, H., Roux, V., Faugere, B., Abegdinni, R., and Birtles, R. J., 1998, Outbreak of epidemic typhus associated with trench fever in Burundi, *Lancet* **352**:353–358.
7. Tarasevich, I., Rydkina, E., and Raoult, D., 1998, Outbreak of epidemic typhus in Russia, *Lancet* **352**:1151.
8. Harris, S., 1992, Japanese biological warfare research on humans: a case study of microbiology and ethics, *Ann. NY Acad. Sci.* **666**:21–49.
9. Alibek, K., 1999, in: *Biohazard*, Random House, New York, pp. 3–319.
10. Sexton, D. J., and Walker, D. H., 1999, Spotted fever group rickettsioses, in: *Tropical Infectious Diseases, Principles, Pathogens, and Practice* (R. L. Guerrant, D. H. Walker, and P. F. Weller, eds.), Churchill Livingstone, Philadelphia, pp. 579–584.
11. Walker, D. H., and Fishbein, D. B., 1991, Epidemiology of rickettsial diseases, *Eur. J. Epidemiol.* **7**:237–245.
12. Miller, J., Engelberg, S., and Broad, W., 2002, *Biological Weapons and America's Secret War Germs*, Simon & Schuster, New York, pp. 165–201.
13. Phillip, C. B., 1959, Some epidemiological consideration in Rocky Mountain spotted fever, *Public Health Rev.* **74**:595–600.
14. Philip, R. N., and Casper, E. A., 1981, Serotypes of spotted fever group rickettsiae isolated from *Dermacentor andersoni* (Stiles) ticks in western Montana, *Am. J. Trop. Med. Hyg.* **30**:230–238.
15. Niebylski, M. L., Peacock, M. G., and Schwan, T. G., 1999, Lethal effect of *Rickettsia rickettsii* on its tick vector (*Dermacentor andersoni*), *Appl. Environ. Microbiol.* **65**:773–778.
16. Silverman, D. J., Boese J. L., and Wisseman, C. L., Jr., 1974, Ultrastructural studies of *Rickettsia prowazeki* from louse midgut cells to feces: search for "dormant" forms, *Infect. Immun.* **10**:257–263.
17. Hatch, T. F., 1961, Distribution and deposition of inhaled particles in respiratory tract, *Bacteriol. Rev.* **25**:237–240.

18. Kenyon, R. H., Kishimoto, R. A., and Hall, W. C., 1979, Exposure of guinea pigs to *Rickettsia rickettsii* by aerosol, nasal, conjunctival, gastric, and subcutaneous routes and protection afforded by an experimental vaccine, *Infect. Immun.* **25**:580–582.
19. Saslaw, S., and Carlisle, H. N., 1966, Aerosol infection of monkeys with *Rickettsia rickettsii, Bateriol. Rev.* **30**:636–644.
20. Gonder, J. C., Kenyon, R. H., and Pedersen, C. E., Jr., 1979, Evaluation of a killed Rocky Mountain spotted fever vaccine in cynomolgus monkeys, *J. Clin. Microbiol.* **10**:719–723.
21. Zinsser, H., 1935, *Rats, Lice and History,* Little Brown and Company, New York, pp. 1–301.
22. Billings, A. N., Rawlings, J. A., and Walker, D. H., 1998, Tick-borne diseases in Texas: a 10-year retrospective examination of cases, *Tex. Med.* **94**:66–76.
23. Walker, D. H., Hudnall, S. D., Szaniawski, W. K., Feng, and H-M., 1999, Monoclonal antibody-based immunohistochemical diagnosis of rickettsialpox: the macrophage is the principal target, *Mod. Pathol.* **12**:529–533.
24. Walker, D. H., and Bouyer, D. H., 2003, Rickettsia in: *Manual of Clinical Microbiology,* 8th Edition (P. R. Murray, E. J. Baron, J. H. Jorgensen, et al., eds.) ASM Press, Washington, D.C., pp. 1005–1014.
25. Anonymous, 2004, Fatal cases of Rocky Mountain spotted fever in family clusters—three states, 2003, *MMWR* **53**:407–410.
26. Kaplan, J. E., and Schonberger, L. B., 1986, The sensitivity of various serologic tests in the diagnosis of Rocky Mountain spotted fever, *Am. J. Top. Med. Hyg.* **35**:840–844.
27. Raoult, D., and Dasch, G. A., 1989, Line blot and western blot immunoassays for diagnosis of Mediterranean spotted fever, *J. Clin. Microbiol.* **27**:2073–2079.
28. Eremeeva, M. E., Balayeva, N. M., and Raoult, D., 1994, Serologic response of patients suffering from primary and recrudescent typhus: comparison of complement fixation reaction, Weil-Felix test, microimmunofluorescence, and immunoblotting, *Clin. Diag. Lab. Immunol.* **1**:318–324.
29. Kelly, D. J., Chan, C. T., Payton, H., Thompson, K., Howard, R., and Dasch, G. A., 1995, Comparative evaluation of a commercial enzyme immunoassay for the detection of human antibody to *Rickettsia typhi, Clin. Diag. Lab. Immunol.* **2**:356–360.
30. Weiss, E., and Dressler, H. R., 1962, Increased resistance to chloramphenicol in *Rickettsia prowazekii* with a note on failure to demonstrate genetic interaction among strains, *J. Bacteriol.* **83**:409–414.
31. Fox, J. P., 1949, The relative infectibility of laboratory animals and chick embryos with rickettsiae of murine or of epidemic typhus, *Am. J. Hyg.* **49**:313–320.
32. Kenyon, R. H., Sammons, L. St. C., and Pedersen, C. E., Jr., 1975, Comparison of three Rocky Mountain spotted fever vaccines, *J. Clin. Microbiol.* **2**:300–304.
33. Oster, C. H., Burke, D. S., Kenyon, R. H., Ascher, M. S., Harber, P., and Pederson, C. E., Jr., 1977, Laboratory-acquired Rocky Mountain spotted fever, the hazard of aerosol transmission, *N. Eng.; J. Med.* **297**:859–863.
34. La Scola, B., and Raoult, D., 1996, Diagnosis of Mediterranean spotted fever by cultivation of *Rickettsia conorii* from blood and skin samples using the centrifugation-shell vial technique and by detection of *R. conorii* in circulating endothelial cells: a 6-year follow-up, *J. Clin. Microbiol.* **34**:722–2727.
35. Labruna, M. B., Whitworth, T., Horta, M. C., Bouyer, D. H., McBride, J. W., Pinter, A., Popov, V., Gennari, S. M., and Walker, D. H., 2004. *Rickettsia* species infecting *Amblyomma cooperi* ticks from an area in the state of Sao Paulo, Brazil, where Brazilian spotted fever is endemic, *J. Clin. Microbiol.* **42**:90–98.
36. Díaz-Montero, C. M., Feng, H-M., Crocquet-Valdes, P. A., and Walker, D. H., 2001, Identification of protective components of two major outer membrane proteins of spotted fever group rickettsiae, *Am. J. Trop. Med. Hyg.* **65**:371–378.
37. Crocquet-Valdes, P. A., Díaz-Montero, C. M., Feng, H-M., Li, H., Barrett, A. D. T., Walker, D. H., 2002, Immunization with a portion of rickettsial outer membrane protein A stimulates protective immunity against spotted fever rickettsiosis, *Vaccine* **20**:979–988.

Index